A Handbook of Spatial Research Paradigms and Methodologies

Volume 2: Clinical and Comparative Studies

edited by

Nigel Foreman and Raphael Gillett
Department of Psychology
University of Leicester, UK

Psychology Press
Taylor & Francis Group
HOVE AND NEW YORK

Copyright © 1998 by Psychology Press

All rights reserved. No part of this book may be reproduced in any form, by photostat, microform, retrieval system, or any other means without the prior written permission of the publisher.

Published 1998 by Psychology Press
27 Church Road, Hove, East Sussex BN3 2FA, UK
711 Third Avenue, New York, NY 10017

First issued in paperback 2015

Psychology Press is an imprint of the Taylor & Francis Group, an informa business

British Library Cataloguing in Publication Data

A catalogue record for this book is available from the British Library.

ISBN 13: 978-1-138-88308-6 (pbk)
ISBN 13: 978-0-8637-7807-0 (hbk)

Typeset by J&L Composition Ltd, Filey, North Yorkshire.

Publisher's Note
The publisher has gone to great lengths to ensure the quality of this reprint but points out that some imperfections in the original may be apparent.

Contents

List of Contributors   ix

General Introduction   1
    Nigel Foreman and Raphael Gillett
    Why this handbook?   1
    Volume 2: Clinical and comparative studies   3
        Human clinical studies   3
        The clinical approach   3
        The neurobehavioural approach   4
        The ethological approach   7
    Symbiosis between human and animal work   8
    References   8

1 Visual-spatial Skill and Standard Psychometric Tests   11
    J. Graham Beaumont
    Introduction   11
    The isolation of visual-spatial skills   12
    Visual-spatial skill and general cognitive ability   13
    Specific psychometric tests of visual-spatial skill   15
        Visual organisation   15
        Visual scanning   17
        Visual-spatial functions   17
        Visual construction   22
        Drawing and copying   23
    Theoretical aspects   24
        Cognitive neuropsychology   24

iv CONTENTS

       Visual-spatial neglect   25
       Internal frames of reference   26
       Contemporary theoretical positions   28
Conclusion   29
References   30

2 A Cognitive Neuropsychological Approach to Spatial Memory Deficits in Brain-damaged Patients   33
Martial Van der Linden and Thierry Meulemans
Introduction   33
Spatial working memory   34
Topographical memory   41
Long-term memory for spatial location   48
Conclusion   52
References   53

3 Animal Spatial Cognition and Exploration   59
Catherine Thinus-Blanc, Etienne Save and Bruno Poucet
Introduction   59
The spatial function of exploration   60
Exploration and problem solving   62
The dishabituation paradigm   66
    Open field with objects   68
    Reaction to a missing stimulus   75
    The dishabituation paradigm for studying hippocampal functions   77
Advantages and pitfalls of paradigms based on exploration   80
Conclusions   83
References   84

4 The Radial Arm Maze: Twenty Years On   87
Nigel Foreman and Irina Ermakova
Introduction   87
Evolution of the radial arm maze   88
Procedures: Practical considerations for the standard continuous version of the task   90
Data collection   91
Measurement of performance: The standard continuous version of the task   92
Containment between choices   95
Performance strategies: How do animals complete the task?   97
    Extramaze vs. intramaze spatial cues   99
    Choice criteria   102
    Olfaction   102
Learning and memory   103
    Spatial vs. operant learning   103
    Memory: Parameters of performance   104

Differential baiting: Working vs. reference memory   106
Influences on RAM performance   110
    Early experience   110
    Sex differences   111
    Age   111
    Stress   112
    Species and strain variation in performance   112
Win-shift and natural foraging   114
Algorithmic response patterning   116
    Scoring of algorithmic response patterns   117
    Predominant algorithms: Influencing factors   124
    Breaking algorithmic response patterns (strategies)   125
    Do algorithms mask poor performance?   126
An aquatic version: The radial water maze   128
Alternative paradigms using a radial structure   130
    The Barnes platform task   130
    Matching and nonmatching to sample   131
Concluding remarks   131
Acknowledgements   132
References   132

5 The Morris Water Maze (is not a Maze)   145
Françoise Schenk

A timely task design   145
The analysis of spatial representations   147
    Which measures must be recorded?   148
    Comparison with a task on solid ground   150
    What do rats actually know: Place memory or alternative strategies?   151
    Critical steps in task acquisition: Platform pre-exposure and instantaneous transfer tests   153
    A role for vestibular and other idiothetic cues   154
    What is the minimum visual information required for place memory?   156
    How critical is the memory of the panorama from the training sector?   158
    Assessing the importance of local cues in place-learning designs   160
    Reversal learning and learning sets   163
    Long-term retention: Memory consolidation or primacy effect?   164
    The ambiguity of stress effects   165
Dissociations between memory components for place learning   167
    Maturational steps   168
    Performance in the Morris task following damage of the hippocampal formation   171
    Multiple aspects of cholinergic functions   173

vi CONTENTS

 Eco-ethological relevance of the Morris task 175
  Is it a valid task for the evaluation of adaptive spatial abilities? 175
 Shall we reject or rather refine our hypotheses? 177
  Do we need to reject the spatial map hypothesis? 177
  Do we need to reconsider which brain structures are involved in spatial memory? 178
  Does it raise some experimental issues? 180
 Acknowledgements 181
 References 181

6 Testing for Spatial Brain Dysfunction in Animals 189
Helen Hodges

 Introduction: Approaches to the assessment of spatial brain dysfunction 189
 Variety of procedures for the assessment of spatial learning and memory 191
 Visuospatial information processing 195
 Cue distribution and salience 196
  Cue removal 197
  Cue stability 198
 Associative vs. cognitive mapping processes 199
 Egocentric spatial information processing 202
 Nonvisual sensory information involved in spatial navigation 203
  Olfactory information 204
  Tactile information and vibrissae movement 205
 Search strategies 207
 Nonsensory factors affecting spatial navigation 209
  Task motivation and stress 209
 Training and testing regimes 211
 Assessment of memory deficits by spatial tasks 213
  Procedural vs. rapid acquisition tasks 213
  Working memory tasks 217
 Working memory and the hippocampus 222
 Assessment of recovery of function in spatial tasks 224
 Environmental enrichment 225
 Intra-cerebral transplantation 226
 Combined effects of enrichment and transplantation 228
 Conclusions 230
 Acknowledgements 230
 References 231

7 Long-distance Travels and Homing: Dispersal, Migrations, Excursions 239
Jacques Bovet

 A preliminary overview and a prefatory note 239
 Observing and recording travels in the field 241

   Capture-Marking-Recapture 241
   Radio-tracking 242
   Miscellaneous opportunistic techniques 245
  The home range 247
   Measuring home range size and shape 247
   Travels within the home range boundaries 250
  Long-distance travels and homing 251
   A hazy history 251
   Homing phenomena and strategies 252
   Spontaneous behaviour 255
   Experiments 259
  Perspectives for the study of mammals 264
  References 264

Author index **270**
Subject index **277**

List of Contributors

J. Graham Beaumont, Department of Clinical Psychology, Royal Hospital for Neuro-disability, West Hill, Putney, London, SW15 3SW, UK
Jacques Bovet, Département de biologie, Université Laval, Québec, QC, Canada G1K 7P4
Irina Ermakova, Institute for Higher Nervous Activity and Neurophysiology, 5a Butlerova Str, Moscow 117485, Russia
Nigel Foreman, Department of Psychology, University of Leicester, University Road, Leicester, LE1 7RH, UK
Raphael Gillett, Department of Psychology, University of Leicester, University Road, Leicester, LE1 7RH, UK
Helen Hodges, Department of Psychology, Institute of Psychiatry, De Crespigny Park, Denmark Hill, London, SE5 8AF, UK
Thierry Meulemans, Service de Neuropsychologie, University of Liège, Boulevard du Rectorat, B33, B-4000 Liège, Belgium
Bruno Poucet, CNRS – Centre de Recherche en Neurosciences Cognitives, 31 chemin Joseph Aiguier, Marseille 13402, cedex 20, France
Etienne Save, CNRS – Centre de Recherche en Neurosciences Cognitives, 31 chemin Joseph Aiguier, Marseille 13402, cedex 20, France
Françoise Schenk, Institute de Psychologie, Universite de Lausanne, Rue du Bugnon 7, CH-1005 Lausanne, Switzerland
Catherine Thinus-Blanc, CNRS – Centre de Recherche en Neurosciences Cognitives, 31 chemin Joseph Aiguier, Marseille 13402, cedex 20, France
Martial Van der Linden, Service de Neuropsychologie, University of Liège, Boulevard du Rectorat, B33, B-4000 Liège, Belgium

General Introduction

Nigel Foreman and Raphael Gillett
University of Leicester, UK

WHY THIS HANDBOOK?

Spatial cognition has been the subject of some of the oldest and fiercest disputes in philosophy, the nature versus nurture controversy often focusing upon the innateness or otherwise of spatial thought (Klein, 1970). Moreover, research into spatial cognition is among the earliest conducted in the name of psychology, influencing the emerging discipline considerably throughout the past century. Indeed, two of the first PhD degrees to be awarded in psychology in the United States went to G. Stanley Hall (Harvard University, 1878) for his thesis on "The Perception of Space", and to Joseph Jastrow (Johns Hopkins University, 1886) for his on "The Perception of Space by Disparate Senses" (Klein, 1970). It is perhaps surprising, therefore, that by comparison with other areas of perception and cognition, good models of spatial cognition have appeared relatively recently. This may reflect the elusive nature of spatial abilities, ambiguity over what is meant by the term "spatial", and the equally ambiguous nature of "spatial cognition", which encompasses processes that are neither classifiable as stimulus nor response. Spatial cognitive skills must be inferred from behaviours, occupying the space between stimuli and the responses made to them. Responses such as reaching toward objects, or judging distances and directions, are made within the context of a spatial framework, but the nature of that framework (or, more accurately, those frameworks) has remained elusive. Perhaps for this reason, until the last 40 years spatial research and its applications have remained rather neglected areas, particularly among psychologists (Siegel and Schadler, 1977). Still now, the teaching of graphicacy and spatial skill in children, the ways in which spatial skills evolve in animals and in humans, and the study of the neural mechanisms that underly these skills—in short, the ways and means by which animals and people organise themselves in relation to their spatial surroundings—remain substantially under-researched.

Spatial cognition is a broad field of enquiry, emerging not only from psychology but from a range of disciplines, including geography and environmental planning at the macro end of the scale (e.g., Abler, Adams and Gould, 1972; Appleyard, 1970; Lynch, 1960; Downs and Stea, 1973), to single cell biology and receptor physiology at the other (e.g., Butelman, 1989; Muller and Kubie, 1987; O'Keefe and Dostrovsky, 1971). Perhaps as a consequence of this diversity of approaches and related disciplines, there is little consistency in what is meant by the term "spatial", and consequently there exists a broad range of paradigms and methodologies that can be employed to examine spatial skill, each measuring disparate aspects of spatial cognition in a wide range of behavioural contexts, in both animals and humans.

The problem of coping with such diversity is the main issue which the present Handbook is set to address. In this volume, authors with specialist knowledge of particular types of experimental paradigms describe their uses, drawbacks, and the results that have emerged from them. They describe techniques that have been applied in attempts to understand how brain damage affects spatial behaviours in humans, how animal studies have shed light on spatial representation in the brain, neurophysiological aspects of spatial cognition, and how animals home and orient themselves within their natural habitats.

In the first volume of the Handbook, authors considered paradigms used in developmental studies to examine the ways in which spatial skills develop in infants and children, how children search strategically among multiple choice locations, and how they navigate in both real and laboratory environments. In addition the first volume was concerned with adults' spatial orientation in built environments and the uses of simulated 3-D (virtual reality) environments in testing spatial skills.

The contributing authors in both volumes are highly respected international authorities in their fields, with many years of experience in using the test paradigms which they describe and evaluate. They have been asked to concentrate specifically on paradigms and test methodologies, and the type and value of the data which each can yield. There is less emphasis than is usual in scientific publications on the theoretical models that provoked the research. These volumes should be of interest to new and old students alike; the student new to spatial research can be brought up-to-speed with a particular range of techniques, made aware of the background and pitfalls of particular approaches, and directed toward useful sources. For seasoned researchers, these volumes provide a rapid scan of the currently available tools that they might consider as alternatives when wishing to answer a particular "spatial" research problem.

The following introduction is intended to provide a broad introduction to some of the terms, concepts and theories that will be encountered throughout the book, although this should be read in conjunction with the introductory chapter of Volume 1.

VOLUME 2: CLINICAL AND COMPARATIVE STUDIES

Human clinical studies

Research described in Volume 1 of this Handbook considered paradigms used in spatial studies with individuals drawn largely from the able-bodied population. Although there is variation in performance on spatial tasks among able-bodied subjects, depending on their sex, age, and experience (see Volume 1 of this Handbook), spatial deficiencies can occur in individuals for particular reasons, for example, as a consequence of their having suffered illness and/or sustained brain damage. Providing a link into Volume 2 is Wilson's final chapter in Volume 1, showing how virtual reality computing could be used to enhance the poor spatial awareness of disabled children. Spatial disability was, in that instance, not necessarily the result of brain damage, and was more likely to have occurred for primarily functional reasons (lack of autonomous exploration by disabled individuals), although in some cases damage to spatial brain structures may have been a primary cause.

Volume 2 moves on to consider specifically the paradigms employed to identify the mechanisms underpinning spatial behaviours, and the parameters of performance, via clinical, neurobehavioural and comparative approaches. These are concerned with spatial representations in the damaged human brain (the clinical approach), comparative studies of nonhuman spatial cognition, brain function and the recovery of spatial brain function (the neurobehavioural approach) and the evolution of spatial processes, addressing the ways in which spatial/navigational skills can be examined in natural conditions in animals inhabiting a variety of environmental niches (the ethological approach).

The clinical approach

As long ago as 1919, Holmes reported patients who could correctly identify objects, but were incapable of reaching accurately toward them, or describing the relative positions of two objects (Holmes, 1919). Following the Second World War, patients with (often widespread) brain damage were first examined using test batteries which included spatial subscales (e.g., Luria, 1973; Beaumont, Chapter 1, this volume), and interest in clinical spatial deficits developed rapidly. Experimentally, Semmes, Weinstein, Ghent and Teuber (1955, 1963) demonstrated, using locomotor mazes and mapping tasks that a variety of spatial deficits could occur, depending upon the site of the damage. Right hemisphere lesions were reported to lead to deficits in mental rotation and global-topographical spatial aspects of a task, while left hemisphere lesions lead to problems in following and describing routes. Ratcliff and Newcombe have shown that when tested on visual-spatial and visual-motor tasks on the one hand, and navigational tasks on the other, patients may succeed on one type of task while failing another (Ratcliff and Newcombe, 1973). The term "spatial" has been used in relation to a number of deficits. As in animals (see later chapters), choice of test paradigm is crucial if the precise nature of lost, spared or recovered spatial abilities is to be identified.

The opening papers cover the types of test typically applied when patients are suspected of having spatial impairments, due to ageing, illness or suffering the direct effects of brain injury. Beaumont, and Van Linden and Meulemans, discuss the use of spatial tests within standard test batteries, and experimental paradigms, respectively, and the ways in which clinical data throw light on underlying spatial impairments and localisation of spatial function within the human brain. In Chapter 1, Beaumont describes tests of extrapersonal space perception, visual object and visual-spatial perception, drawing and copying, and construction tasks found in the standard test batteries. Beaumont concludes that, with the possible exception of visual-spatial neglect, relatively little work has been devoted to cognitive aspects of visual-spatial function and forsees a future expansion in the range of test materials that will become available for assessment purposes. Van der Linden and Meulemans, in Chapter 2, discuss functional dissociations in visual-spatial short-term memory, in topographical spatial orientation, and in long-term memory for spatial locations, particularly emphasising studies of orientation in locomotor space. Examples are illustrated of possible dissociations among "spatial" deficits, and the caution that is required in interpreting deficits as having separate or related underlying cognitive bases.

The neurobehavioural approach

It is only in recent decades that it has been possible to study physiological measures as indicators of underlying spatial processes. Russell (1932) was perhaps one of the first to address this issue, since earlier studies of animal spatial behaviour had tended to concentrate on the matter of animal "intelligence" rather than spatial skill *per se*. Russell was interested in the way that rats could judge the effort required to leap between two raised platforms. By measuring the force exerted and varying the interplatform distance, he was able to demonstrate a systematic relationship between force and distance. Early studies were especially concerned with the innateness or otherwise of spatial skill. Lashley and Russell (1934) showed that the relationship between leap distance and exerted force could also be observed in rats reared in darkness, indicating that this was an innate ability in rats, and not one in which vision needed early education by movement and touch. Later, Held and Hein (1963) showed that some spatial behaviours related to distance estimation, such as paw placement, were deficient in animals raised with restricted visual-motor conditions.

A thread can be traced through the development of maze methodologies, also originating with Lashley and coworkers in the early part of the century. Many studies of animal learning incorporated spatial components, for example, the Hebb–Williams mazes, which required animals to find routes to food locations via a series of interconnected alleyways. The speed with which animals negotiated the maze to reach the goal was taken as a measure of learning (Hebb, 1949). Such tasks could be solved via the use of many alternative cues and strategies, of which spatial memory was only one element.

Among the first psychologists to study animals' discrimination of places was E.C. Tolman, who published "Cognitive maps in animals and man" in 1948. Tolman had developed in the company of conventional S-R (stimulus-response) behaviourists, who attempted to explain all behaviour in terms of laws of reinforcement. Conventionally, rewards and punishments respectively increased and decreased the likelihood of behaviours immediately preceding them. Tolman's assertion that animals (and, indeed, people) placed in novel environements acquire spatial "map"-like representations (without being rewarded for doing so; hence so-called "latent learning") seriously conflicted with the established wisdom of the time. Tolman was accused of leaving animals at runway choice points " . . . buried deep in thought". Such a "cognitivist" viewpoint as Tolman's appeared to ascribe a high level of intelligence to animals. A similar theoretical divide occurred in Russia during the 1940s between the behaviourist reflex psychologists Pavlov and Bechterev, and the Georgian Beritov (Beritashvilli), a psychologist in Tolman's tradition, whose work also showed that animals use spatial maps or representations (*predstavlenniya*) to locate places where they have obtained food, even when the food dishes are later removed (Joravsky, 1989).

To some extent, the traditional positions of cognitivists and behaviourists have become compromised within more recent models that allow the coexistence and availability of a variety of spatial strategies, applicable according to task requirements (Thinus-Blanc, 1996).

A major development in spatial theorising occurred in 1978 with the publication of *The hippocampus as a cognitive map*, by O'Keefe and Nadel. These authors took Tolman's work as a starting point, but identified a particular neural structure, the forebrain hippocampus, as the seat of cognitive spatial mapping ("cognitive", because these are nonmetric mental maps, and "spatial" because they define places and the relationships among them.) The O'Keefe and Nadel model is discussed from various perspectives in the current volume (Chapters 3, 4, 5 and 6). The model was based not only on behavioural data but a range of electrophysiological findings, often collected from animals in the course of performing spatial tasks (O'Keefe and Dostrovsky, 1971). Studies of cellular correlates of spatial behaviour are beyond the scope of Volumes 1 and 2 of this Handbook.

The O'Keefe and Nadel model was largely tested with rodents, and more recent work has shown that cells in the monkey hippocampus have firing patterns that relate more closely to particular bodily movements and facing direction than to position in space *per se* (O'Mara, Rolls, Berthoz and Kesner, 1994). The model has been elaborated rather than replaced, in so far as neural structures other than hippocampus, such as frontal and parietal cortices, have also been shown to play important roles in spatially organised behaviour (Poucet, 1993), different forms of spatial representation and strategy may be engaged, according to task requirements (Thinus-Blanc, 1996), and aspects of the environment other than landmark arrays, such as room geometry (Cheng, 1986; Gallistel, 1990) may be used for orientation purposes. It follows from this recognition of a diversity of possible spatial task solutions, that methodology is once again extremely important in determining the ways in which spatial skill should be tested.

Not all studies of spatial ability require that rewards are presented (as Tolman had realised) and many sensitive tests of spatial ability can take advantage of rodents' strong innate tendency to explore open field arenas and to investigate encountered objects via approaching and contacting them. The behaviour of animals in response to selective spatial changes in object locations can provide powerful evidence for spatial memory, using rapid paradigms that require no deprivation or training. Test paradigms that take advantage of rats' reactions to spatial change are discussed by Thinus-Blanc, Save and Poucet (Chapter 3, this volume). These authors discuss the relationship between exploration and the acquisition of spatial knowledge, a theme that recurs frequently throughout later chapters.

Olton's (1977) report of the Radial Arm Maze (RAM) revolutionised the field of spatial research and still provides a common laboratory test that can be used to examine aspects of spatial memory in animals, and quantify spatial impairments. Foreman and Ermakova (Chapter 4, this volume), describe the development of the various forms of the test over the past 20 years and discuss the measurement and significance of the various strategies that can be used to perform the task. This chapter concentrates upon parameters of performance.

In 1980, Morris described an alternative testing paradigm, which also had a substantial impact on spatial research. This consisted of a circular bath of opaque water in which animals had to swim toward an invisible, submerged platform on to which they could escape (see Schenk, Chapter 5, this volume). Instead of requiring successive responses to several spatial locations, the water task (Schenk insists it is not a maze) required just one swim response to locate the platform. The task was intended to provide a purer test than the RAM of specifically spatial impairments, since animals could arguably only use the layout of ambient room cues to locate the platform. Testing in water prevented the use of odour cues or trails.

Both the RAM (in its many versions) and the water maze, among others, have been put to use in recent years in the investigation of loss of spatial function (deficient across-trial, or within-trial memory, for example) that occur following brain damage, drug treatments, or ageing, but also to examine recovery of function that can occur following treatments improving deficient brain function. The neurophysiological techniques have appeared only recently but provide important indicators of the future role of spatial paradigms in cognitive spatial research. They are reviewed in Chapter 6 by Hodges, who considers both the theoretical and practical aspects of data interpretation using several maze configurations and paradigms.

The ethological approach

Clearly, species' use of space will differ. Animals living in burrows, or emerging into the environment in semi-darkness will not have a visually differentiated environment, and navigation may require a combination of strategies used sequentially or simultaneously, thus studies of spatial cue use in laboratory tasks will not always replicate the types of situation that animals encounter in their natural environments. This is especially so when long-distance travelling is involved among visually disconnected environments. Most of the tasks described in Chapters 3 to 6 involve environments in which the entire test space was visible from any part of the apparatus. In contrast, Bovet in Chapter 7, describes a range of techniques that can be used to investigate large-scale spatial orientation, including that involved in migratory travels, patrolling in home ranges and homing from distant sites. Each of these situations requires quite different approaches to measurement and data interpretation, as Bovet illustrates.

SYMBIOSIS BETWEEN HUMAN AND ANIMAL WORK

Although the two volumes of this Handbook address different areas of research, there are several areas in which linkage, overlap or symbiosis exists between them. Historically, animal and human areas of spatial research have developed in parallel (cf., Tolman, 1948). Within the human domain, children's spatial development (see Volume 1 of this Handbook) can provide an interesting source of information for several reasons, particularly because the child probably attempts to solve spatial tasks without employing the sophisticated verbal mediation that might be used by adults. Moreover, the representations that children acquire might be more fragile than those of adults, making them more vulnerable to behavioural interventions.

There are numerous studies showing that children's spatial behaviour changes systematically in the early years (reviewed in Volume 1 of this Handbook), and sometimes these changes have been related very directly to underlying neurology (Douglas, 1975; Harris, 1975; Wertlieb and Rose, 1979), although extrapolation between humans and animals requires cautious interpretation (cf., Diamond, 1988). Similar methodological issues apply equally to animal and human research, for example, the analysis of item and order information in radial search paradigms (Dale, 1987), the quantification of choice behaviour in multiple choice tasks (see Foreman and Ermakova, Chapter 4, this Volume) and definitions of "representation" in child studies (see Liben, Chapter 1 of Volume 1) and in animal studies (Roitblat, 1982). Furthermore, much animal work is, directly or indirectly, aimed at understanding human spatial disorders that characterise Korsakov's amnesia, Parkinson's disease, Alzheimer's dementia and other conditions involving disruption of spatial behaviour (Olton, 1982; see Hodges, Chapter 6 in this volume).

REFERENCES

Abler, R., Adams, J. S., & Gould, P. (1972). *Spatial organization: The geographer's view of the world.* London: Prentice Hall.

Appleyard, D. (1970). Styles and methods of structuring a city. *Environment and Behavior, 1,* 131–156.

Butelman, E.R. (1989). A novel NMDA (*n*-methyl-*d*-aspartate) antagonist, MK-801, impairs performance in a hippocampal-dependent spatial learning task. *Pharmacology, Biochemistry and Behavior, 34,* 13–16.

Cheng, K. (1986). A purely geometric module in the rat's spatial representation. *Cognition, 23,* 149–178.

Dale, R.H. (1987). Similarities between human and animal spatial memory: Item and order information. *Animal Learning and Behavior, 15,* 293–300.

Diamond, A. (1988). Abilities and neural mechanisms underlying AB performance. *Child Development, 59,* 523–527.

Douglas, R.J. (1975). The development of hippocampal function. In R. Isaacson & K. Pribram (Eds.), *The hippocampus*. New York: Plenum.

Downs, R.M., & Stea, D. (Eds.) (1973). *Image and environment: Cognitive mapping and spatial behavior*. London: Edward Arnold.

Gallistel, C.R. (1990). *The organization of learning*. Cambridge, MA: Bradford Books/MIT Press.

Harris, L.J. (1975). Neurophysiological factors in the development of spatial skills. In J. Eliot & N.J. Salkind (Eds.), *Children's spatial development*. Springfield, IL.: Charles C. Thomas.

Hebb, D.O. (1949). *The organization of behavior*. New York: Wiley.

Held, R., & Hein, A. (1963). Movement-produced stimulation in the development of visually guided behavior. *Journal of Comparative and Physiological Psychology, 56*, 872–876.

Holmes, G. (1919). Disturbances of visual orientation. *British Journal of Ophthalmology, 2*, 449–468.

Joravsky, D. (1989). *Russian psychology: A critical history*. Cambridge, UK: Blackwell.

Klein, D.B. (1970). *A history of scientific psychology: Its origins and philosophical backgrounds*. London: Routledge & Kegan Paul.

Lashley, K.S., & Russell, J.T. (1934). The mechanism of vision: A preliminary test of innate organization. *Journal of Genetic Psychology, 45*, 136–144.

Luria, A.R. (1973). *The working brain: An introduction to neuropsychology*. Harmondsworth, UK: Penguin.

Lynch, K. (1960). *The image of the city*. Cambridge, MA: MIT Press.

Muller, R.U., & Kubie, J.L. (1987). The effects of changes in the environment on the spatial firing of hippovcampal complex-spike cells. *Journal of Neuroscience, 7*, 1951–1968.

O'Keefe, J., & Dostrovsky. J. (1971). The hippocampus as a spatial map. Preliminary evidence from unit activity in the freely moving rat. *Brain Research, 34*, 171–175.

O'Keefe, J., & Nadel, L. (1978). *The hippocampus as a cognitive map*. London: Oxford University Press.

Olton, D.S. (1977). Spatial memory. *Scientific American, 236*, 82–98.

O'Mara, S.M., Rolls, E.T., Berthoz, A., & Kesner, R.P. (1994). Neurons responding to whole body motion in the primate hippocampus. *Journal of Neuroscience, 14*, 6511–6523.

Poucet, B. (1993). Spatial cognitive maps in animals: New hypotheses on their structure and neural mechanisms. *Psychological Review, 100*, 163–182.

Ratcliff, G., & Newcombe, F. (1973). Spatial orientation in man: Effects of left, right, and bilateral cerebral lesions. *Journal of Neurology, Neurosurgery and Psychiatry, 36*, 448–454.

Roitblat, H.L. (1982). The meaning of representation in animal memory. *Behavioral and Brain Sciences, 5*, 353–406.

Russell, J.T. (1932). Depth discrimination in the rat. *Journal of Genetic Psychology, 40*, 136–159.

Semmes, J., Weinstein, S., Ghent, L., & Tauber, H-L. (1955). Spatial orientation: I. Analysis of locus of lesion. *Journal of Psychology, 39*, 227–244.

Semmes, J., Weinstein, S., Ghent, L., & Tauber, H-L. (1963). Impaired orientation in personal and extrapersonal space. *Brain, 86*, 747–772.

Siegel, A.W., & Schadler, M. (1977). The development of young children's representations of their classrooms. *Child Development, 48*, 388–394.

Thinus-Blanc, C. (1996). *Animal spatial cognition: Behavioral and neural approaches*. London: World Scientific.

Tolman, E.C. (1948). Cognitive maps in rats and men. *Psychological Review, 55*, 189–208..

Wertlieb, D., & Rose, D. (1979). Maturation of maze behavior in preschool children. *Developmental Psychology, 15*, 478–479.

1 Visual-spatial Skill and Standard Psychometric Tests

J. Graham Beaumont
Royal Hospital for Neuro-disability, London, UK

INTRODUCTION

Neuropsychology, in both its experimental and clinical aspects, has provided a powerful research context for cognitive psychology in recent years. The result has been a forum in which cognitive psychological models can be developed and tested, and which in turn have informed and enhanced approaches to the resolution of clinical problems.

The description and analysis of cognitive processes and their interrelationship may be accomplished through a variety of paradigms, but an important approach which has figured throughout the history of psychology is through observing the dissociation of independent processes which may accompany abnormal states. For example, the patient who may be able to correctly identify objects from drawings, and match pairs of representations of objects presented in various orientations, but is nonetheless unable to indicate the normal upright orientation of the object (and hence makes rotational errors in drawing and copying), demonstrates that detection of object orientation relies upon a process which is independent of the process required for object recognition.

Such cases of dissociation between functional elements are widely and frequently documented in the neuropsychological literature, and provide critical evidence which must be accounted for in any model of the relevant psychological function. The resultant models, which therefore account for this evidence, can be employed in a clinical setting to provide the psychological descriptions of the general deficits suffered by the patient, and be used to guide interventions appropriate to the rehabilitation and management of the patient's problems. By extension, not only single case studies, but also group studies of patients with cerebral lesions classified by localisation, aetiology and other relevant factors, may be employed to establish functional dissociations at a more general level.

While visual-spatial performance has naturally been among the functional topics addressed by neuropsychologists over the past century, it has been less extensively studied than certain other aspects of cognitive behaviour. The reasons are simple: it is an area of function about which the patient is less likely to complain, and the impact of visual-spatial deficits on everyday life are (unless they are profound) less disabling than certain other deficits. It is about problems of memory and language that patients primarily complain and which disrupt their lives and, perhaps as a consequence, these are the areas which have received greater attention in neuropsychological research.

With the possible exceptions of unilateral neglect (see below), and visual-constructive ability (for theoretical reasons), visual-spatial abilities have been less studied, less extensively modelled, are less frequently the object of interventions, and are less well understood. As a consequence, there are fewer procedures available, and of a less sophisticated standard, than may be found in other neuropsychological functional areas. This is perhaps surprising considering the sophistication of psychophysics and perception, both within the area of experimental psychology, and as the focus of advanced neurosciences research.

Nevertheless, certain assessment techniques have been developed which will be the topic of this chapter. They broadly fall into two classes: (1) discrete instruments which historically have more in common with behavioural neurology than with experimental psychology; and (2) more recent developments which have sought to introduce a more systematic approach to the general domain of visual-spatial function. The latter herald a resurgence of interest in visual-spatial function which is to be welcomed. Before these classes of instrument are discussed, two more general topics need to be addressed.

THE ISOLATION OF VISUAL-SPATIAL SKILLS

One of the principal difficulties in assessing visual-spatial skill (or any other mental ability) is not so much the measurement of the ability as it is expressed, but in being sure that any decrement in performance is specifically attributable to a deficit in the ability being assessed. A trivial example will suffice (trivial conceptually, if not in practice): a patient fails to complete test items presented in the left visual field, while successfully completing those in the right visual field. Before attributing this to a unilateral right hemispatial neglect, it is important to ensure that the patient has normal visual sensory function in the affected visual half field; that is, that there is no left hemianopia. If this is not the case, then the patient's poor performance is attributable to a failure to see the relevant stimuli, rather than a neglect of them.

It is therefore generally impossible to consider the assessment of visual-spatial function without also considering the assessment of functions upon which that function depends for satisfactory performance. Such functions may be considered to be more "peripheral" or at a "lower level" in neurological terms, or be considered to be prior to visual-spatial function in terms of the processing sequence of a cognitive psychological model. (The ability also to make an effective response based upon the satisfactory completion of the visual-spatial operation must also be intact, but is less acutely a problem than establishing the prerequisites for visual-spatial performance.)

In short, a subject must be able adequately to sense the material employed in any task, and to perform any essential perceptual operations upon which the visual-spatial task depends. While it is the object of any assessment to ensure that this is the case, it is of course not always possible to establish these facts and much doubt remains with regard to a variety of reports in the literature as to whether they can be regarded as purely visual-spatial phenomena. Indeed, it is often an assumption, even a matter of belief, that certain phenomena relate to a failure in visual-spatial functioning rather than to a more fundamental perceptual deficit.

Nevertheless, the models employed by neuropsychologists generally embody an independent class of visual-spatial processes, which do receive some support both from the neuropsychological literature and from psychology more generally (see elsewhere in this Volume). On the hypothesis that this is valid, the attempt continues to isolate cases of pure visual-spatial deficits, uncontaminated by deficits in other functionally related skills.

The assessment of these functionally related skills, which include visual acuity (in its multi-dimensional aspects: location, size, brightness, contrast, movement, depth, temporal distribution and stimulus complexity), visual recognition, visual organisation, visual inference, colour perception and recognition, and visual scanning, is a topic which deserves more space than is available here, but general reviews of the available test instruments can be found in Beaumont and Davidoff (1992) and in Lezak (1995). Discussions of the neuropsychological theories and data relating to these areas can be found in Beaumont, Kenealy and Rogers (1996), and Heilman and Valenstein (1993).

VISUAL-SPATIAL SKILL AND GENERAL COGNITIVE ABILITY

It is important to recognise that visual-spatial skill has been considered to be a fundamental component of general intellectual ability by many researchers in individual differences. There is no opportunity here to review adequately the relevant aspects of the continuing debate about the structure of intelligence, but historically, prominence has been given to the role of spatial abilities (by which is generally meant *visual* spatial functions).

The Hierarchical Theory of Burt and Vernon, which developed out of Spearman's Two-factor Theory (general ability: g; and specific factors: s), retained the concept of general ability (g), but interposed two group abilities between the general factor and the specific abilities. These group abilities were a verbal, educational factor (v: ed) and a kinaesthetic, motor factor (k: m).

The most widely used and respected individual assessment of intelligence currently available, the Wechsler Adult Intelligence Scale–Revised (WAIS-R) is a product arising out of these theories of the nature of intelligence in combination with the existing traditions of intelligence assessment at the time of its initial development as the Wechsler Bellevue Intelligence Scale in 1939. This, and subsequent psychometric analysis of its properties, has resulted in three intelligence quotients (IQs) being routinely calculated when the WAIS-R is administered: a Verbal IQ, a Performance IQ, and the Full Scale IQ. While the Performance IQ is far from a pure measure of visual-spatial ability, it "reflects the efficiency and integrity of the individual's perceptual organisation, including nonverbal reasoning skills, the ability to employ visual images in thinking, and the ability to process visual material" (WAIS-R Analysis Worksheet, 1985, The Psychological Corporation).

Within the subtests comprising the Performance scale, two of the five subtests have a particular bearing on visual-spatial performance: Block Design ("essentially a measure of ability to handle spatial relations", *op. cit.*) and Object Assembly ("A sense of space relations, visual-motor coordination, and persistence are among the qualities measured by this subtest", *op. cit.*).

The important fact to note is that the standardisation data (given in the WAIS-R Manual: Wechsler, 1981) indicate that the Performance score correlates with the Full Scale score at $r = 0.91$ ($N = 1880$), while the Block Design and Object Assembly subtests have correlations respectively of 0.70 and 0.62 with the Performance score, and 0.68 and 0.57 with the Full Scale score. These results underline the importance of visual-spatial skills as a component element of general intellectual skills.

More recent analyses of the concept of intelligence have resulted in a greater diversity of views of intelligence, but both Thurstone's Multi-Factor Theory (Primary Mental Abilities) includes spatial ability as one of seven component mental abilities, as does Gardner's Theory of Multiple Intelligences (Gardner, 1983).

Before leaving this issue, it is also worth noting that the single test which is generally regarded as providing the purest measure of g (begging for the present whether *g* actually exists) is the Ravens Progressive Matrices. The various versions of this test are described as assessing "observation and clear-thinking", but it is relevant to note that while successful completion of the items relies on abilities beyond purely spatial functions, an ability to appreciate spatial relationships and details of spatial configuration (in the visual modality) is crucial to performance on the test. The fact that the test items are displayed in a matrix format is itself evidence of this essential aspect of the test.

The point is simply that visual-spatial function should be understood not simply as a component among others in the constellation of abilities which we regard as "intelligent", but that visual-spatial ability plays a critical role in many aspects of intelligent behaviour and, conversely, that many tests designed to assess general cognitive function may be employed to shed light on the efficiency of visual-spatial processes.

SPECIFIC PSYCHOMETRIC TESTS OF VISUAL-SPATIAL SKILL

Visual organisation

Beyond the level of basic visual processes, visual organisation might be considered the next higher level of visual processing, and may be impaired in association with, or as a component of, visual-spatial skills. Although ambiguous stimuli (such as the Rorschach plates) are used in North America to assess visual organisation, this function is normally assessed through the presentation of either incomplete or fragmented figures.

The most widely used incomplete visual stimuli appear in the Picture Completion subtest of the WAIS-R, but this test will only reveal the most severe impairments, performance on this task being relatively resistant to the effects of perceptual dysfunction, perhaps because the stimuli are so highly structured. The same is true of the very similar Mutilated Pictures subtest at level VI of the Stanford–Binet test.

Other tests which use incomplete pictures, and have been specifically designed to assess visual organisation are the Street Completion Test (Street, 1931), the Mooney Closure Test ("Mooney Figures"; Mooney & Ferguson, 1951) and the Gestalt Completion Test (Ekstrom, French, Harman & Dermen, 1976). However, while these tests discriminate between clinical patients and controls, normal subjects show relatively low intercorrelations among these tests (Wasserstein, 1986). Wasserstein constructed a new test out of items drawn from these previous tests which in some respects may be regarded as a superior tests although the range of norms available is limited.

An alternative, and more widely used, test involving similar figures is the Gollin Incomplete Figures Test (Gollin, 1960; a computer-based version is described by Foreman & Hemmings, 1987). There are 20 sets of stimuli in this test, each set being a progressively more complete representation of the object from barely suggestive through to the complete drawing. Subjects begin with the least complete and progress through the set until the object is correctly identified. Warrington and Rabin (1970) examined the nature of this test and found that while it was not highly sensitive to right posterior cerebral lesions (those usually associated with deficits of visual-spatial skill) successful performance on the test depended on more than the extraction of perceptual features.

Fragmented, rather than incomplete, figures feature in two other well-known tests: the Hooper Visual Organisation Test (Hooper, 1958) and the Revised Minnesota Paper Form Board Test (Likert & Quasha, 1970). The Hooper was originally devised for the identification of organic impairment, for which purpose, like so many other tests it is ineffective, but it is able to detect specific deficits. The test items simply involve line drawings of objects which have been sectioned into a number of pieces which are displayed without reference to their original orientation in the figure. Interestingly, patients with right frontal lesions may be impaired on performance of this task, particularly on items in which the reduced cognitive flexibility of these patients leads them to base a response on the incorrect interpretation of a distinct fragment which they are unable to reject in favour of alternative hypotheses. A version of the Hooper test and a guide to normative data available are provided by Spreen and Strauss (1991). The Revised Minnesota Paper Form Board Test appears to be unacceptably influenced by subjects' cognitive strategies (Lezak, 1995). The Object Assembly subtest of the WAIS-R is, of course, another example of a test based upon a fragmented figure.

Another aspect of visual organisation is visual inference which has traditionally been assessed by figure-ground tasks. The most familiar of these must be the Gottschaldt Embedded Figures Test, and such figures have been employed by clinical neuropsychologists since the time of the First World War, and figured in Luria's neuropsychological battery. However, these tests have tended to be employed in a behavioural-neurological context rather than a strictly psychometric one. However, the Visual Closure subtest of the Illinois Test of Psycholinguistic Ability (ITPA; Kirk, McCarthy & Kirk, 1968) is a more formal version of such a test, and the Southern California Figure-Ground Perceptual Test (Ayres, 1966) provides normative data for children. Spreen and Strauss (1991) also provide a version of an Embedded Figures Test with normative data for both adults and children.

The most widely used test of visual inference is, however, the Closure Flexibility (Concealed Figures) Test (Thurstone & Jeffrey, 1982) which is close to the original concept of Gottschaldt, but prepared in a multiple-choice version. The task is simple: the subject must select the one among a number of more complex figures which includes the simpler unembedded stimulus.

All of these tests, which are based upon incomplete, mutilated or embedded figures rely on some degree of visual synthesis; it is matter of debate whether this is independent of visual organisation and visual inference.

Visual scanning

Disordered visual scanning may be an important aspect of failures of visual-spatial skill, and is often underrated as a significant clinical disability. Beyond the analysis of ocular movements in association with complex stimuli, there are three formal psychometric approaches to its assessment.

Elithorn's Perceptual Maze Test (Elithorn, Jones, Kerr & Lee, 1964) had a vogue but is not now widely used. It was conceived as a nonverbal measure of intelligence, rather in the manner of the Raven's Progressive Matrices, although it has proved less successful than the latter test. The test materials are typically V-shaped lattices with points placed at various intersections on the lattice. The subject is required to trace from the stem to the top of the lattice passing through as many of the points as possible but without reversing direction at any intersection. The more points placed on the lattice, the more difficult the task. Not all the functions involved in performing this task are, of course, perceptual but the task discriminates between patients with left and right hemisphere cerebral lesions, and patients may fail on the task because of a visual-spatial deficit.

A visual search task, which involves searching for small chequered stimulus matrices within a set of larger matrices, forms part of the Repeatable-Cognitive-Perceptual-Motor Battery and is, incidentally, sensitive to the effects of medication in epileptic patients (Lewis & Kupke, 1977).

Visual tracking may be assessed by the Talland Line Tracing Task (Talland, 1965) which has a format familiar to anyone who has seen a child's puzzle book in which the child is asked to identify which hand holds the string attached to which balloon. Difficulty is manipulated by varying the number of lines involved. Normal subjects can complete this task without difficulty; failure on the task would normally suggest some failure of visual scanning.

Visual-spatial functions

Visual-spatial functions have been more directly assessed in a number of ways. Many of the tests available rely upon a behavioural neurology approach (i.e. they involve tasks upon which all normal individuals may be confidently expected to succeed), although recently attempts have been made to adopt a more psychometric approach to the assessment of these functions.

Benton's neuropsychological tests

Benton, Hamsher, Varney and Spreen (1983) have published a formalised series of standard neuropsychological procedures, more properly known as *Contributions to neuropsychological assessment*. Although firmly derived from the behavioural neurology tradition, psychometric norms are provided for the tests included in this collection, variously derived from normal adults and children, and from brain-damaged clinical populations. Relevant to the present context, the tests include an assessment of Right-Left Orientation which includes orientation with respect to one's own body; orientation to a confronting person; and orientation which combines the subject's own body and that of the confronting person.

There is also the well-known Judgment of Line Orientation task. In this task the subject is required to match line segments with a multiple choice response array of radiating lines drawn at intervals of 18 degrees. Difficulty is manipulated by including as stimuli only partial lines, more than one stimulus line, and whether line segments are near, middle or distal. The test has good reliability and a range of norms with neuropsychological groups.

Besides a test of Three-Dimensional Block Construction (see below), there are also tests of Visual Form Discrimination and Finger Localisation. Finger localisation may seem an unusual ability to discuss in the present context, but abnormalities of finger localisation (finger agnosia) have long been known to be associated with other visual-spatial disturbances, as for example in the Gerstmann syndrome (begging questions of the validity of this syndrome). Benton et al. (1983) provide a systematic examination of this ability.

Personal and extrapersonal orientation

One aspect of visual-spatial function is the ability to maintain both personal and extrapersonal orientation. Personal orientation includes both body orientation and special aspects such as right-left orientation. Body orientation is normally an element of the neurological examination, but the Personal Orientation Test (Semmes, Weinstein, Ghent & Teuber, 1963; Weinstein, 1964) provides a more extended and formal approach to its assessment. It is similar in structure to Benton's test of Right-Left Orientation in that the subject is required to touch own body parts named by the examiner, name own body parts touched by the examiner, touch parts which the examiner names on the examiner's body, imitate touches made by the examiner, and make touches indicated by schematic diagrams. While there is clearly a tactual component in sections of this task, the later stages all involve visual perceptual components, although problems of neglect, apraxic difficulties and body schema disturbances may all contribute to poor performance on the test.

Right-left orientation has already been indicated as included among Benton's tests, but another formal scheme for its assessment is to be found in Heilman and Valenstein (1993). This scheme also allows for the particular contribution of visual perceptual components to be assessed in that elements of the task are performed without visual guidance. Spreen and Strauss (1991) provide two forms of assessment for right-left orientation, Benton's form, and also a form devised by Culver (1969) in which hands and feet are presented in various orientations and postures to be labelled as "left" or "right".

There are few formal tests for the assessment of extrapersonal orientation. Money (1976) has published the *Standardised Road Map Test of Direction Sense* (a computer-based version is described by Beaumont & French, 1987). In this test the subject watches as a route is traced around a schematic street plan. The route is then revisited and at each intersection the subject must decide whether the route turns to the left or to the right with respect to the current direction of travel. An advantage of the test is that it is possible to distinguish between extrapersonal orientation (turns of the route) and personal orientation (the turns considered by reference to the subject's own body). The test was, however, developed in a developmental context, and most of the available norms are for children.

Visual Object and Space Perception Battery (VOSP)

The VOSP was first published in 1991 by Warrington and James, and has the object of providing an assessment of the relevant areas of performance through tasks which not only require simple responses, but were also designed to focus on one component of visual perception while minimising the contribution of other cognitive skills. In this, the test battery is relatively successful.

After an initial screening test which assesses visual sensory efficiency and allows subjects for whom the test would be inappropriate to be excluded, there are four tests which examine visual object perception, and a further four which address space perception. The visual object subtests are Incomplete Letters, Silhouettes, Object Decision and Progressive Silhouettes.

The subtests which assess visual space perception are, first, Dot Counting in which a number of dots arranged on a white card are to be counted, providing a test of simple spatial scanning in addition to being a screening tests for single point localisation. Second, Position Discrimination assesses the ability of the subject to perceive the relative position of objects in two-dimensional space. The task is, when presented with two outline squares one of which contains a dot in the centre and one of which has the dot slightly off-centre, to identify which square contains the dot in the centre. In the third subtest, Number Location, there are again two outline squares, in one of which are placed a number of randomly placed digits. A single dot appears in the second square, and the subject is required to report the digit which is in the location which corresponds to the location of the dot. This task is therefore a more demanding task of spatial location. The final subtest is Cube Analysis. The stimuli are graphic representations of a number of cubes in a three-dimensional arrangement and the subject has to indicate how many cubes are represented; a task which has a long history in the psychology of individual differences. The task is inferred to require spatial analysis of the two-dimensional representation of three-dimensional space, and is clearly the most complex in the battery.

Standardisation data are presented for 350 normal subjects aged between 20 and 69, as well as validation data in a neuropsychological context for 256 patients with unilateral cerebral lesions. The battery is proving valuable in clinical practice, partly because it is quick to administer and is tolerated well by patients; certain subtests are designed in a multiple choice format, and are therefore available for assessing clients with communication difficulties or relatively severe cognitive impairment.

Behavioural Inattention Test (BIT)

Another clinical test, the Behavioural Inattention Test (Wilson, Cockburn & Halligan, 1987) provides an equally attractive assessment for various aspects of unilateral spatial neglect. The test opens with six "conventional" subtests of neglect, which represent slightly formalised versions of many of the procedures employed in clinical practice by neuropsychologists, such as line cancellation, line bisection, figure copying and drawing. What follow are nine novel "behavioural" subtests which test practical aspects of neglect in an ecologically valid context. These subtests include picture scanning, telephone dialling, menu reading, address and sentence copying, and map navigation. Formal materials, procedure and scoring criteria are provided for each subtest, and the tasks are generally acceptable to patients, nor too demanding for the neurologically impaired.

Normative data were obtained from 50 control subjects and from 80 brain damaged subjects with unilateral cerebrovascular accidents, and from these data cut-off scores for "acceptable" performance are provided. The manual also provides estimates of reliability and validity, both of which are high.

Rivermead Perceptual Assessment Battery (RPAB)

Whiting, Lincoln, Bhavnani and Cockburn published the RPAB in 1985, and it provides a structured assessment of a range of perceptual functions. Although the battery as a whole only provides an indication of the presence of "perceptual deficit", several of the subtests can be employed to assess visual-spatial aspects of perceptual function. These include: Sequencing-Pictures, in which cards have to be arranged in terms of their appropriate position in a sequence; Body Image, two tests closely similar to the Object Assembly subtest of the WAIS-R; Right/Left Copying-Shapes-Words, which assesses neglect to the left or right of the midline; 3D Copying, a simple constructional task (see below); Cube Copying, a similar task employing a two-dimensional representation as the stimulus; and Cancellation, a conventional neglect task.

Although there is a good amount of technical data available in the test manual, and information from a number of clinical criterion groups, the data on normals are limited to a study involving only 69 subjects. However, the test does have the advantage that it is possible to account for level of intelligence (as assessed by the Mill Hill Vocabulary Test). This is a useful battery, but it is relatively weak on aspects of perception which are specifically visual-spatial, and contains tasks which are generally available elsewhere in alternative formats.

Trail Making Test

The Trail Making Test (see Spreen & Strauss, 1991) also has a long history in neuropsychology, being developed in 1944 and still used clinically in a variety of contexts. It requires (in Part A) the connection in sequence, by drawing, of 25 numbers in circles arranged randomly on a page, and then (Part B) the connection of 25 numbers and letters in alternating sequence. The task is timed. Although the test involves not only visual-spatial functions, but also attention, mental flexibility and motor functions, it may be useful as a test of visual-spatial search and spatial organisation.

Versions are available for both children and adults, and the test is quick and easy to administer. There is a variety of normative data, including large groups of normal subjects, which are provided by Spreen and Strauss (1991).

Standardised Analysis of Visual-Spatial Perception (VS)

A recent and promising development has been the creation of the VS test by Kerkhoff and Marquardt (1996). This is a computer-based battery which as a consequence permits much greater control over and manipulation of relevant task parameters. Normative data are contained within the package and the results can be presented in a variety of graphical formats.

The clinical tests contained within VS are: subjective orientation of axes; judgment of line orientation; judgment of length and distance; line and interval bisection; matching of rectangular forms; and copying of point localisation. A large number of the stimulus and task variables are open to modification, and subsets of the tests to form screening examinations are available. The normative data were collected from 36 (screening tests) and 40 normal control subjects, and the reliability has been established in a clinical group of 20 brain-damaged patients. The greater attention paid to psychophysical aspects of the assessment situation, and the improved control over stimulus and task variables, make this an attractive new test battery.

Visual construction

Constructional functions have regularly featured in perceptual assessments throughout the history of neuropsychology, and are most sensibly considered in terms of: two-dimensional construction; three-dimensional construction; and drawing tasks.

Two dimensional construction

Tests of two-dimensional construction have generally been based upon Kohs Blocks, and the best known and most widely used of the resulting versions is the Block Design subtest of the WAIS-R (see above). However, it is worth noting that the original version of Koh's Blocks includes items of greater difficulty than those contained in the WAIS-R, which may be useful in certain contexts (Arthur, 1947).

The one alternative to tasks involving block arrangement is the Stick Construction Task (Butters & Barton, 1970). This task requires both normal and rotated copying and is quick and easy to administer.

Three dimensional construction

The most commonly employed test is Benton's Three-Dimensional Block Construction Test, which is one of the elements of the battery of Benton's Neuropsychological Tests discussed above. There are, helpfully, two equivalent forms to the test, with three items in each: a model pyramid of 6 blocks; an 8-block, four-level construction; and a 15-block four-level construction. The subject must copy the model presented with the materials provided, which are laid out in a standard format before the subject. The models are available, ready prepared, in the published form of the test.

Various features of defective performance may be scored, and norms are available for adults, children and a variety of clinical groups.

The only other similar test appears to be the two relevant subtests (Tower; Level II; Bridge: Level III) of the 1960 Revision of the Stanford–Binet, which are similar in nature, but much easier; it would require grossly abnormal performance for these tests to be failed by an adult.

Drawing and copying

Tests of drawing and copying are normally less well formalised within neuropsychological assessment, although a number of scoring systems have at various times been proposed. However, it has to be said that analysis of graphic constructions is a difficult business, and adequate reliability is rarely attained in the use of these procedures.

Drawing tasks frequently require the subject to draw a person, a house, a tree, or a bicycle (there is a formal Draw-a-Person, and a House-Tree-Person, Test). Lezak (1995) gives a useful scoring protocol and norms for drawings of a bicycle, and similar procedures for Clock Drawing are to be found in Spreen and Strauss (1991). The Behavioural Inattention Test, BIT (see above) also incorporates formal procedures for drawing of simple objects.

Copying tests are available in more standardised forms, the most widely known being the Bender Gestalt Test (which has applications in personality as well as in cognitive psychology and neuropsychology). There are various procedures which relate to the original Bender Gestalt of which Hutt's (1977) revision is probably the one to be recommended. Lezak (1995), however, gives data which relates to Hain's (1964) procedure which may be more useful in a neuropsychological context.

The Bender Gestalt figures have also been carried over into other tests: the Background Interference Procedure developed by Canter (see Heaton, Baade & Johnson, 1978) involves making both normal copies and then further copies against a background of bold curved lines; the Minnesota Percepto-Diagnostic Test (Fuller & Laird, 1963) makes use of certain Bender Gestalt figures, as does the Benton Visual Retention Test: Copy Administration (Benton, 1974).

The one popular alternative to the Bender Gestalt Test is the Complex Figure Test: Copy Administration (also known as the Rey Figure, or the Rey–Osterreith Figure; Osterreith, 1944) and a parallel version was published by Taylor in 1979. Its scoring procedure is more reliable, and norms are available for both normals and clinical groups, making it probably the best of the copying tests.

THEORETICAL ASPECTS

Although the preceding sections present a rather indigestible catalogue of relevant tests, it is hoped that the information may be of value to researchers seeking to identify procedures which they might employ, and that the brief comments given may assist them in selecting among the available tests. The list given is not, of course, exhaustive.

While space precludes an adequate discussion of the theoretical underpinnings derived from the cognitive psychology and neuropsychology of visual-spatial performance, a number of critical current issues will be briefly addressed below.

Cognitive neuropsychology

Cognitive neuropsychological approaches have dominated the development of neuropsychology in recent years, and have therefore come to have an impact on formal approaches to the assessment of visual-spatial function as of other areas of function. Cognitive neuropsychological approaches depend on the application of explicit, normally information processing but sometimes connectionist, cognitive models to the analysis of both normal and abnormal cognitive function. The model assists in the analysis and decomposition of functional elements of cognitive performance which may then be assessed as operating in a normal or abnormal fashion.

Although the cognitive neuropsychological approach is not inherently incompatible with formal psychometric approaches to the assessment of functional competence, it is in practice rare for formal psychometric procedures to map on to the specific functional elements of a cognitive neuropsychological model, and in consequence formal tests rarely feature in this form of analysis. Because the approach is generally based around single case studies, and an individual-based approach, there is little stimulus for the development of more formal psychometric procedures related to this analytic approach. Indeed, while cognitive neuropsychology has both been greatly enriched by evidence from clinical neuropsychology research and been a major contributor to neuropsychological knowledge, its impact on the routine practices adopted by clinical neuropsychologists has been less than might have been anticipated. Cognitive neuropsychological investigations are generally too lengthy and expensive to be feasible within a routine clinical context, although it has to be recognised that the evidence such an investigation could, in principle, provide would be of considerable value in planning the rehabilitation and management of clients.

Cognitive neuropsychological approaches have generally been applied within a limited number of functional areas where well-articulated models of performance can be derived from cognitive psychology: single word recognition, reading, spelling and arithmetical functions. Some work has also been undertaken on certain perceptual functions including basic visual processes, object recognition and (especially) face recognition, but there appears to be very little work to date on visual-spatial functions.

Visual-spatial neglect

Visual-spatial neglect is currently one of the most active areas of neuropsychological research. Unilateral neglect is characterised by an unawareness and subsequent failure to orient or respond to events in the contralesional hemispace, which cannot be explained by primary sensory or motor deficits. A fundamental issue, which is still hotly debated, is whether this form of neglect is an attentional or a representational disorder; whatever the nature of the disorder, it is a dysfunction of visual-spatial performance. Excellent reviews are to be found in Robertson and Marshall (1993) and McCarthy (1996).

Representational accounts of neglect depend on such evidence as the findings of Bisiach and Luzzatti (1978). Their patients were asked to describe the scene of the Piazza del Duomo in Milan, a place thoroughly familiar to their subjects, as viewed from one end of the piazza. As these patients had unilateral neglect, their account omitted significant features on the imagined left side of the scene. However, when asked to describe the scene as if they were standing at the opposite end of the piazza, they again omitted landmarks on the imagined left side of the scene, reporting features which had been omitted on the first occasion, but omitted features previously reported. As this demonstration is independent of any sensory input, it is argued to be powerful evidence in favour of a defective ability to generate internal representations of the external visual world. Subsequent research, including more formal laboratory experiments, has supported such a conclusion.

Attentional theories have, however, had more prominence in recent discussions. Most develop from Heilman's hypothesis (see Heilman & Watson, 1977) that the right hemisphere attends to both lateral halves of space, while the left hemisphere attends only to the right. This principle operates in conjunction with a unilateral decrease in arousal affecting orienting responses to the contralateral half of space. Subsequent developments by Posner (Posner & Dehane, 1994) in relation to his general theory of attentional processes, and Mesulam (1985) have elaborated on this general conception. Mesulam specifically proposes that three areas of the cortex (inferior parietal lobe, frontal eye field, cingulate gyrus) are involved in a system which is responsible for the distribution of attention upward from reticular structures. There is, similarly, empirical support for these hypotheses.

A further aspect of this debate has been the demonstration that patients may have some covert semantic awareness of neglected aspects of the stimulus array. For example, in Marshall and Halligan's classic study (1988), a patient with left sided neglect who reported two drawings of a house as identical despite the presence of bright red flames issuing from the left side of one of the houses, nevertheless consistently selected the house without the flames as the house in which she would choose to live. Similar demonstrations have followed employing other kinds of stimuli and a rather more formal paradigm.

A detailed discussion of the neglect phenomenon is inappropriate here. The point that should be made, however, is that the assessment of visual-spatial function is more complex than one might be tempted to suppose. Behind the objective performance (and the stimuli and task demands of the procedure) lie a variety of more fundamental psychological processes which are critical to successful performance. Internal representations are a psychological phenomenon only indirectly determined by the external physical work, and attentional factors also play a part in the perception of the material presented in the course of the test. As studies in other areas of visual function have elegantly shown (cf. blindsight; Cowey & Stoerig, 1991), explicit report of conscious experience, or simple behavioural responses to cognitive tasks, is not necessarily a valid indicator of the degree of semantic covert access which may be available in relation to visual-spatial material.

Internal frames of reference

The importance of internal spatial frames of reference, independent from objective person-orientated and objective external space, has already been alluded to. However, its significance has been emphasised by recent lines of research.

The first centres around a phenomenon first described by Gazzaniga and Ladavas (1987) which relates to lateral differences in relation to the two visual half fields. It has now long been established that performance asymmetries may be demonstrated contingent upon a variety of stimuli lateralised in their presentation to two visual half fields. This procedure, the divided visual field paradigm, has been associated with a quite enormous literature, although the fashion for visual half field studies has now largely passed. Tachistoscopic presentation is normally employed in these studies, although so-called "free field" effects may also be observed, although these effects are somewhat weaker. The standard, core effect is that verbal material will be responded to faster, or more accurately, when presented to the right of fixation, than if it is presented to the left. Although a variety of factors play a part in generating this effect, including attentional variables and response compatibility, it is accepted that structural variables (the more direct projection of the right visual field to the left cerebral hemisphere which is the location of language processing) are the most important.

The divided visual field paradigm has therefore always been regarded as depending upon isolation of the relevant stimuli to the lateral visual half fields. The control condition, of presenting stimuli across the midline above and below the point of fixation, does not result in performance differences similar to those observed with lateralised presentation.

Gazzaniga and Ladavas (1987) undertook a most elegant manipulation. They administered the standard divided visual field paradigm (although in this case used very simple stimuli: the appearance of an x within a stimulus box; and the experiment involves clinical groups), and observed the expected asymmetry. They then required subjects to tilt their heads 90 degrees to either the right or the left, and repeat the task. The expectation was that the performance asymmetry would disappear as, although the stimuli still appear on the left or right of the projection screen, the stimuli now fall above or below the fixation point across the midline with respect to retinal projection. The remarkable finding was that the performance asymmetry was not entirely abolished, and this has been supported in later studies.. Gazzaniga and Ladavas discuss their results in relation to previous hypotheses about the dissociation of retinotopic and gravitational frames of reference. The important conclusions are that lateralised visual field effects may be, at least partly, determined by gravitational frames of reference (so further diminishing the role of structural factors), and that there is a clear dissociation between these two forms of reference. This is taken into account in no current psychometric assessments of visual-spatial function, and this must be a deficiency in terms of our current models of visual-spatial orientation.

A recent paper by Pellegrino, Frassinetti and Basso (1995) has contributed to this theme, but within the context of visual-spatial neglect. As they point out, in conventional studies of neglect, viewer-centred and object-centred frames of reference are confounded. It had already been observed that neglect patients might demonstrate neglect with respect to objects or to the whole visual array: presented with two flowers side-by-side to be copied, some neglect patients would omit to copy the whole flower to the left, while other patients would neglect to copy the left half of both flowers. Pellegrino et al. extended this distinction, introducing a manipulation essentially similar to that employed by Gazzaniga and Ladavas· (1987). They employed chimeric figures, in which the stimulus is composed of the left and right halves of two different complete stimuli, joined at the vertical midline to form a chimeric figure. They required their patient, who had severe left visual field neglect, to view the stimuli in the normal orientation, or when the stimulus display was rotated 90 degrees to the left or right. Their patient continued to neglect the left half of the stimulus even when it was rotated so that the neglected half fell in the upper or lower visual fields. The neglect was demonstrated to be related to object-centred space rather than to egocentric space.

A related observation, although the link may not be immediately apparent, was published by Turnbull, Laws and McCarthy in 1995. They were interested in the ability to recognise objects independently of observer viewpoint. Previous explanations had distinguished between orientation-dependent accounts which take account of the viewer's position, and orientation-independent accounts which rely on an analysis of the component elements of the object with reference to the object's principal axis of elongation. Turnbull et al. studied a patient in whom the knowledge of the normal upright position of objects was disrupted. This resulted in the patient making drawings and copies which were rotated with respect to the normal upright, and committing errors in an orientation-matching task. This deficit was independent of any problem with object recognition, and establishes a dissociation between object knowledge and object orientation. This observation has important implications for cognitive theories of object recognition and for the description of neuropsychological processes involved in this ability, but it is also relevant to the assessment of visual-spatial function in that it again demonstrates the importance of internal frames of reference, and the independence of orientational information from other aspects of object evaluation and description. Psychometric approaches to the evaluation of visual-spatial function have yet to take account of such factors.

Contemporary theoretical positions

As these brief notes will have suggested, psychometric approaches to the assessment of visual-spatial abilities have failed to take account of recent developments in our general understanding of the cognitive bases of these abilities. A discussion of such recent developments is outwith the remit of this chapter, but some pointers to the relevant literature may be useful.

Apart from the material to be found within other contributions to this volume, and the general neuropsychology texts identified above, the most comprehensive account specifically devoted to disorders of space exploration and cognition has been provided by De Renzi (1982). This volume is not exclusively concerned with visual functions, but vision being such a salient modality for humans, the greater part of the book addresses visual-spatial functions. De Renzi addresses not only spatial perception, but also emphasises the importance of space exploration (which includes attention;

visual-spatial function being an active exploratory and adaptive process) and what he terms, rather neatly, spatial thought.

A rather more general discussion of the "brain bases" of spatial cognition, and their development, is to be found in the book edited by Stiles-Davis, Kritchevsky and Bellugi (1988). This is a more theoretically oriented volume, although a great deal of empirical evidence is produced in discussing the various theoretical aspects and it provides a useful source for contemporary accounts of spatial thought including, an aspect not included in this chapter, mental imagery.

However, the most stimulating book relevant to these issues is the book edited by Eilan, McCarthy and Brewer (1993) which is both an exciting and a challenging read. It brings together contributions by both psychologists and philosophers who are interested in spatial representation, and not surprisingly explores the theme of how our understanding of the physical world interacts with the representations which we construct of that external world. Philosophical and psychological approaches enlighten each other. Here are to be found analyses of the role of frames of reference, and of spatial representations in the respective modalities, together with their relationship to action and the central interaction of "what" and "where" information within the human cognitive system. It can be strongly recommended to those interested in psychological aspects of visual-spatial performance.

CONCLUSION

This chapter has attempted to describe, in rather breathless detail, the most commonly employed instruments for the psychometric assessment of visual-spatial function. The context has been necessarily rather neuropsychological, for this has been the main arena for the development and application of these instruments. The term "psychometric" has also been interpreted rather loosely, for an embarrassingly small proportion of these instruments can withstand careful scrutiny of their formal psychometric merits. Nonetheless, many of the procedures described have been demonstrated to be of practical clinical utility and are in regular effective use for the analysis of human visual-spatial performance.

At the same time, an attempt has been made to set these assessment instruments in the general context of the evaluation of individual differences within psychology, and to address the problem of the isolation of purely visual-spatial abilities from related areas of psychological function.

Finally, a hint has been given of the very rich vein of contemporary research which is enhancing our understanding of visual-spatial processing, yet which is not reflected (yet) in the available psychometric assessment procedures. Formal tests always lag well behind conceptual development, and this is an enduring problem for the applied psychologist. The construction of good psychometric tests is an expensive and laborious business (and not always potentially rewarding), and applied psychologists must frequently search outside the formal published tests for procedures which will enable them to make assessments which reflect contemporary theoretical knowledge.

It is also the case that, the influence of cognitive psychology being so strong, that visual-spatial aspects of psychological function have been given less attention than other, more symbolic, areas of cognitive processing (as well as, it should be recognised, more basic visual processes). It is hard to imagine that this imbalance will not be redressed in the coming years, and a corresponding expansion of procedures for the assessment of visual-spatial function should be the welcome result.

REFERENCES

Arthur, G.A. (1947). *A Point Scale of Performance Tests: Revised Form II*. New York: Psychological Corporation.

Ayres, A.J. (1966). *Southern California Figure-Ground Visual Perception Test: Manual*. Los Angeles, CA: Western Psychological Services.

Beaumont, J.G., & Davidoff, J.B. (1992). Assessment of visuo-perceptive dysfunction. In J.R. Crawford, D.A. Parker, & W.W. McKinlay (Eds.), *A handbook of neuropsychological assessment*. Hove, UK: Lawrence Erlbaum Associates.

Beaumont, J.G., & French, C.C. (1987) A clinical field study of eight automated psychometric procedures: The Leicester/DHSS project. *International Journal of Man-Machine Studies, 26*, 661–682.

Beaumont, J.G., Kenealy, P.M., & Rogers, M.J.C. (Eds.) (1996). *The Blackwell dictionary of neuropsychology*. Oxford: Blackwell.

Benton, A.L. (1974). *The Revised Visual Retention Test*. (4th ed.). New York: Psychological Corporation.

Benton, A.L., Hamsher, K. de S., Varney, N.R., & Spreen, O. (1983). *Contributions to neuropsychological assessment*. New York: Oxford University Press.

Bisiach, E., & Luzzatti, C. (1978). Unilateral neglect of representational space. *Cortex, 14*, 129–133.

Butters, N., & Barton, M. (1970). Effect of parietal lobe damage on the performance of reversible opeartions in space. *Neuropsychologia, 8*, 205–214.

Cowey, A., & Stoerig, P. (1991). The neurobiology of blindsight. *Trends in Neuroscience, 14*, 140–145.

Culver, C.M. (1969). Test of right-left discrimination. *Perceptual and Motor Skills, 29*, 863–867.

De Renzi, E. (1982). *Disorders of space exploration and cognition*. Chichester, UK: Wiley.

Eilan, N., McCarthy, R., & Brewer, B. (Eds.) (1993). *Spatial representation: Problems in philosophy and psychology*. Oxford: Blackwell.

Ekstrom, R.B., French, J.W., Harman, H.H., & Dermen, D. (1976). *Manual for kit of factor-referenced cognitive tests*. Princeton, NJ: Educational Testing Service.

Elithorn, A., Jones, D., Kerr, M., & Lee, D. (1964). The effects of the variation of two physical parameters on empirical difficulty in a perceptual maze test. *British Journal of Psychology, 55*, 31–37.

Foreman, N.P., & Hemmings, R. (1987). The Gollin incomplete figures test: A flexible, computerised version. *Perception, 16*, 543–548.

Fuller, G.B., & Laird, J.T. (1963). The Minnesota Percepto-Diagnostic Test. *Journal of Clinical Psychology*, (Suppl. 16).

Gardner, H. (1983). *Frames of mind: The theory of multiple intelligences*. London: Heinemann.

Gazzaniga, M.S., & Ladavas, E. (1987). Disturbances in spatial attention following lesion or disconnection of the right parietal lobe. In M. Jeannerod (Ed.), *Neurophysiological and neuropsychological aspects of spatial neglect*. Amsterdam: Elsevier.

Gollin, E.S. (1960). Developmental studies of visual recognition of incomplete objects. *Perceptual and Motor Skills, 11*, 289–298.

Hain, J.D. (1964). The Bender Gestalt Test: A scoring method for identifying brain damage. *Journal of Consulting Psychology, 28*, 34–40.

Heaton, R.K., Baade, L.E., & Johnson, K.L. (1978). Neuropsychological test results associated with psychiatric disorders in adults. *Psychological Bulletin, 85*, 141–162.

Heilman, K.M., & Valenstein, E. (Eds.) (1993). *Clinical neuropsychology* (3rd ed.). New York: Oxford University Press.

Heilman, K.M., & Watson, R.T. (1977). The neglect syndrome—a unilateral defect of the orienting response. In S. Harnad, R.W. Doty, L. Goldstein, J. Jaynes, & G. Krauthamer (Eds.). *Lateralization in the nervous system*. New York: Academic Press.

Hooper, H.E. (1958). *The Hooper Visual Organization Test: Manual*. Los Angeles, CA: Western Psychological Services.

Hutt, M.L. (1977). *The Hutt Adaptation of the Bender Gestalt Test* (3rd ed.). New York: Grune & Stratton.

Kerkhoff, G., & Marquardt, C. (1996). *Standardized analysis of visual-spatial perception after brain damage*. Unpublished manuscript, Clinical Neuropsychology Research Group, EKN, Munich, Germany.

Kirk, S.A., McCarthy, J.J., & Kirk, W.D. (1968). *Illinois Test of Psycholinguistic Abilities: Examiners Manual* (rev. ed.). Urbana, IL: University of Illinois Press.

Lewis, R., & Kupke, T. (1977). *The Lafayette Clinic repeatable neuropsychological test battery: Its development and research applications*. Paper presented at the Southeastern Psychological Association, Hollywood, FL.

Lezak, M.D. (1995). *Neuropsychological assessment* (3rd ed.). New York: Oxford University Press.

Likert, R., & Quasha, W.H. (1970). *The Revised Minnesota Paper Form Board Test: Manual*. New York: Psychological Corporation.

McCarthy, M.M. (1996). Neglect. In J.G. Beaumont, P.M. Kenealy, & M.J.C. Rogers (Eds.), *Blackwell dictionary of neuropsychology*. Oxford: Blackwell.

Marshall, J.C., & Halligan, P.W. (1988). Blindsight and insight into visuospatial neglect. *Nature, 336*, 766–767.

Mesulam, M.M. (1985). Attention, confusional states, and neglect. In M.M. Mesulam (Ed.), *Principles of behavioral neurology*. Philadelphia, PA: F.A. Davies.

Money, J. (1976). *A Standardized Road Map Test of Direction Sense: Manual*. San Rafael, CA: Academic Therapy.

Mooney, C.M., & Ferguson, G.A. (1951). A new closure test. *Canadian Journal of Psychology, 5*, 129–133.

Osterreith, P.A. (1944). Le test de copie d'une figure complexe. *Archives de Psychologie, 30*, 206–356.

Pellegrino, G. di, Frassinetti, F., & Basso, G. (1995). Coordinate frames for naming misoriented chimerics: A case study of visuo-spatial neglect. *Cortex, 31*, 767–777.

Posner, M.I., & Dehane, S. (1994). Attentional networks. *Trends in Neuroscience, 17*, 75–79.

Robertson, I.H., & Marshall, J.C. (1993). *Unilateral neglect: Clinical and experimental studies.* Hove, UK: Lawrence Erlbaum Associates

Semmes, J., Weinstein, S., Ghent, L., & Teuber, H.-L. (1963). Correlates of impaired orientation in personal and extrapersonal space. *Brain, 86,* 747–772.

Spreen, O., & Strauss, E. (1991). *A compendium of neuropsychological tests: Administration, norms and commentary.* New York: Oxford University Press.

Stiles-Davis, J., Kritchevsky, M., & Bellugi, U. (Eds.) (1988). *Spatial cognition: Brain bases and development.* Hillsdale, NJ: Lawrence Erlbaum Associates.

Street, R.F. (1931). *A gestalt completion test.* Contributions to Education, N. 481. New York: Bureau of Publications, Teachers College, Columbia University.

Talland, G.A. (1965). *Deranged memory.* New York: Academic Press.

Taylor, L.B. (1979). Psychological assessment of neurosurgical patients. In T. Rasmussen & R. Marino (Eds.), *Functional neurosurgery.* New York: Raven.

Thurstone, L.L., & Jeffrey, T.E. (1982). *Closure flexibility (Concealed figures).* Park Ridge, IL: London House Press.

Turnbull, O.H., Laws, K.R., & McCarthy, R.A. (1995). Object recognition without knowledge of object orientation. *Cortex, 31,* 387–395.

Warrington, E.K., & James, M. (1991). *The Visual Object and Space Perception Battery.* Bury St Edmunds: Thames Valley Test Company.

Warrington, E.K., & Rabin, P. (1970). Perceptual matching in patients with cerebral lesions. *Neuropsychologia, 8,* 475–487.

Wasserstein, J. (1986). *Differentiation of perceptual closure: Implications for right hemisphere functions.* Unpublished doctoral dissertation. New York: City University of New York.

Wechsler, D. (1981) *WAIS-R: Wechsler Adult Intelligence Scale–Revised: Manual.* San Antonio, TX: The Psychological Corporation.

Weinstein, S. (1964). Deficits concomitant with aphasia or lesions of either cerebral hemisphere. *Cortex, 1,* 151–169.

Whiting, S., Lincoln, N., Bhavnani, G., & Cockburn, J. (1985). *RPAB: Rivermead Perceptual Assessment Battery: Manual.* Windsor: NFER-Nelson.

Wilson, B.A., Cockburn, J., & Halligan, P. (1987). *The Behavioural Inattention Test.* Bury St Edmunds: Thames Valley Test Company.

2
A Cognitive Neuropsychological Approach to Spatial Memory Deficits in Brain-damaged Patients

Martial Van der Linden and Thierry Meulemans
Neuropsychology Unit, University of Liège, Belgium

INTRODUCTION

Cognitive neuropsychology is a discipline which is based upon a two-way relationship between cognitive theory and neuropsychological data (see Seron, 1993). On the one hand, cognitive neuropsychology uses models of normal cognitive function in order to understand cognitive deficits in brain-damaged patients. On the other hand, data collected in brain-damaged subjects can be used to test, constrain, or develop models of normal cognitive processing. Over the last decade, cognitive neuropsychology has significantly contributed to the development of cognitive theory in several domains of cognition, such as reading (Coltheart, 1985), spoken language (Coltheart, Sartory, & Job, 1987), objects or face recognition (Humphreys & Riddoch, 1987; Young, Newcombe, de Haan, Small, & Hay, 1993) and memory (Van der Linden, 1994a). These contributions have mainly come through the demonstration of dissociations between individual patients (Shallice, 1988). More specifically, a multitude of inferences concerning the modular structure of the information-processing systems underlying particular tasks have been made by demonstrating that some patients may have selective problems with one task but not with another, while other patients show the opposite profile.

The contributions of cognitive neuropsychological research to the domain of spatial memory have been more limited. However, cognitive neuropsychology has produced some evidence of dissociations suggesting that spatial memory is underlain by a set of specialized mechanisms and also that several varieties of spatial memory can be distinguished. In particular, neuropsychological studies have shown that spatial memory deficits can occur selectively, that is to say, in the context of normal memory for nonspatial information. In addition, several neuropsychological studies have revealed that some abilities of spatial memory can be severely impaired while others are relatively spared (for reviews, see Schacter & Nadel, 1991; De Renzi, 1982).

The main purpose of the present chapter is to describe those neuropsychological studies and to point out some of the theoretical and methodological problems they are confronted with. Three domains of spatial memory deficits will be considered: (1) spatial working memory; (2) topographical memory; and (3) long-term memory for spatial locations.

SPATIAL WORKING MEMORY

Working memory refers to a limited capacity system which is responsible for the processing and temporary storage of information while cognitive tasks are performed. Baddeley's model represents the most extensively investigated theoretical account of working memory (Baddeley & Hitch, 1974; Baddeley, 1986). This model comprises a modality-free controlling central executive which is aided by a number of subsidiary slave systems ensuring temporary maintenance of information. Two such systems have been more deeply explored: the phonological loop and the visuospatial sketchpad. The phonological loop system is specialized in processing verbal material and is composed of two subsystems: a phonological store and an articulatory rehearsal process. The visuospatial sketchpad system is assumed to hold and to manipulate visuospatial images.

The core of the working memory model is the central executive. The central executive is assumed to be an attentional control system responsible for strategy selection and for the control and coordination of the various processes involved in short-term storage and more general processing tasks. An important characteristic of this system is that its resources are limited and divided into different processing and storage functions. Baddeley (1986) has suggested that the Supervisory Attentional System (SAS) component of the attentional control of action model proposed by Norman and Shallice (1986) might be an adequate approximation of the central executive system. According to Shallice (1988), a dysfunctioning of the SAS could plausibly account for the cognitive deficits following frontal-lobe lesions. More recently, Shallice (1994) also suggested that the function of SAS could be fractionated into several components.

Convincing evidence for the existence of the different components included in the working memory model comes from the study of brain-damaged patients with specific short-term memory impairments (for reviews, see Della Sala & Logie, 1993; Gathercole, 1994; Van der Linden, 1994a; Vallar & Papagno, 1995). Some patients with a deficit of auditory short-term memory but a normal visuospatial memory have been reported.

This deficit was attributed either to a selective impairment of the phonological store (e.g., Vallar & Baddeley, 1984) or to a disturbance of the articulatory rehearsal process (e.g., Belleville, Peretz, & Arguin, 1992). Other patients with a specific impairment of the visuospatial sketchpad but with a normal phonological loop system have also been described (De Renzi & Nichelli, 1975; Hanley, Pearson, & Young, 1990; Hanley, Young, & Pearson, 1991). Finally, a specific central executive deficit has also been documented (Van der Linden, Coyette, & Seron, 1992; Baddeley, Logie, Bressi, Della Sala, & Spinnler, 1986; Baddeley, Bressi, Della Sala, Logie, & Spinnler, 1991).

The deficits of the visuospatial sketchpad have received much less attention than the deficits of the phonological loop. De Renzi and Nichelli (1975) retrospectively described two cases of patients (selected among a group of 32 patients with right hemisphere damage) with a selective deficit of visuospatial span on the block tapping test but with a normal performance on digit span (see also De Renzi, Faglioni, & Previdi, 1977a). In the block tapping test, devised by Corsi (see Milner, 1971), nine cubes are arranged on a wooden board. The examiner taps some of the blocks in sequences of increasing length, and the patient is required to tap out exactly the same sequence immediately afterwards. In the digit span test, the patient has to reproduce (verbally or by a pointing procedure) sequences of digits. Neither of these patients showed topographical disorientation or deficit in long-term visuospatial memory. This observation of a visuospatial short-term memory deficit without a corresponding visuospatial long-term memory impairment suggests a parallel organization of short-term and long-term memory systems.

Several group studies suggested that Alzheimer patients showed a more severe impairment on the visuospatial block tapping test than on the auditory digit span (e.g., Cantone, Orsini, Grossi, & De Michele, 1978). However, this dissociation was not observed in other studies (e.g., Spinnler, Della Sala, Bandera, & Baddeley, 1988). Ergis, Van der Linden, Boller, Degos, and Deweer (1995) also examined visuospatial short-term memory in Alzheimer patients by means of the visuospatial pattern span test devised by Wilson (1993). Contrary to the block tapping test, the pattern test does not involve sequencing, but rather consists in memorizing spatial positions (filled cells in matrix patterns) which are presented simultaneously. This task also limits the possibilities of verbal encoding and it allows both visuospatial short-term and long-term memory to be assessed with the same material. Group analyses showed that Alzheimer patients were impaired in both visuospatial short-term and long-term memory. However, the examination of the individual profiles revealed the existence of dissociations among patients, between short-term and long-term visuospatial memory, and also between visuospatial and verbal short-term memory (assessed with the digit span test). Similar dissociations have also been observed in Alzheimer patients by Baddeley, Della Sala, and Spinnler (1991). Ergis et al. (1995) found significant correlations between the performance of Alzheimer patients on the visuospatial pattern task and a test of visual attention (the Gottschaldt's Hidden Figure Test; Gottschaldt, 1926, 1929) which suggest that the short-term memory deficit could be due to an impairment affecting visuospatial coding abilities. However, Trojano, Chiacchio, De Luca, and Grossi (1994) found that the defective performance observed in Alzheimer's disease on a pattern span test similar to that used by Ergis et al. (1995) was not the consequence of impaired basic visuospatial coding or motor output abilities.

Wilson, Wiedmann, Hadley, and Brooks (1989) compared the performance of patients with closed head injuries on the pattern span test, on a sequential visuospatial test similar to the block tapping test, and on the digit span test. They observed that patients were significantly impaired on the pattern but not on the sequential visuospatial or digit span tests. In addition, values of pattern span in both patients and control subjects were consistently above those observed on the sequential test (see also Wilson, 1993, for similar results obtained with normal subjects in various European centres). A deficit on a simultaneous pattern span test has also been found by Pigott and Milner (1994) in patients with right frontal lesions. The authors suggested that this deficit was due to a difficulty in actively encoding complex visual stimuli in a condition of temporal constraint.

In a detailed single case study, Hanley et al. (1990; 1991) described a right brain-damaged patient (E.L.D.), who showed a defective span on the block tapping test and on a short-term memory task for unfamiliar faces. She was also impaired on the Brooks (1967) matrix task (which requires recalling sentences of increasing length, describing the relative location of numbers in a 4 × 4 matrix). Furthermore, E.L.D.'s performance was defective on mental rotation and in using visual imagery mnemonics, which suggest that the spatial manipulation of mental images may require temporary storage in a visuospatial short-term memory system. Her visuospatial long-term memory for unfamiliar faces and objects (but also for unfamiliar voices) was poor. In addition, she also encountered problems in learning routes in unfamiliar areas, such as the way back to her new flat. By contrast, this patient was unimpaired in her ability to recall letter sequences with auditory and with visual presentation. The retrieval of visual imagery from long-term memory was also intact (e.g., indicating the characteristic colour of an object, or making a size comparison). It could be argued from these data that E.L.D. did not show a specific deficit of the visuospatial working memory system but rather suffered from a more general impairment of nonverbal memory (see Vallar & Papagno, 1995). However, according to Hanley et al. (1991), this pattern of results could also suggest that the visuospatial sketchpad plays a role in the long-term acquisition of new visuospatial information, similar to that played by the phonological loop in the acquisition of new vocabulary (see Gathercole & Baddeley, 1993). In another single case study, Farah, Hammond, Levine, and Calvanio (1988) described a patient (L.H.), who was unimpaired on tests of visuospatial sketchpad functioning (specifically the Brooks matrix task) and spatial manipulation of images (e.g., a mental rotation task), but was impaired in the retrieval of visual imagery from long-term memory. These data suggest that visual and spatial imagery represent distinct functional processing systems and therefore that a subdivision of the sketchpad between a visual and a spatial component is likely to be necessary.

More recently, Morris and Morton (1995) reported the case of a patient (M.G.), who showed a normal visuospatial short-term memory (assessed by the block tapping test and the Brooks matrix task) but significant impairments in mental rotation, detected by using a variety of tasks, as well as in visual long-term memory. These findings argue in favour of a dissociation between the visuospatial sketchpad component and the processes responsible for mental transformation.

Finally, in a group study, Warrington and Rabin (1971) reported that patients with posterior left hemisphere lesions showed a poor performance on a visual, rather than a visuospatial, test. In this test, the patients were visually presented sequences of letters, digits, or straight and curved lines for brief periods of 50msec to 160msec and were required to report as many of the presented items as possible. However, they obtained normal results when the letter sequences were approximations to words and on auditory digit span tests. By contrast, the results of patients with right hemisphere lesions were normal on the visual test. It should also be reminded that a defective span for visuospatial locations, as assessed by the block tapping test, has been observed in patients with right brain lesions (e.g., De Renzi & Nichelli, 1975). Ross (1980) also described two patients presenting a specific deficit of visual short-term memory. These patients were unable to reproduce or recognize drawings or patterns when a brief delay was interposed between presentation and test. However, verbal short-term memory and long-term visual memory were normal in both patients. Other patients were reported showing the converse pattern of deficits. These patients showing a severe verbal short-term memory deficit could retain visually presented verbal items more than auditorily presented items, contrary to normal subjects (e.g., the patient K.F. described by Shallice & Warrington, 1970, and Warrington & Shallice, 1972; the patient P.V. reported by Vallar & Baddeley, 1984, and Basso, Spinnler, Vallar, & Zanobio, 1982). This dissociation has been interpreted by suggesting that the patients could use a visual short-term memory system storing the visual form of verbal items while their phonologically based, verbal short-term memory system was defective. Finally, some authors have suggested the existence of selective short-term memory deficits affecting specifically colours (Davidoff & Ostergaard, 1984) or tactile information (Ross, 1980).

With regard to the physiological substrate of the visuospatial sketchpad, the only detailed single case (E.L.D.) who was described as having a specific visuospatial short-term memory deficit suffered from an extensive lesion in the frontotemporal region of the right hemisphere (Hanley et al., 1991). When group studies are considered, it appears that the most frequent lesion associated with a visuospatial short-term memory deficit involves the posterior parietal lobe near its junction with the occipital lobe (e.g., Warrington & James, 1967; De Renzi & Nichelli, 1975; De Renzi et al., 1977a). However, some other data also suggest that patients with frontal cortex lesions have a deficit of spatial working memory (Pigott & Milner, 1994). In a recent positron emission tomography (PET) activation study, Perani et al. (1993) observed, in Alzheimer patients, significant correlations between visuospatial span performance and metabolic activity in the right frontal and parietal regions. It should also be noted that the upper part of the left occipital lobe has been implicated in the deficits affecting short-term storage of visually presented verbal material (Kinsbourne & Warrington, 1962) and visual nonverbal material (Warrington & Rabin, 1971). To summarize, the posterior-inferior parietal and frontal association regions of the right hemisphere seem to be the anatomical correlates of the visuospatial short-term memory system. However, it also appears that the posterior region of the left hemisphere is implicated in short-term storage of visual verbal and nonverbal stimuli. Finally, a recent PET study (Smith & Jonides, 1995; Smith et al., 1995) showed that a spatial and a visual working memory task led to different patterns of activation: the spatial task led to activation in the right hemisphere occipital, parietal, and prefrontal regions, whereas the object task resulted in activation in left hemisphere parietal and prefrontal regions.

If the visuospatial short-term storage seems to be associated with different brain regions, the evidence concerning the physiological substrate of the phonological loop component is more consistent. Neuroimaging studies of patients with a selective deficit of the phonological loop suggest that this component is localized in the left hemisphere. More specifically, in a review of all patients for whom sufficient details were available, Della Sala and Logie (1993; see also Shallice & Vallar, 1990) found that the common denominator seems to be the inferior part of the parietal lobe, close to the junction with the superior, posterior temporal lobe. In the few cases where the lesion was described in more detail, it involved the supra-marginal gyrus of the inferior parietal lobe. The PET study carried out by Perani et al. (1993) on Alzheimer patients reported that verbal short-term memory was associated with the temporal, parietal, and frontal areas of the left hemisphere. Finally, a recent PET activation study (Paulesu, Frith, & Frackowiak, 1993) confirmed this finding, localizing the phonological store in the left supramarginal gyrus while the articulatory rehearsal process was localized in Broca's area (see also Salmon et al., 1996, for similar results). Finally, concerning the localization of the central executive, some authors have proposed that executive processes involve the frontal lobes (e.g., Shallice, 1988). In recent PET studies, Petrides and colleagues (Petrides, Alivisatos, Evans, & Meyer, 1993; Petrides, Alivisatos, Meyer, & Evans, 1993) and Salmon et al. (1996) observed an activation of the mid-dorsolateral frontal cortex (areas 46 and 9) during the execution of random generation and updating memory tasks which are considered to place significant demands on the central executive (see Baddeley, 1986; Van der Linden, Brédart, & Beerten, 1994). However, the frontal localization of the central executive functions has been questioned and some researchers consider that executive functioning relies on a network distributed between anterior and posterior areas of the brain (e.g., Fuster, 1993; Morris, 1994).

Although neuropsychological explorations of the visuospatial sketchpad have been far from extensive, they have provided very convincing evidence that this system operates independently of the phonological loop. However, the contributions of neuropsychological studies to the identification of the precise processes and mechanisms involved in this subsystem have been rather limited for both methodological and theoretical reasons. First, most studies which examined visuospatial sketchpad deficits used a group methodology. Many neuropsychologists have rejected this group approach arguing that patients do not have identical cerebral dysfunction and so cannot be grouped meaningfully in terms of their cognitive deficits (Caramazza, 1986). A good illustration of this problem is the exploration of the visuospatial sketchpad deficits in Alzheimer's disease. In fact, several studies recently demonstrated the extreme heterogeneity of cognitive deficits of Alzheimer patients in different domains, such as language, memory or even face processing (see Van der Linden, 1994b). These variations do not seem to be the consequence of random fluctuation since no examples of "nonsense syndromes" were found with respect to cognitive theory and modelling (e.g., Baddeley et al., 1991a,b; Della Sala, Muggia, Spinnler, & Zuffi, 1995). In that context, the nature of the visuospatial sketchpad impairment may vary considerably among Alzheimer patients. This does not mean that the examination of Alzheimer patients cannot help to model the functioning of the sketchpad. On the contrary, the fine-grained dissociations caused by the disease could be used to test or develop the available sketchpad model.

Another reason explaining why so few detailed studies have been devoted to the exploration of the visuospatial sketchpad is that for a long time the concept of visuospatial sketchpad has been underspecified.

Some evidence from studies of normal subjects (e.g., Baddeley & Lieberman, 1980) but also of brain-damaged patients (see Hanley et al., 1991; Farah et al., 1988) suggested that the system has both a visual and a spatial component. This distinction within the sketchpad has frequently been assimilated to the distinction between the "what" and "where" systems serving respectively in the identification of visual inputs and the location of the inputs (Ungerleider & Mishkin, 1982).

More recently, Logie (1995) proposed a model of the visuospatial system comprising a visual temporary store which is subject to decay and to interference from new incoming information and a spatial temporary store which can be used to plan movement and also to rehearse the contents of the visual store. Visual input reaches both stores via the activation of long-term memory representations of the visual form of stimuli or the spatial information about a dynamic scene. The sketchpad is linked to visual imagery in that it provides temporary storage of information that can be used by the central executive for a particular mental imagery task.

Some aspects of this model are loosely defined or seem inconsistent with recent data. Careful investigation of brain-damaged patients might certainly contribute to clarifying those problematic points. For example, it remains to determine the precise relationships between, on the one hand, the passive visual store and the active "spatial" store and, on the other hand, the independent "where" and "what" systems that process respectively visual (colour and shape) and spatial (locations) information. Another difficult problem concerns the nature of the processes involved in short-term memory for order information. In some spatial tasks, such as in the block tapping test, presentation of spatial information is sequential and serial recall is required, contrary to the pattern span test (Wilson, 1993) in which presentation is simultaneous and recall is nonserial. Whether the block tapping and the pattern span tests tap distinct abilities should be examined from a neuropsychological point of view. Evidence for such a distinction would be provided by a double dissociation of block tapping span and pattern span in brain-damaged patients.

It also appears that the separation of the sketchpad from the central executive system has been less clearly established than for the phonological loop system (Barton, Matthews, Farmer, & Belyavin, 1995). In that context, it should be noted that the central executive resources are involved in a substantial number of tasks used to assess the sketchpad, particularly in the Brooks matrix task (Salway & Logie, 1995). Simpler sketchpad tasks should be designed, with central executive demands reduced to a minimum (see Barton et al., 1995). Another question concerns the processes involved in the short-term storage of the visual form of verbal items: are these processes distinct from the sketchpad processes or are they similar to those implicated in visuospatial storage? Finally, the observation of a patient

(M.G.; Morris & Morton, 1995) showing a deficit in mental rotation and visuospatial long-term memory in the absence of a sketchpad deficit argues against a part of Logie's (1995) view. Indeed, according to Logie, the sketchpad is needed to store the image which is being manipulated by the central executive, so that a deficit of mental rotation cannot occur independently from a sketchpad deficit. Morris and Morton's data suggest the existence of complex dynamic relations between the sketchpad, the central executive, the visuospatial long-term memory and mental imagery processes. The exploration of other patients showing mental manipulation deficits may help to disentangle these relationships.

According to Young et al. (1993), there exist some limitations which could make the dissociations observed within a group or between single cases difficult to interpret. Further studies exploring sketchpad deficits should take into account these limitations. A first problem concerns the risk that a dissociation happened by chance. This problem is particularly evident when single case examples are drawn from a group study. Consequently, a sound statistical design is required to account for such a risk. Another problem concerns the dangers of equating tests and abilities. For example, in a number of studies, a deficit of the visuospatial sketchpad and of the phonological loop has been inferred from the performance on two tests, for example, the block tapping test and the digit span test. The danger of this approach is that a patient may fail a certain test for a number of reasons, apart from an impairment of the explored function. It is possible to avoid this problem by ensuring that the different tests are strictly comparable in terms of demands and response requirements. Another solution is to choose two very different tests of the same ability, and to conclude that a patient is impaired on a specific ability only if a poor performance is observed on both tests. A third problem is that it is sometimes difficult to compare directly two separate reports of patients because different tests have been used to assess the specific abilities. Finally, it could be that patients apply idiosyncratic strategies to compensate for their functional deficit, so that they can achieve normal levels of performance on accuracy scores. According to Young et al., inasmuch as these unusual strategies tend to be time-consuming, a way to solve this problem is to measure response latency as well as accuracy scores. This should also reveal whether patients will trade speed for accuracy in tasks which are sensitive to their deficit.

TOPOGRAPHICAL MEMORY

Over the last 100 years, neuropsychologists have described many cases of brain-damaged patients whose main deficit was the inability to remember or follow well-known itineraries, or learn new ones (see De Renzi, 1982).

These case studies have provided some important dissociations. One of the most interesting findings is that topographical disorders do not always co-occur with other types of visual problems. For example, Levine, Warach, and Farah (1985) reported the case of a patient (patient 2) who suffered from severe topographical deficits but who was not impaired with regard to recognizing faces and had an intact knowledge of objects and colours (see also the patient Mr Smith, described by Hanley & Davies, 1995). By contrast, Farah et al. (1988) and Levine et al. (1985) described a patient (L.H.), who was unable to recognize familiar faces and to remember the colour, shape and relative size of objects, but who could describe familiar routes, and travelled around his city without getting lost. These data suggest that specific processes are involved in topographical orientation.

Another important issue concerns the nature of the functional impairment that is responsible for the topographical disorders. Classically, topographical disorders have been classified into two main categories: topographical agnosia (Patterson & Zangwill, 1945; Pallis, 1955; Landis, Cummings, Benson, & Palmer, 1986) and topographical amnesia (De Renzi, Faglioni, & Villa, 1977b; Whiteley & Warrington, 1978; Habib & Sirigu, 1987; Bottini, Cappa, Geminiani, & Sterzi, 1990). Topographical agnosia has been described as a perceptual defect, in which patients are no longer able to identify the environmental landmarks (well-known places and buildings) or to appreciate how environmental stimuli are related to each other in space. Topographical amnesia has been defined as a deficit in memory for spatial relationships or landmarks.

Quite obviously, this classification is too crude in view of the various processes which underlie topographical recognition and exploration. Indeed, the ability to travel through an environment in a purposeful way involves the interaction of many cognitive components which could be specifically affected by a brain lesion (see Byrne, 1982; Riddoch & Humphreys, 1989). First, a subject who wants to move from one location to another should be able to recognize objects and to appreciate the spatial relations between them. In particular, he needs an integrated viewpoint-dependent representation of the world, that is to say a representation which preserves information about the positions of objects relative to the viewer. Numerous visual processes underlie these abilities, and impairments to these processes could particularly produce failure to recognize visual landmarks and routes and to interact with the environment appropriately (e.g., difficulties in reaching objects). In addition to these visual processes, stored memories for previously experienced landmarks and for routes ("spatial maps") are also necessary. Byrne (1982) distinguishes two types of stored maps that could apply to the same environment. A network map represents routes as a network of strings, and locations along them as nodes. Nodes identify a physical location but may also contain instructions for a change of direction. This representation is based on our route knowledge: it develops from our locomotor experiences and is bound by contextual aspects. Vector maps encode horizontal information about directions and distances. They are isomorphic with the real world as viewed from above. In that theoretical perspective, topographical orientation disorders could affect either network-map or vector-map representations: some patients would be unable to follow or describe familiar routes but could appreciate relative locations in large-scale space, whereas others would manifest disorders in direction and compass bearings but not impairments in their route-finding ability. In addition, the ability to update one's current position and bearings on a route and to plan future movements requires some form of spatial working memory as well as central executive resources. A disorder affecting the "updating position" and "maintaining orientation" skills would appear when a patient proves to be able to describe and draw routes that he cannot follow. Inversely, a patient with intact "updating position" and "maintaining orientation" abilities might be able to find his way with a printed map, even though he had lost all representations of familiar areas. Finally, another useful skill is the ability to maintain orientation and position with respect to a printed map.

The approach which attempts to link topographical disorders to a model of normal navigation has not really been addressed hitherto. Indeed, most studies which have described patients with topographical disorders were not founded upon a detailed theoretical model of the processes normally involved in the recognition and exploration of the environment. In two very stimulating papers, Byrne (1982) and Riddoch and Humphreys (1989) have tried to reclassify the patients described in the literature in terms of a disruption to particular processes in the normal model. These analyses are confronted with all the problems arising from retrospective studies but, nevertheless, they clearly revealed that topographical disorders could arise in a number of different ways and also that the patients can be classified on the basis of the normal framework.

In most studies on topographical disorders, the deficit has not been explored by means of formal tasks providing a quantitative measure, but it has only been clinically described. However, there exist two laboratory tasks which have been frequently used in the past for exploring topographical orientation in brain-damaged patients: the road-map test (Money, Alexander, & Walker, 1965) and the map-reading test (Semmes, Weinstein, Ghent, & Teuber, 1955). The road-map test consists of a schematic outline map of several city blocks in which a route has been drawn including 32 right or left turns, some when the route is going down (towards the bottom of the sheet) and some while it is going up (towards the top of the sheet). The patients are asked to imagine a walk along the route and they must indicate at each corner whether they would be going to the right or to the left, without turning the map. The map-reading test consists of nine coloured discs laid out in a 3 × 3 arrangement on the floor of a large room, and of a series of maps. The patients are required to walk from disc to disc following a route drawn on the map. The direction "north" is indicated on the wall facing the patients at the beginning of each trial and is also written on the map. The orientation of the map relative to the patients is kept constant but it changes relative to the room as the patients progress along the route. Consequently, in order to interpret correctly the directions indicated by the map, the patients must realize correct mental rotation of their body. Both tests have little to do with the task of remembering the spatial knowledge of a familiar route, but could be used to assess the ability to keep a correct orientation and position with respect to a printed map. However, the problem is that they can be realized by adopting various strategies, thus recruiting different abilities (De Renzi, 1982). For example, in the map-reading test, a patient may rely on visuospatial information, while another may verbally code the directional indications of the map. As a matter of fact, in order to investigate a patient with topographical disorders, the cognitive neuropsychologists will have to design tasks which specifically tap the different cognitive components involved in the acquisition of new routes as well as in the ability to remember or follow well-known itineraries. In fact, investigation of topographical deficits is particularly difficult because it has to be conducted, to a great extent, outside the laboratory.

As an illustration of this cognitive approach in the exploration of topographical orientation, we will describe the tasks we have devised to investigate network-map and vector-map representations in a right brain-damaged patient who showed very marked topographical disorders (Van der Linden & Seron, 1987). As noted by Byrne (1982), tasks designed to tap these mental representations should not require actual route-finding since this would presuppose an intact ability to encode current location and bearings.

Considering first the network representation evaluation, the patient was presented with the following task.

(1) *Verbal description of routes* (outside routes).
(2) *Direct questions about routes*: "If you go from point A to point B, do you get to X or Y first?" and "If you go from point A to point B, do you have to go through point C?"
(3) *Verbal presentation of false routes* (i.e., routes including errors in the succession of elements or irrelevant elements). In the first situation, the subject was asked questions such as: "If you go from this place to this place, will you have to go through X, yes or no? Will you have to go through Y, yes or no?" In the second situation, we gave the

subject a very detailed account of a route, including irrelevant elements, erroneously located components, and reversed left-right turns.
(4) *Presentation of falsified maps*, including erroneous locations of buildings, landmarks, etc., along a route.

For the vector-map representation, the following tests were administered:

(5) *Direction estimates*: the subject pointed to places, buildings, etc., in the real world or on slides. In the actual situation, the subject was inside the hospital facing the main exit in such a way that he could see the streets and the nearby landmarks. In the slide situation, the subject faced the screen and imagined that he was the photographer. He then tried to recognize the site and was asked to point to different places as quickly as possible so that he could not resort to a network-map representation.
(6) *Distance estimates*: in the comparison method, the subject answered as rapidly as possible questions such as: "As the crow flies, is it further to go from here to the University or to go from here to the shopping centre?" A direct estimation of short distances was the other method. We asked for estimates of the length in metres of several streets in the city of Liège and in the subject's own village.
(7) *Location of different sites on sketch maps* without route representations. For instance, we showed the subject a sketch map of Liège with a central point representing the Market Square; the subject then indicated as quickly as possible different locations. A new map was supplied for each site so that the subject could not rely on route knowledge. After this, the subject tried to locate Belgian cities on a blank Belgium map.
(8) *Presentation of falsified maps* including road distortions and orientation errors.
(9) *Finally*, the subject drew maps and plans as viewed from above, that is, isomorphic to reality.

The results showed that the tasks intended to evaluate network-map representation were well performed, whereas tasks aimed at vector-map evaluation gave poor results. Among the latter, only the distance estimation task (with a comparison method) gave good results, but travel-time estimate strategy may have been used by the subject. Although these results seem to support Byrne's distinction between vector-map representations and network-map representations, this analysis can only be regarded as tentative. Other subjects must be found fitting this pattern as well as the opposite one. The evidence of the opposite dissociation might even assume greater significance. Indeed, we could consider an ordered sequence of transition from network-map to vector-map and, by so doing, relate the observed disorder to the severity of the lesion. Some damage at the network-map level might be caused by a less severe lesion than the one disturbing the vector-map level (a more abstract level). As a matter of fact, several cases which have been described in the literature seem to show the same (Zangwill, 1951) or the opposite (Spalding & Zangwill, 1950; Whiteley & Warrington, 1978) dissociation as seen in our subject. Furthermore, several variables should be systematically explored and, specifically, patterns of error occurrence: Does the subject make errors at random or does he reorganize his representation on the wrong bases? Such an orientation might help us distinguish between an impairment of access to the representation or an impairment of representation itself.

From a more general point of view, the study of topographical impairments by means of theoretically oriented tasks should contribute to test, extend or even modify the theoretical apparatus built up with normal subjects. For example, it should be particularly interesting to approach the role of working memory in topographical orientation. In a recent paper, Hanley and Davies (1995) reported the case of a patient (Mr Smith) who showed severe topographical problems along with very poor performance on immediate tests of visuospatial processing, such as mental rotation, and on tests of visuospatial short-term memory, such as the block tapping test. According to the authors, it may be that his spatial short-term memory impairment makes it no longer possible for him to make use of an intact long-term spatial knowledge. Alternatively, it may be that the short-term and long-term spatial memory reflect distinct functional deficits. Further information concerning the role of the visuospatial sketchpad in topographical difficulties will be provided by other studies of patients such as Mr Smith. A reduction of the capacity of the central executive could also affect the process of maintaining one's current position on a route, as well as updating and planning future movements. Indeed, Böök and Gärling (1980) have shown that keeping track of one's current position within a full-size maze does require conscious attention and is disrupted by a concurrent task. Recently, Riddoch and Humpreys (1995) presented the cases of three patients who all experience, among other problems, difficulties in route finding which seem to be underlain by working memory (central executive) impairments.

We have already mentioned in the section, Spatial working memory, some of the precautions that a neuropsychologist must take before interpreting a dissociation observed in the performance of a brain-damaged patient. One of them is to determine whether patients adopt a strategy to compensate for their deficit. Recently, Clarke, Assal, and De Tribolet (1993) described a patient with severe topographical disorientation who used compensatory strategies such as relying on detail-by-detail analysis of buildings for recognition and on identification of landmarks and memorizing their sequences for finding routes. A patient (E.L.D.), described by Hanley et al. (1990) also learnt a set of verbal instructions which will enabled her to find her way.

The investigation of topographical orientation raises other relatively specific methodological problems. One of these problems is the great inconsistency of results obtained in normal subjects in distance estimation tests, angle width tests, location estimation, etc. (e.g., Giraudo & Pailhous, 1994). According to Giraudo and Pailhous, the fluctuation of responses appears to be a fundamental element of spatial memory. These fluctuations provide evidence of the intrinsic flexibility of the spatial system but also show that this flexibility is a base for the plasticity of the system. However that may be, this inconsistency of results will have to be considered in future experiments. More specifically, it suggests that spatial processes cannot be studied with only one production per subject.

Another problem in some topographical tasks is to distinguish spatial knowledge and representation from the enabling skills (Byrne, 1982) needed to demonstrate it in the experimental situation. For example, inability to draw a map may not be due to limited topographical knowledge but to constructional apraxia. In the same vein, the inability to follow a formerly familiar route may be attributed to unilateral neglect rather than to a topographical deficit. In this context, Morrow, Ratcliff, and Johnston (1985) were able to demonstrate deficits in spatial representations which are not confounded by enabling skills. They compared the performance of patients with right hemisphere lesions and control subjects on three spatial tasks: (1) imagining a map of the United States and verbally estimating distances between all possible pairs of nine cities; (2) locating these cities on an outline map of the United States; and (3) verbally estimating distances between all possible pairs of nine symbols which were placed on a page in the same arrangement as the nine cities. The patients performed significantly worse than the controls on the first task, and this deficit cannot be attributed to a failure to estimate verbally distance or to an ignorance of the city locations. At another level, the study of relations between hemineglect and topographical disorders seems particularly important (e.g., Tromp, Dinkla, & Mulder, 1995). Patients with unilateral neglect may experience problems in negotiating their environment because of their difficulty in initiating action to their neglected side of space or even in visualizing a side of space (Bisiach, Bulgarelli, Sterzi, & Vallar, 1983). The contribution of executive skills and planning deficits should also be explored (e.g., Boyd & Sautter, 1993).

LONG-TERM MEMORY FOR SPATIAL LOCATION

Long-term memory for spatial locations has been investigated extensively within human neuropsychology in the last fifteen years. For example, Smith and Milner (1981) found that patients with right, but not left, hippocampal excisions were impaired in their ability to remember locations of objects placed on a grid, whereas they later found that patients with frontal excisions performed this task normally (Smith & Milner, 1984). However, the patients whom Smith and Milner explored had a material-specific memory deficit and it was difficult to generalize their results to amnesic patients with a more severe and global impairment.

Global amnesia is classically defined as a deficit in the acquisition of new information (anterograde amnesia) accompanied by a deficit in recall and recognition of information acquired before the lesion was sustained (retrograde amnesia). The importance of the deficit may vary from one patient to another but, even in the most severe cases, intellectual abilities and short-term memory may be preserved. Hirst and Volpe (1984) examined the spatial memory performance of a group of global amnesic patients using the same tasks as Smith and Milner in both incidental and intentional spatial memory conditions. In the incidental spatial memory condition, subjects were shown the objects placed on a grid and told to remember them. Two minutes later, they were asked to recall the objects and then to place them on the grid in the positions they had previously occupied. In the intentional condition, with a new set of objects and locations, they were told to remember both objects and positions. The amnesics were severely impaired in spatial memory as well as in object recall but, unlike the controls, their spatial memory improved under the intentional condition. The absence of a memory difference under incidental and intentional instructions is one of the criteria used by Hasher and Zacks (1979) to indicate that encoding is automatic. From this perspective, Hirst and Volpe (1984) interpreted their results by suggesting that amnesic patients encode spatial information with effort, whereas normal people encode it automatically. However, several studies (Kovner, Dopkins, & Goldmeyer, 1988; Smith, 1988, in a study conducted on the amnesic patient H.M.; MacAndrew & Jones, 1993) showed that amnesic patients did not improve following intentional spatial encoding instructions.

More recently, a series of studies investigated spatial memory in the framework of the context-memory deficit hypothesis (Mayes, Meudell, & Pickering, 1985). This hypothesis states that amnesics have a severe and selective memory deficit for independent contextual information, and that this primary deficit causes a secondary deficit in memory for target information. Independent context includes the spatial, temporal, and other information that is incidental to what is being intentionally learned. This contextual view rests on the postulates that independent context must be processed differently from target information and by somewhat different brain structures (including the limbic-diencephalic areas that are lesioned in amnesia), and that associating target items to their contextual markers must be essential for target recognition and recall.

The context-memory hypothesis predicts that amnesics are more impaired at remembering contextual information than they are recognizing the target information to which they attended during the learning experience. In order to demonstrate the existence of a disproportionate deficit in contextual information, it is crucial to respect a main methodological constraint (see Mayes, Meudell, & MacDonald, 1991). Recognition performance for target information must be matched in the control and patient groups, so as to be able to show that patients are more impaired in tasks assessing contextual information memory than in tasks assessing target information memory. This is the only way to establish that a contextual memory impairment represents the fundamental deficit in patients and is the cause, rather than the consequence, of a globally poor memory. Matching has been usually obtained by giving patients more opportunity to learn the material or by testing them after shorter delays. The logic of this matching procedure is provided by the two-component theories of recognition memory (Mandler, 1980; Jacoby & Dallas, 1981) which postulate that recognition involves conscious recollection and/or familiarity, the former but not the latter, requiring the availability of contextual information. The view is that amnesics cannot use contextual information to recollect normally stored target information and that they have to rely on the less efficient familiarity process. This explains why amnesics can only reach a recognition performance comparable to control subjects when they are given more opportunity to learn or when they are tested after a shorter delay than normal subjects.

Evidence for a disproportionate context-memory deficit in amnesics has been found for memory of temporal order information (e.g., Kopelman, 1989; Squire, 1982), memory for source of information (Schacter, Harbluk, & McLachlan, 1984; Shimamura & Squire, 1987) and memory for the sensory modality via which information is presented (Pickering, Mayes, & Fairbairn, 1989). Several studies have also explored memory for spatial information but they provided contradictory results. Some of them found a disproportionate spatial memory deficit in amnesia (Hirst & Volpe, 1984; Mayes et al., 1991; Shoqeirat & Mayes, 1991) while others (Backer Cave & Squire, 1991; MacAndrew & Jones, 1993) reported that amnesics were impaired with regard to spatial memory tasks relative to control subjects, matched in target memory.

Indeed, there exists considerable variability in spatial memory performance among amnesics. Typically, studies explored amnesics with a variety of aetiologies: Korsakoff patients who have lesions in the midline diencephalon, post-encephalitic patients who have lesions in the medial temporal lobe region, and patients suffering from a rupture of an anterior communicating artery aneurysm and who have lesions in the cholinergic basal forebrain. In view of the studies on animals suggesting that the hippocampus plays a role in allocentric (viewpoint-independent) spatial memory (see Poucet, 1993), it could be that only hippocampal patients show a spatial memory deficit. In the Mayes et al. (1991) study, there was no strong evidence that the amnesic subgroups differed in degree of disproportionate spatial memory impairment. However, the size of their subgroups was small and so they had limited statistical power to detect any differences between subgroups. After a close inspection of the Backer Cave and Squire (1991) data, Pickering (1993) identified several elements suggesting that hippocampal patients perform more poorly on the spatial memory task than their diencephalic counterparts. Finally, Pickering (1993) reported unpublished observations confirming that some patients have deficits specific to spatial memory which are significantly more marked than those observed in other amnesics. However, it remains to determine clearly whether damage to the hippocampus underlies these differences.

Some of the discordances concerning spatial memory performance in amnesic patients could be due to methodological differences between studies (e.g., the delay between study and test; the way spatial memory is tested) or even to methodological inadequacies of certain studies (e.g., ceiling effects). The procedure used in the various studies was generally adapted from the spatial memory task developed by Smith and Milner (1981). In this task, subjects are presented with a list of items (words, little objects or nameable shapes and letters), each item being placed on a particular location of a grid. Encoding of spatial locations is incidental, subjects being told to remember the words. Testing is divided into two parts, one for targets and one for locations.

In fact, this task raises several methodological and theoretical problems. A first difficulty concerns the matching procedure in which recognition performance is matched across groups by varying the delay or the amount of learning opportunity given to subjects. Indeed, it cannot be excluded that these experimental manipulations influence target information more than contextual information memory (Shoqeirat & Mayes, 1991). A second problem concerns the sensitivity of the tasks which are used for comparing target and contextual information memory. It is essential that these two forms of memory are not confused with recall and recognition memory, inasmuch as these tasks are differentially sensitive to experimental and pathological situations. In other words, it is crucial that memory for target information and memory for contextual information are assessed by the same type of tasks, for instance recognition tasks.

Another methodological problem concerns the way spatial memory is tested. In a recent study, we investigated context memory for spatial information in schizophrenic patients (Rizzo, Danion, Van der Linden, Grangé, & Rohmer, 1996). A modified version of the spatial memory task developed by Shoqeirat and Mayes (1991) and Smith and Milner (1981) was used. Two different spatial location tasks were given: a location memory task and a relocation memory task. These tasks were intended to contrast a nonassociative form of spatial memory, that is, spatial information memory *per se*, and an associative form of spatial memory (i.e., memory for the association between target and contextual information). The location task consisted in determining which of three words was in a particular spatial location, and the relocation task consisted in determining which of three locations was occupied by a particular word. The main finding of this study was that schizophrenic patients were more impaired with regard to a location memory task than to target recognition and, to a lesser extent, relocation tasks. This result cannot be explained as an artefact arising from testing recognition memory for target and context information at different delays. Nor can the result be explained in terms of scaling effects or differences in task difficulty. The reasons why patients were disproportionately impaired in the location memory task, but not in the relocation task, deserve comment. Although both tasks investigated the ability to associate target and spatial information, they differed in one major aspect. Whereas in the former, subjects could only perform on the basis of their associative spatial memory, in the latter, they could also perform on the basis of a nonassociative spatial memory: their response could have been based on recognition memory for the locations which were, or were not, previously occupied by words, irrespective of which particular words occupied particular locations. The fact that schizophrenic patients were not disproportionately impaired in the relocation task, in conditions where their associative spatial memory was defective, strongly suggests that they used a nonassociative spatial memory in the relocation task to compensate for their associative spatial memory deficit. This interpretation is further supported by the demonstration of a significant correlation between relocation performance and the visuospatial pattern span (Wilson, 1993) in the schizophrenic group, and between relocation performance and the visual memory score of the Wechsler Memory Scale–Revised in the control group; these correlations provide evidence that performance in the relocation task relied heavily on visuospatial information processing. Taken together, these results suggest that spatial information memory *per se* was not impaired in schizophrenic patients, but rather that the ability to bind target and spatial information was defective. More generally, they also indicate that the way spatial memory is tested is a crucial aspect of the procedure which may or may not lead to a defective performance.

From a more general point of view, the context-memory deficit hypothesis will be validated only if some predictions are rigorously tested. First, this hypothesis predicts that the degree to which amnesic patients show disproportionately severe deficit for context memory will correlate significantly with the severity of the memory deficit (Shoqeirat & Mayes, 1991). Furthermore, inasmuch as we could consider that free recall depends on associating target information to its contextual markers to a greater extent than recognition, another prediction is that a selective impairment of contextual memory should disrupt free recall more than recognition. Finally, it is absolutely necessary to demonstrate that the spatial context deficit is an essential feature of amnesia and is not due to other brain lesions (in particular, frontal lobe lesions) which are incidental to their amnesia. In that perspective, Smith and Milner (1984) found that frontal patients showed normal performance on a spatial memory task. In addition, Shoqeirat and Mayes (1991) showed that the selective spatial memory observed in amnesic patients was unrelated to frontal lobe dysfunction.

Thus, it appears that the investigation of spatial memory in the framework of the context-memory hypothesis of amnesia is a complex process requiring a lot of methodological and theoretical controls. Future research should be conducted to compare systematically amnesics with various aetiologies, as well as various experimental paradigms, by using procedures which respect essential methodological constraints. Further research will also have to give a clear account of why contextual (spatial) memory should be disproportionately impaired in amnesics. More particularly, it should explore whether the problem is located at the encoding, consolidation or retrieval phase and whether it is a spatial problem *per se* or rather a difficulty to bind spatial information and targets (see Mayes, 1992, for a detailed discussion of the context-memory hypothesis with regard to other interpretations of amnesia).

CONCLUSION

The cognitive neuropsychological approach to spatial memory disorders has been much less productive than in other domains of cognitive function, such as reading or object and face recognition. Nevertheless, it has already revealed theoretically interesting dissociations in the performance of brain-damaged patients. Thus, in the domain of visuospatial working memory, the functional independence of the visuospatial sketchpad and the phonological loop seems largely supported by neuropsychological evidence. Similarly, different patterns of topographical memory deficits have been found, indicating that topographical disorientation is not a homogeneous syndrome. Finally, exploration of long-term memory for spatial locations in amnesic patients has raised fundamental questions concerning the role of contextual information in target memory.

We believe that further progress in the understanding of the spatial memory deficits will be achieved by research centred on individual cases and conducted within the framework of theoretical models. In addition, the detailed investigation of patients with spatial memory impairments will contribute to the testing or modification of the theoretical models of normal spatial processing.

However, this cognitive neuropsychological approach is confronted with numerous methodological difficulties and it has to take into account several limitations which could render the dissociations observed between patients difficult to interpret.

Finally, it is most likely that the cognitive approach is the best way for clinical neuropsychologists to devise efficient re-educative strategies. In that perspective, we recently elaborated a re-education project which proposes different re-education strategies adapted for each type of topographical disorder (Van der Linden & Seron, 1991). Future studies will tell the validity of this project.

REFERENCES

Backer Cave, C., & Squire, L.R. (1991). Equivalent impairment of spatial and nonspatial memory following damage to the human hippocampus. *Hippocampus, 1*, 329–340.

Baddeley, A.D. (1986). *Working memory.* Oxford: Oxford University Press.

Baddeley, A.D., Bressi, S., Della Sala, S., Logie, R., & Spinnler, H. (1991a). The decline of working memory in Alzheimer's disease. A longitudinal study. *Brain, 114*, 2521–2542.

Baddeley, A.D., Della Sala, S., & Spinnler, H. (1991b). The two-component hypothesis of memory deficit in Alzheimer's disease. *Journal of Clinical and Experimental Neuropsychology, 13*, 372–380.

Baddeley, A.D., & Hitch, G. (1974). Working memory. In G.H. Bower (Ed.), *The psychology of learning and motivation* (Vol. VIII). New York: Academic Press.

Baddeley, A.D., & Lieberman, K. (1980). Spatial working memory. In R.S. Nickerson (Ed.), *Attention and performance* (Vol. VIII). Hillsdale, NJ: Lawrence Erlbaum Associates.

Baddeley, A.D., Logie, R.H., Bressi, S., Della Sala, S., & Spinnler, H. (1986). Dementia and working memory. *Quarterly Journal of Experimental Psychology, 38A*, 603–618.

Barton, A., Matthews, B., Farmer, E., & Belyavin, M. (1995). Revealing the basic properties of the visuospatial sketchpad: The use of complete spatial arrays. *Acta Psychologica, 89*, 197–216.

Basso, A., Spinnler, H., Vallar, G., & Zanobio, M.E. (1982). Left hemisphere damage and selective impairment of auditorily-verbal short-term memory. *Neuropsychologia, 20*, 263–274.

Belleville, S., Peretz, I., & Arguin, H. (1992). Contribution of articulatory rehearsal to short-term memory: Evidence from a selective disruption. *Brain and Language, 43*, 713–746.

Bisiach, E., Bulgarelli, E., Sterzi, R., & Vallar, G. (1983). Line bisection and cognitive plasticity of unilateral neglect of space. *Brain and Cognition, 2*, 32–38.

Böök, A., & Gärling, T. (1980). Processing of information about location during locomotion:

Effects of a concurrent task and locomotion patterns. *Scandinavian Journal of Psychology, 21*, 185–192.

Bottini, G., Cappa, S., Geminiani, G., & Sterzi, R. (1990). Topographic disorientation: A case report. *Neuropsychologia, 28*, 309–312.

Boyd, T.M., & Sautter, S.W. (1993). Route-finding: A measure of everyday executive functioning in the head-injury adult. *Applied Cognitive Psychology, 7*, 171–181.

Brooks, L.R. (1967). The suppression of visualisation by reading. *Quarterly Journal of Experimental Psychology, 19*, 289–299.

Byrne, R.W. (1982). Geographical knowledge and orientation. In A.W. Ellis (Ed.), *Normality and pathology in cognitive functions*. London: Academic Press.

Cantone, G., Orsini, A., Grossi, D., & De Michele, G. (1978). Verbal and spatial memory span in dementia. *Acta Neurologica, 33*, 175–183.

Caramazza, A. (1986). On drawing inferences about the structure of normal cognitive systems from the analysis of patterns of impaired performance: The case for single-patients studies. *Brain and Cognition, 5*, 41–66.

Clarke, S., Assal, G., & De Tribolet, N. (1993). Left hemisphere strategies in visual recognition topographical orientation and time planning. *Neuropsychologia, 31*, 99–113.

Coltheart, M. (1985). Cognitive neuropsychology and the study of reading. In M.I. Posner & O.S.M. Marin (Eds.), *Attention and performance* (Vol. XI). Hillsdale, NJ: Lawrence Erlbaum Associates.

Coltheart, M., Sartori, G., & Job, R. (1987). *The cognitive neuropsychology of language*. Hillsdale, NJ: Lawrence Erlbaum Associates.

Davidoff, J.B., & Ostergaard, A.L. (1984). Colour anomia resulting from weakened short-term colour memory. *Brain, 107*, 415–431.

De Renzi, E. (1982). *Disorders of space exploration and cognition*. New York: Wiley.

De Renzi, E., Faglioni, P., & Previdi, P. (1977a). Spatial memory and hemispheric locus of lesion. *Cortex, 13*, 424–433.

De Renzi, E., Faglioni, P., & Villa, P. (1977b). Topographical amnesia. *Journal of Neurology, Neurosurgery and Psychiatry, 40*, 498–505.

De Renzi, E, & Nichelli, P. (1975). Verbal and nonverbal short-term memory impairment following hemispheric damage. *Cortex, 11*, 341–354.

Della Sala, S., & Logie, R.H. (1993). When working memory does not work: The role of working memory in neuropsychology. In F. Boller & H. Spinnler (Eds.), *Handbook of neuropsychology* (Vol. 8). Amsterdam: Elsevier.

Della Sala, S., Muggia, S., Spinnler, H., & Zuffi, M. (1995). Cognitive modelling of face processing: Evidence from Alzheimer patients. *Neuropsychologia, 33*, 675–687.

Ergis, A-M., Van der Linden, M., Boller, F., Degos, J-D., & Deweer, B. (1995). Mémoire visuo–spatiale à court et à long terme dans la maladie d'Alzheimer débutante. *Neuropsychologia Latina, 1*, 18–25.

Farah, M.J., Hammond, K.M., Levine, D.N., & Calvanio, R. (1988). Visual and spatial mental imagery: Dissociable systems of representation. *Cognitive Psychology, 20*, 439–492.

Fuster, J.M. (1993). Frontal lobes. *Current Opinions in Neurobiology, 3*, 160–165.

Gathercole, S.E. (1994). Neuropsychology and working memory. *Neuropsychology, 8*, 494–505.

Gathercole, S.E., & Baddeley, A.D. (1993). *Working memory and language*. Hillsdale, NJ: Lawrence Erlbaum Associates.

Giraudo, M.-D., & Pailhous, J. (1994). Distortions and fluctuations in topographic memory. *Memory and Cognition, 22*, 14–26.

Gottschaldt, K. (1926; 1929). Über den Einfluss der Erfahrung auf die Wahrnehmung von Figuren. *Psychologische Forschung, 8*, 261–317; *12*, 1–87.

Habib, M., & Sirigu, A. (1987). Pure topographical disorientation: A definition and anatomical basis. *Cortex*, *23*, 73–85.
Hanley, J.R., & Davies, A.D.M. (1995). Lost in your own house. In R. Campbell & M. Conway (Eds.), *Broken memories. Case studies in memory impairment*. Oxford: Blackwell.
Hanley, J.R., Pearson, N.A., & Young, A.W. (1990). Impaired memory for new visual forms. *Brain*, *113*, 1131–1148.
Hanley, J.R., Young, A.W., & Pearson, N.A. (1991). Impairment of the visuo-spatial sketch pad. *Quarterly Journal of Experimental Psychology*, *43A*, 101–125.
Hasher, L., & Zacks, R.T. (1979). Automatic and effortful processes in memory. *Journal of Experimental Psychology: General*, *108*, 356–388.
Hirst, W., & Volpe, B.T. (1984). Encoding of spatial relations with amnesia. *Neuropsychologia*, *5*, 631–634.
Humphreys, G.W., & Riddoch, M.J. (1987). *Visual object processing: A cognitive neuropsychological approach*. London: Lawrence Erlbaum Associates.
Jacoby, L.L., & Dallas, M. (1981). On the relationship between autobiographical memory and perceptual learning. *Journal of Experimental Psychology: General*, *110*, 306–340.
Kinsbourne, M., & Warrington, E.K. (1962). A disorder of simultaneous form perception. *Brain*, *85*, 461–486.
Kopelman, M.D. (1989). Remote and autobiographical memory, temporal context memory, and frontal atrophy in Korsakoff and Alzheimer patients. *Neuropsychologia*, *27*, 437–460.
Kovner, R., Dopkins, S., & Goldmeier, E. (1988). The effects of instructional set on amnesic recognition memory performance. *Cortex*, *24*, 477–483.
Landis, T., Cummings, J.L., Benson, F., & Palmer, E.P. (1986). Loss of topographical familiarity. An environmental agnosia. *Archives of Neurology*, *43*, 132–136.
Levine, D.N., Warach, J., & Farah, M.J. (1985). Two visual systems in mental imagery: dissociation of 'what' and 'where' in imagery disorders due to bilateral posterior cerebral lesions. *Neurology*, *35*, 1010–1018.
Logie, R.H. (1995). *Visuo-spatial working memory*. Hillsdale, NJ: Lawrence Erlbaum Associates.
MacAndrew, S.B.G., & Jones, G. (1993). Spatial memory in amnesia: Evidence from Korsakoff patients. *Cortex*, *29*, 235–249.
Mandler, G. (1980). Recognizing: The judgement of previous occurrence. *Psychological Review*, *87*, 252–271.
Mayes, A.R. (1992). Automatic memory processes in amnesia: How are they mediated? In A.D. Milner & M.Rugg (Eds.), *The neuropsychology of consciousness*. London: Academic Press.
Mayes, A.R., Meudell, P.R., & MacDonald, C. (1991). Disproportionate intentional spatial-memory impairments in amnesia. *Neuropsychologia*, *29*, 771–784.
Mayes, A.R., Meudell, P.R., & Pickering, A. (1985). Is organic amnesia caused by a selective deficit in remembering contextual information? *Cortex*, *21*, 167–202.
Milner, B. (1971). Interhemispheric differences in the localization of psychological processes in man. *British Medical Bulletin*, *27*, 272–277.
Money, J., Alexander, D., & Walker, H.T. (1965). *A standardized road-map test of direction sense*. Baltimore, MD: The Johns Hopkins Press.
Morris, R.G. (1994). Working memory in Alzheimer-type dementia. *Neuropsychology*, *8*, 544–554.
Morris, R.G., & Morton, N. (1995). Not knowing which way to turn: A specific image transformation impairment dissociated from working memory functioning. In R. Campbell & M. Conway (Eds.), *Broken memories. Case studies in memory impairment*. Oxford: Blackwell.
Morrow, L., Ratcliff, G., & Johnston, C.S. (1985). Externalizing spatial knowledge in patients with right hemisphere lesions. *Cognitive Neuropsychology*, *2*, 265–273.

Norman, D.A., & Shallice, T. (1986). Attention to action: Willed and automatic control of behavior. In R.J. Davidson, G.E. Schwarts, & D. Shapiro (Eds.), *Consciousness and self-regulation. Advances in research and theory* (Vol. 4). New York: Plenum.

Pallis, C. (1955). Impaired identification of faces and places with agnosia for colors: Report of a case due to cerebral embolism. *Journal of Neurology, Neurosurgery and Psychiatry, 18,* 218-224.

Patterson, A., & Zangwill, O.L. (1945). A case of topographical disorientation associated with a unilateral cerebral lesion. *Brain, 68,* 188-211.

Paulesu, E., Frith, C.D., & Frackowiak, R.S.J. (1993). The neural correlates of the verbal component of working memory. *Nature, 362,* 342-345.

Perani, D., Bressi, S., Cappa, S.F., Vallar, G., Alberoni, M., Grassi, F., Caltagirone, C., Cipolotti, L., Franceschi, M., Lenzi, G., & Fazio, F. (1993). Evidence for multiple memory systems in the human brain: A (18F) FDG PET metabolic study. *Brain, 116,* 903-919.

Petrides, M., Alivisatos, B., Evans, A.C., & Meyer, E. (1993). Dissociation of human mid-dorsolateral from posterior dorsolateral frontal cortex in memory processing. *Proceedings of the National Academy of Science, USA, 90,* 873-877.

Petrides, M., Alivisatos, B., Meyer, E., & Evans, A.C. (1993). Functional activation of the human frontal cortex during the performance of verbal working memory tasks. *Proceedings of the National Academy of Science, USA, 90,* 878-882.

Pickering, A.D. (1993). The hippocampus and space: Are we flogging a dead (sea-)horse? *Hippocampus, 3,* 113-114.

Pickering, A.D., Mayes, A.R., & Fairbairn, A.F. (1989). Amnesia and memory for modality information. *Neuropsychologia, 27,* 1249-1259.

Pigott, S., & Milner, B. (1994). Capacity of visual short-term memory after unilateral frontal or anterior temporal-lobe resection. *Neuropsychologia, 32,* 969-981.

Poucet, B. (1993). Spatial cognitive maps in animals: New hypotheses on their structure and neural mechanisms. *Psychological Review, 100,* 163-182.

Riddoch, M.J., & Humphreys, G.W. (1989). Finding the way around topographical impairments. In J.W. Brown (Ed.), *Neuropsychology of visual perception.* Hillsdale, NJ: Lawrence Erlbaum Associates.

Riddoch, M.J., & Humphreys, G.W. (1995). 17 + 14 = 41? Three cases of working memory impairment. In R. Campbell & M. Conway (Eds.), *Broken memories. Case studies in memory impairment.* Oxford: Blackwell.

Rizzo, L., Danion, J.M., Van der Linden, M., Grangé, D., & Rohmer, G. (1996). Impairment of memory for spatial context in schizophrenia. *Neuropsychology, 10,* 376-384.

Ross, E.D. (1980). Sensory-specific and fractional disorders of recent memory in man: II. Unilateral loss of tactile recent memory. *Archives of Neurology, 37,* 267-272.

Salmon, E., Van der Linden, M., Collette, F., Delfiore, G., Maquet, P., Degueldre, C., Luxen, A., & Franck, G. (1996). Regional brain activity during working memory tasks. *Brain, 119,* 1617-1625.

Salway, A.F., & Logie, R.H. (1995). Visuospatial working memory, movement control and executive demands. *British Journal of Psychology, 86,* 253-269.

Schacter, D.L., Harbluk, J.L., & McLachlan, D.R. (1984). Retrieval without recollection: An experimental analysis of source amnesia. *Journal of Verbal Learning and Verbal Behavior, 23,* 593-611.

Schacter, D.L., & Nadel, L. (1991). Varieties of spatial memory: A problem for cognitive neuroscience. In R.G. Lister & H.J. Weingartner (Eds.), *Perspectives on cognitive neuroscience.* New York: Oxford University Press.

Semmes, J., Weinstein, S., Ghent, L., & Teuber, H.L. (1955). Spatial orientation: 1. Analysis by locus of lesion. *Journal of Psychology, 39,* 227-244.

Seron, X. (1993). *La neuropsychologie cognitive.* Paris: Presses Universitaires de France.
Shallice, T. (1988). *From neuropsychology to mental structure.* Cambridge: Cambridge University Press.
Shallice, T. (1994). Multiple levels of control processes. In C. Umilta & M. Moscovitch (Eds.), *Attention and Performance: Vol. XV. Conscious and nonconscious information processing.* Cambridge, MA: Bradford/MIT Press.
Shallice, T., & Vallar, G. (1990). The impairment of auditory-verbal short-term storage. In G. Vallar & T. Shallice (Eds.), *Neuropsychological impairments of short-term memory.* Cambridge: Cambridge University Press.
Shallice, T., & Warrington, E.K. (1970). Independent functioning of verbal memory stores: A neuropsychological study. *Quarterly Journal of Experimental Psychology, 22,* 261–273.
Shimamura, A.P., & Squire, L.R. (1987). A neuropsychological study of fact memory and source amnesia. *Journal of Experimental Psychology: Learning, Memory, and Cognition, 13,* 464–473.
Shoqeirat, M.A., & Mayes, A.R. (1991). Disproportionate incidental spatial-memory and recall deficits in amnesia. *Neuropsychologia, 29,* 749–769.
Smith, E.E., & Jonides, J. (1995). Working memory in humans: Neuropsychological evidence. In M.S. Gazzaniga (Ed.), *The cognitive neurosciences.* Cambridge, MA: Bradford/MIT Press.
Smith, E.E., Jonides, J., Koeppe, R.A., Awh, E., Schumacher, E.H., & Minoshima, S. (1995). Spatial versus object working memory: PET investigations. *Journal of Cognitive Neurosciences, 7,* 337–356.
Smith, M.L. (1988). Recall of spatial location by the amnesic patient H.M. *Brain and Cognition, 7,* 178–183.
Smith, M.L., & Milner, B. (1981). The role of the right hippocampus in the recall of spatial location. *Neuropsychologia, 19,* 781–793.
Smith, M.L., & Milner, B. (1984). Differential effects of frontal-lobe lesions on cognitive estimation and spatial memory. *Neuropsychologia, 22,* 697–705.
Spalding, J.M.K., & Zangwill, O.L. (1950). Disturbance of number form in a case of brain injury. *Journal of Neurology, Neurosurgery and Psychiatry, 13,* 24–29.
Spinnler, H., Della Sala, S., Bandera, R., & Baddeley, A.D. (1988). Dementia, aging, and the structure of memory. *Cognitive Neuropsychology, 5,* 193–211.
Squire, L.R. (1982). Comparison between forms of amnesia: Some deficits are unique to Korsakoff syndrome. *Journal of Experimental Psychology: Learning, Memory, and Cognition, 8,* 560–571.
Trojano, L., Chiacchio, L., De Luca, G., & Grossi, D. (1994). Exploring visuospatial short-term memory defect in Alzheimer's disease. *Journal of Clinical and Experimental Neuropsychology, 16,* 911–915.
Tromp, E., Dinkla, A., & Mulder, T. (1995). Walking through doorways: An analysis of navigation skills in patients with neglect. *Neuropsychological Rehabilitation, 5,* 319–331.
Ungerleider, S., & Mishkin, M. (1982). Two cortical visual systems. In D.J. Ingle, R.J.W. Mansfield, & M.A. Goodale (Eds.), *The analysis of visual behavior.* Cambridge, MA: MIT Press.
Vallar, G., & Baddeley, A.D. (1984). Fractionation of working memory: Neuropsychological evidence for a phonological short-term store. *Journal of Verbal Learning and Verbal Behavior, 23,* 151–161.
Vallar, G., & Papagno, C. (1995). Neuropsychological impairments of short-term memory. In A.D. Baddeley, B.A. Wilson, & F.N. Watts (Eds.), *Handbook of memory disorders.* Chichester, UK: Wiley.
Van der Linden, M. (1994a). Neuropsychologie de la mémoire. In X. Seron & M. Jeannerod (Eds.), *Traité de neuropsychologie humaine.* Bruxelles: Mardaga.

Van der Linden, M. (1994b). Neuropsychologie des démences. In X. Seron & M. Jeannerod (Eds.), *Traité de neuropsychologie humaine*. Bruxelles: Mardaga.

Van der Linden, M., Brédart, S., & Beerten, A. (1994). Age-related differences in updating working memory. *British Journal of Psychology, 85,* 145–152.

Van der Linden, M., Coyette, F., & Seron, X. (1992). Selective impairment of the Central Executive component of working memory: A single case. *Cognitive Neuropsychology, 9,* 301–326.

Van der Linden, M., & Seron, X. (1987). A case of dissociation in topographical disorders: The selective breakdown of vector-map representation. In P. Ellen & C. Thinus-Blanc (Eds.), *Cognitive processes and spatial orientation in animal and man*, (Vol. II). Boston, MA: Martinus Nijhoff.

Van der Linden, M., & Seron, X. (1991). I disturbi dell'orientamento topographico. Proposte per un progetto terapeutico. In D. Grossi (Ed.), *La riabilitazione dei disturbi della cognizione spaziale*. Milano: Masson.

Warrington, E.K., & James, M. (1967). Disorders of visual perception in patients with localized cerebral lesions. *Neuropsychologia, 5,* 253–266.

Warrington, E.K., & Rabin, P. (1971). Visual span of apprehension in patients with unilateral cerebral lesions. *Quarterly Journal of Experimental Psychology, 23,* 423–431.

Warrington, E.K., & Shallice, T. (1972). Neuropsychological evidence of visual storage in short-term memory tasks. *Quarterly Journal of Experimental Psychology, 24,* 30–40.

Whiteley, A.M., & Warrington, E.K. (1978). Selective impairment of topographical memory: A single case study. *Journal of Neurology, Neurosurgery and Psychiatry, 41,* 575–578.

Wilson, J.T.L. (1993). Visual short-term memory. In F.J. Stachowiack (Ed.), *Development in the assessment and rehabilitation of brain-damaged patients*. Tübingen: Gunter Narr.

Wilson, J.T.L., Wiedmann, K.D., Hadley, D.M., & Brooks, D.N. (1989). The relationship between visual memory function and lesions detected by magnetic resonance imaging after closed head injury. *Neuropsychology, 3,* 255–265.

Young, A.W., Newcombe, F., de Haan, E.H.F., Small, M., & Hay, D.C. (1993). Face perception after brain injury. Selective impairments affecting identity and expression. *Brain, 116,* 941–959.

Zangwill, O.L. (1951). Discussion on parietal lobe syndromes. *Proceedings of the Royal Society of Medicine, 44,* 343–346.

3
Animal Spatial Cognition and Exploration

Catherine Thinus-Blanc, Etienne Save, and Bruno Poucet
Laboratory of Cognitive Neurosciences, Marseille, France

INTRODUCTION

Maze learning has long been psychologists' favourite method of investigating spatial cognition in animals. In such situations, animals are required to learn by trial and error the location of a place where a reward is hidden or where they can escape from a potential danger. The experimenter exerts strong control over the situation since he or she decides the outcome (reward, lack of reward, punishment) of a given trial as a function of the subject's behaviour (i.e., whether an error is committed or a correct response is made). Mazes have become extremely popular and represent powerful tools for investigating the behavioural and brain mechanisms of spatial cognition in animals. This is particularly the case of the radial maze (Chapter 4) and the Morris water maze (Chapter 5.) However, in addition to the considerable advantages that justify their extensive use, such experimental designs also have their drawbacks and limits. For instance, task repetition may lead to the implementation of stereotyped behaviour and/or alternate strategies that correspond, for the animal, to an optimal balance between the amount of reward obtained and the minimizing of expended cognitive effort. In other words, the behaviour observed by the experimenter does not always correspond to the subject's utmost capabilities. Therefore, in addition to experiments based on maze learning, it may be valuable to conduct behavioural tests which do not have those drawbacks arising from the repetitive aspects of the task.

In this regard, experiments that involve the study of exploratory activity (i.e., the behaviours which are observed when an animal is exposed to an unfamiliar situation) are of interest because they may yield some properties of animal spatial cognition, using quite different methodologies. They usually take place within a relatively short period of time and they rely on animals' exhibiting spontaneous behaviours, with no specific task to complete and with minimal intervention on the part of the experimenter.

There are two distinct ways of examining how exploration results in spatial knowledge: the first is by manipulating the amount or quality of exploration that an animal experiences and then look at performance using a spatial test. This allows us to assess the extent to which previous manipulation of exploration affects the representation on which such spatial performance is based. The second way of using exploration to investigate spatial cognition is similar to the first, except that instead of examining performance level in spatial tasks following exploration, it is based on the study of the time course of exploratory activity (habituation) and on its renewal (dishabituation) following spatial modifications that are made to the initial situation/stimulus configuration. The acquisition phase and testing both represent parts of the exploratory process.

In this chapter, we first provide a presentation of exploratory activity and of its meaning in relation to spatial knowledge. Then, experiments are presented which rely on the two above-mentioned paradigms for using exploration as an index of spatial knowledge, and their advantages and limits are discussed.

THE SPATIAL FUNCTION OF EXPLORATION

Exploratory activity is displayed by most mammalian species in the presence of novelty. This reaction can take several forms depending both on the nature of the new event and on the species under study. For instance, a rat or a hamster placed in a new environment containing objects displays a feverish investigatory activity, which finds expression in multiple contacts with the objects, sniffing, stopping, rearing, etc. All of these easily observable activities appear to be different manifestations of some common process. They are induced by curiosity and aimed at reducing the uncertainty related to novelty (Berlyne, 1966). One feature of this activity is its decrease over time, called "habituation". Depending on the complexity of the situation, its degree of novelty, and the species being tested, the time course of habituation varies.

The first experimental situation more or less related to exploratory activity was the open field (an empty field, the floor of which is divided into squares/zones), devised by Hall (1934). It has been used to determine the effects of an unfamiliar environment upon emotional behaviour in the rat. In spite of its interest in certain fields of research, the original, unmodified version of the open field arena does not provide much information related to spatial knowledge.

In contrast, Berlyne's seminal experiment (1950), which consisted of replacing one of the three familiar objects contained in a cage by a new one, is the basis of the dishabituation experiments that are reported in this chapter. Berlyne had observed that rats displayed investigative behaviors that are selectively directed toward the unfamiliar object. These reactions appear to be modulated by various factors, such as the level of familiarity of the context in which the novel object is presented and the rearing conditions (e.g., whether enriched or impoverished) (Cowan, 1976).

A well-known classic test, the spontaneous alternation task, based on curiosity for an unfamiliar part of a maze, is a starting point for examining the link between exploratory activity and spatial cognition. On consecutive trials, a rat is placed at the start of a double choice T- or Y-shaped apparatus and allowed to choose (the left or right alley). Following the choice it is replaced at the starting point and allowed to choose again. On the second trial, there is a 70–80% likelihood that the animal will choose the arm which was not visited during the first trial. This spontaneous alternation phenomenon is considered to be a form of exploratory activity since it reflects the animal's tendency to investigate an unknown part of the environment. This means that familiar and unfamiliar locations are correctly identified and differentiated from one another. The tendency to alternate increases if the subject is kept in the goal-box chosen on the first trial for a while (Glanzer, 1953). In addition, this behaviour does not rely on the alternation of different motor responses (turn left, turn right) but on visiting distinct places. When two paths lead to the same location, the alternation rate decreases (Sutherland, 1957).

Theoretical developments about the spatial function of exploration, for example Piaget's theory which emphasized the important role of exploratory activity (1937) as well as Berlyne's ideas about curiosity and reactions to novelty (1960), have fostered the idea that investigatory behaviours are part of the general process of knowledge. Berlyne's key idea (1960) is the notion that previously stored information regarding some characteristics of an object or of a situation serves as a reference for novelty detection. Novelty *per se* does not exist. Rather, it is defined by reference to familiarity, and reciprocally, familiarity to novelty. The relativity of these two opposite features has substantial methodological and theoretical implications; methodological, since the dishabituation paradigm is now extensively used in many studies in young children and animals. The main theoretical consequence of Berlyne's pioneer developments is that exploratory activity has a far greater role than simply "reducing a primary motivation" because it corresponds to actual cognitive processing and elaboration. This view is shared by other authors, for example, Renner (1988) and Thorpe (1963).

O'Keefe and Nadel (1978) go even further in the role they attribute to exploratory activity since they argue that one of its functions is to update "cognitive spatial maps". The less familiar a situation, the greater the intensity of exploration. When confronted with a partly or totally novel situation, there is a systematic screening of the stored memories of familiar environments. The detection of a discrepancy or mismatch between what is known/expected and what is currently perceived is the factor which triggers exploratory activity. This can be aimed either at constructing a map of a totally new situation or at updating an already available internal model. Accordingly, habituation would correspond to the increasing matching between the two (stored, and currently perceived) situations being compared. The Russian physiologist, Evgeniye Nikolaievitch Sokolov (1960, 1963a,b) had developed similar reasoning about orienting responses, which form one of the important constituent elements of exploratory activity. Gray (1982) and O'Keefe and Nadel (1978) have put forward the hypothesis that the forebrain hippocampus could perform this comparison function. Indeed, the fact that the hippocampus is strongly implicated in both exploration and spatial memory is a weighty argument for their reciprocal relationship.

EXPLORATION AND PROBLEM SOLVING

In this section, we give some examples of experiments that illustrate the beneficial effects of exploration on the performance scores of animals solving spatial tasks.

The Maier three-table test has been successfully used since Maier's pioneer experiments (1932). Three platforms are linked to each other by a runway system (Fig. 3.1). Each daily session begins with a period of exploration. In Maier's experiments, however, exploration was not entirely free but was guided by the experimenter so that it was unidirectional and fragmented; one day the rats ran X–Y several times, finding food at Y; the following day, they ran Y–Z, the food being at Z, and the third day Z–X was run with food at X. During the test, the rats were shown food on one of the platforms, X for example, and they were allowed to eat a small part of it. They were then released from Y. The task for the animals was to return directly to the table where they had just been fed by running a segment (Y–X) of the apparatus that they had not traversed before. A screen with a door at the front of each table prevented the animals from seeing from one table to another. Unlike a control group which had been allowed to freely explore each leg of the apparatus, rats that had experienced only unidirectional and fragmented exploration failed to choose the shorter path during the first test trial but ran along the longer one in the direction they were familiar with.

In a follow-up experiment, the segments X–Z and Z–Y could be explored in both directions and in a second pre-test phase rats ran X–Y–Z for food.

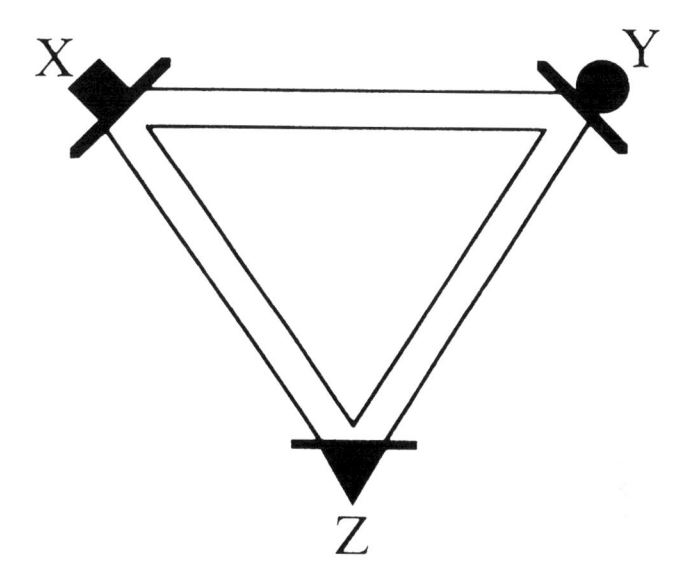

FIG. 3.1 The Maier three-table maze. A screen with a door is placed before each platform in order to prevent the rats from seeing one platform while standing on another.

Only one leg (X–Y) was run unidirectionally. During the tests, rats were fed at X and placed at Y (Fig. 3.1). Therefore, the optimal choice corresponded to the path Y–X which had not been explored in that direction before. Under these conditions, a large number of rats immediately chose the shorter path, unlike the previous experiment.

This elegant experiment is a demonstration of the functional role of exploratory activity in the setting up of spatial relationships; it also illustrates some properties of various spatial behaviours. When rats are denied bi-directional exploration, they learn a route (i.e., a sequence of places). The term "route" is used here in a broad sense. It merely means that the structure of the maze itself determines the entire trajectory once the initial choice during the test has been made. The behaviour is dependent on the sequence in which information has been collected. In contrast, rats provided with bi-directional exploration appear to set up spatial relationships between places regardless of the temporal features of the acquisition phase. Even though bi-directional exploration did not take place on the segment of the maze involved in the test, the spatial representation of the whole situation in a map-like form would have allowed the inference of the direction of the goal (i.e., of the shorter path hitherto not experienced in that direction).

In a modified version of Maier's three-table task with a different runway configuration (Fig. 3.2), Ellen, Soteres and Wages (1984) obtained evidence

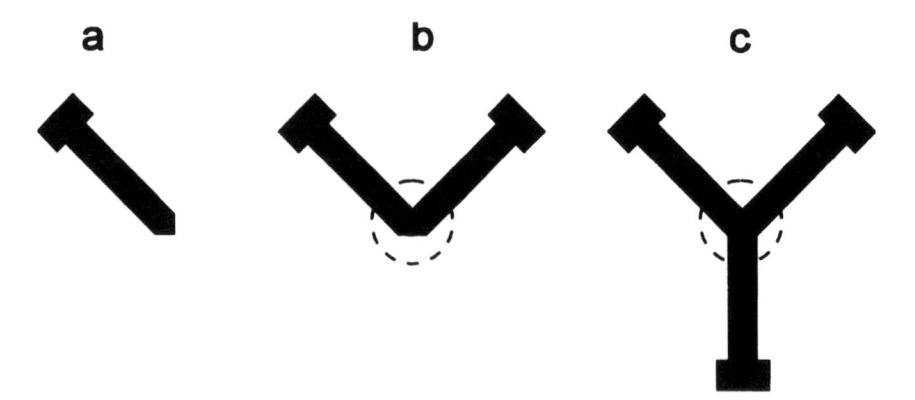

FIG. 3.2 A version of the Maier three-table maze and the various segmentations of the maze used for exploration. (Reproduced from Thinus-Blanc, 1996, with kind permission of World Scientific, Singapore.)

that a "piecemeal" exploration of the apparatus can lead to successful performance depending on the segments which have been separately explored. For instance, no rat that explored one table and runway per day (Fig. 3.2a) was able to solve the task, whereas 60% of the rats that explored two tables and their interconnecting runways (Fig. 3.2b) did succeed. All rats that explored the entire apparatus (Fig. 3.2c) on each exploratory day were able to solve the task. Since the amount of exploration was the same for all groups, it is likely that differences in performance levels resulted from the partitioning of previous exploration. In another experiment, Ellen, Parko, Wages, Doherty and Herrmann (1982) demonstrated that exploration of either the tables or the runways alone led to unsuccessful performance. From these two experiments, it appears that the runway connections, as well as the runways themselves, must have been investigated to ensure problem-solving success: animals need continuous exploration of basic elements of the apparatus (e.g., two tables and the runway linking them). Indeed, the three basic elements (the runways) do have one part in common, corresponding to their point of junction. It can be assumed that it is this partial overlap which allows the reorganization of the fragmented information into a whole. It is also from this central point that panoramic views involving the three possible goals could be perceived. Therefore, this central choice point may have served as a reference for the spatial organization of the whole situation.

Indeed, results obtained by Poucet, Bolson and Herrmann (1990) using a modified version of the maze, support this interpretation. In their experiment, interconnecting systems with longer runways could be added to the initial apparatus consisting of the three tables and direct runways (Fig. 3.3). The primary aim of this experiment was to test the deficits of brain-

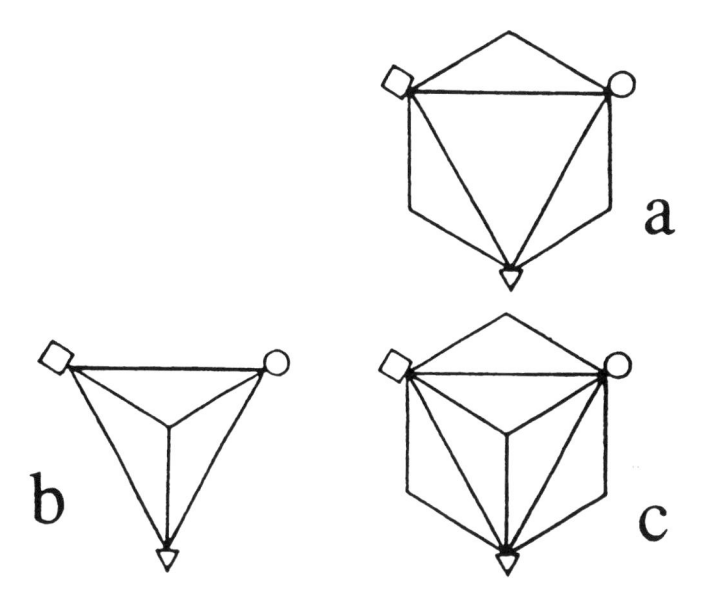

FIG. 3.3 Complex versions of the Maier three-table maze with various interconnecting runways. (Adapted from Poucet et al., 1990.)

damaged animals. However, the observation of intact rats' strategies is of particular interest with regard to their adaptation to the increased complexity of the situation. The groups tested in conditions a and b (Fig. 3.3) consistently chose the most direct route between tables. However, in the more complex situation, c, normal rats demonstrated a strong preference for the inner route (leading to the central choice point), rather than for the direct route. The main difference between conditions b (which also provided an inner runway pattern) and c is that the latter situation provided a much more complex runway pattern as compared with the former condition in terms of the number of possible routes. Poucet et al. (1990, p.380) attribute the preference for the inner runway to the advantages that it provided: "First it allowed delaying choice until animals were closer to either goal table than was possible by the other available routes. Second, it permitted animals to self-correct at the choice-point by gathering additional relevant information at that place. Lastly, in the configuration c, the number of alternatives at the choice-point was reduced to two routes, whereas five routes were available at the start in the same condition".

When the rats became more familiar with the problem, their preferences switched from the inner route to the direct route. Together with the experiments discussed above, which examined the effects of a "piecemeal" exploratory experience, these data demonstrate that some places and their interrelations are of particular importance with regard to spatial organization.

One of these places in the Y-shaped three-table apparatus appears to be the central choice point, arguably because it offers various views of all three goal locations as well as views of the visual background.

Although using different procedures, other experiments have also shown that animals are able to solve spatial tasks provided that they have previously been allowed adequate exploration of the situation. In an experiment with hamsters (Chapuis, Durup & Thinus-Blanc, 1987), two subspaces each made up of two platforms connected by an elevated runway, were explored separately (see Fig. 3.4a). Pieces of food were placed on each of the platforms. In a second phase, half the subjects explored the pathway AB (Fig. 3.4b) connecting the two subspaces. All the animals were finally given a shortcut test for two trials. They were shown the food at D and they were then placed at C (Fig. 3.4c). They had three possibilities: they could take the longer familiar pathway CBAD, or one of the two new shorter ones, CAD or CD. By comparison with the control group, which did not undergo the second exploratory phase and which chose the longer familiar path significantly more often that would be expected by chance, the experimental group displayed a preference for the unfamiliar shortcut CD (Fig. 3.5).

THE DISHABITUATION PARADIGM

The second approach which relies on measures of exploratory activity takes the evolution of this behaviour as an index of spatial knowledge. It consists of studying the decrease in exploratory activity over time (habituation) and its reactivation following a change (dishabituation). In this procedure, animals are first allowed to explore an open field containing one or several objects. The number as well as the duration of contacts with these objects is recorded. Although these measures of exploratory activity are not exhaustive, they are valid indices of object investigation which can be easily contrasted with more diffuse locomotor activity (Buhot, Soffié & Poucet, 1989; Buhot, Rage & Segu, 1989). After habituation of exploration has occurred, some spatial relationships between the elements of the situation are modified (e.g., by displacing one object). What is new is the spatial arrangement of the objects but not their individual characteristics since they have all been previously investigated. If the animals react to the change (e.g., by displaying a renewal of exploratory activity), it can then be assumed that this detection of spatial novelty relied upon a comparison being made between the current perceived arrangement and a representation or stored "internal model" of the initial situation.

Several species have been tested using such a procedure. Hamsters (Poucet, Chapuis, Durup & Thinus-Blanc, 1986; Thinus-Blanc et al., 1987), mice (Ammassari-Teule, Tozzi, Rossi-Arnaud, Save & Thinus-Blanc,

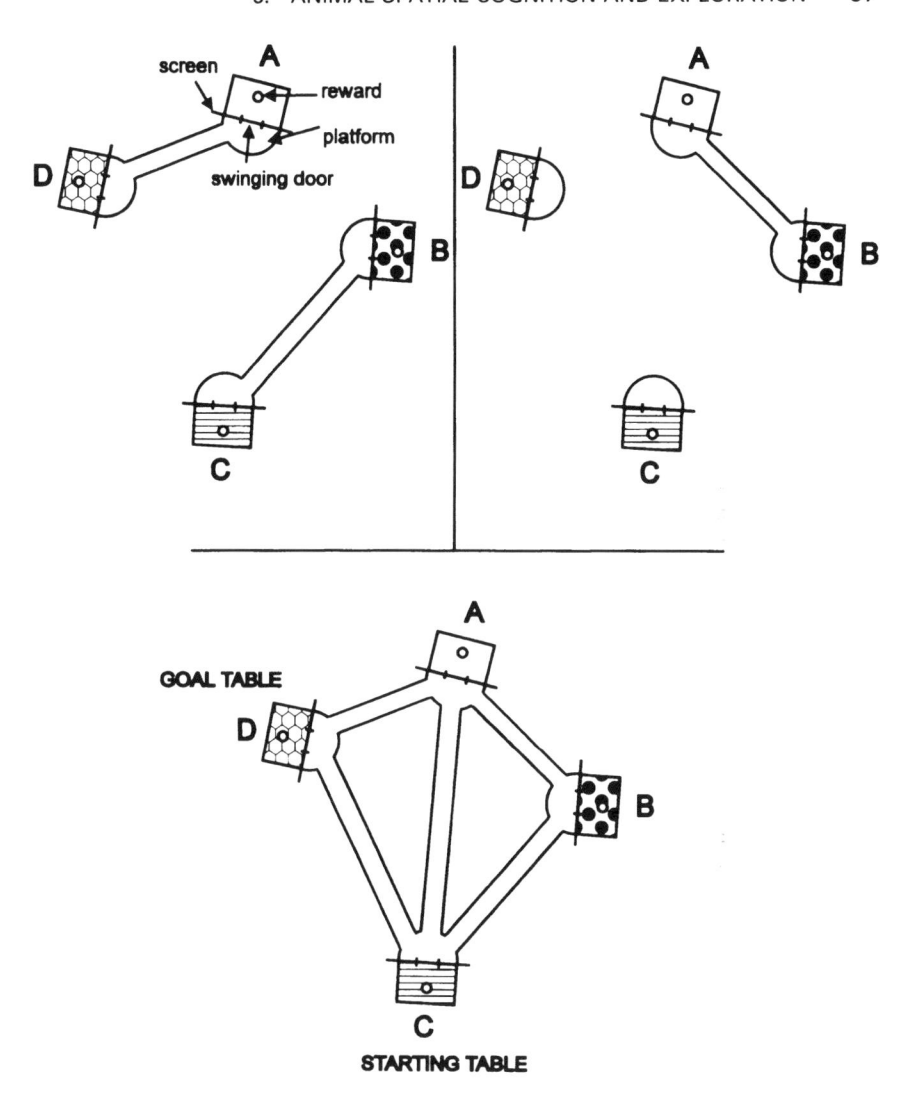

FIG. 3.4 Schematic representation of the apparatus for the three phases of the shortcut experiment. (Reproduced from Chapuis et al. 1987, with kind permission of The Psychonomic Society.)

1995; Misslin, 1983), mongolian gerbils (Cheal, 1978; Thinus-Blanc & Ingle, 1985; Wilz & Bolton, 1971), rats (Poucet, 1989; Lukaszewska, 1978) and monkeys (Menzel & Menzel, 1979; Joubert & Vauclair, 1986). All species display renewed exploratory reactions to a spatial change, which provide evidence that the animals have created an internal model of the initial situation. In addition, this dishabituation paradigm allows us to

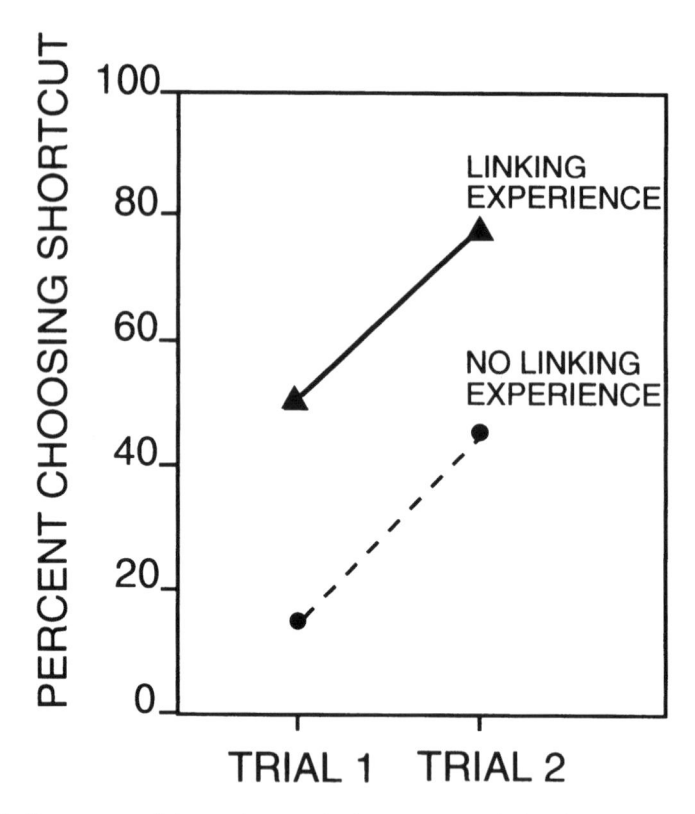

FIG. 3.5 Percentages of chosen shortcuts in the two groups during the test trials. (Reproduced from Chapuis et al., 1987, with kind permission of The Psychonomic Society.)

glimpse some of the spatial features that are spontaneously encoded during exploration, since not all kinds of rearrangement induce similar reactions.

Open field with objects

In a series of experiments (Poucet et al., 1986 ; Thinus-Blanc, 1988; Thinus-Blanc et al., 1987), hamsters were given three 15–minute sessions of exploration separated by several hours. The apparatus was a circular open-field containing four different objects, forming a square shape, located within a curtained environment, visually homogeneous except for a single striped patterned cue. The aim of using a curtained environment was to induce animals to focus their attention on the objects contained in the experimental field by preventing them from being distracted by irrelevant distal cues. The number and duration of their contacts with the objects were recorded.

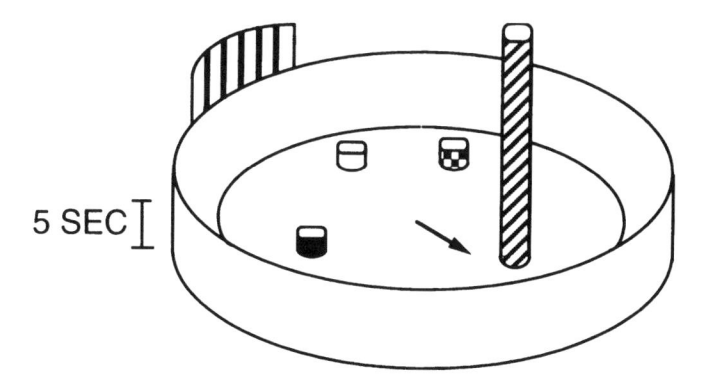

FIG. 3.6 Example of a deformation of the initial square arrangement which induces a selective re-exploration of the displaced objects. The "height" of the objects represents the difference in exploratory activity between the test session and the last session of habituation. (Adapted from Thinus-Blanc, 1988.)

After habituation had occurred from the first to the second session, a spatial change was made to the initial situation by displacing one or several object(s). The changes affected specific classes of spatial information, such as the overall geometric structure provided by the object set, the topological relationships between the objects, their absolute locations, and so on. For the control groups, the arrangement remained unchanged during the third session. Three types of behaviour were seen. (1) Re-exploration selectively directed towards the actually displaced object. This occurred when one object was set apart from the other ones (Fig. 3.6). The same pattern of results was obtained when the objects were identical. (2) Re-exploration directed towards all of the objects whether they had been displaced or not. Such nonselective behaviour was observed when one object was put closer to the others (Fig. 3.7) even with large distances between the objects. Hamsters reacted as if the situation were completely new. (3) No re-exploration was observed, as was the case when the four objects were displaced equally so as to form a larger square than the initial one. The reverse modification, consisting of reducing the size of the initial square, was no more effective in eliciting re-exploration. It is unlikely that such a massive change was not detected by the animals which had been able to detect far more subtle modifications in other tests. Therefore, a process of generalization may have occurred, preventing the animals from reacting to a change which modified neither the shape of the arrangement nor the topological relationships between the objects.

The specific conclusions which can be drawn from these experiments are the following. First, hamsters appear to encode spontaneously geometrical relationships since they always react, selectively or not, to a change bearing

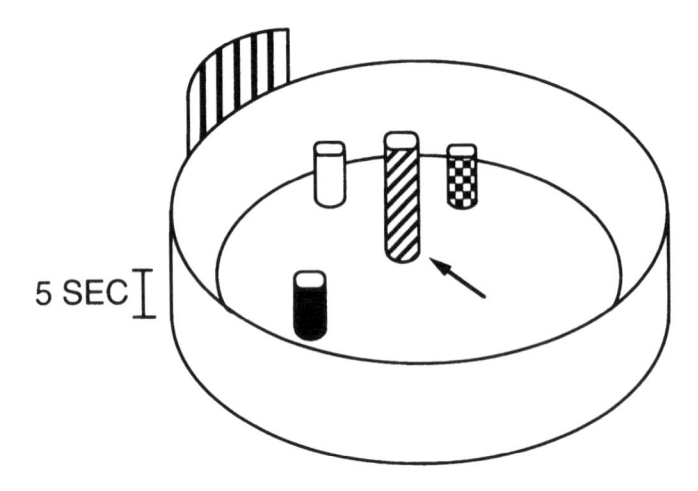

FIG. 3.7 Example of a deformation of the initial square arrangement which induces a diffuse re-exploration of all the objects. (Adapted from Thinus-Blanc, 1988.)

on this parameter. This finding is convergent with Cheng and Gallistel's data (1984) in rats. Second, absolute distances between objects do not appear to be an important spatial parameter, at least within the range of modifications used in this experiment. However, the level of familiarization and the requirements of the task may be important modulating factors in terms of the abilities displayed by the animals in evaluating distances. Finally, a third conclusion is that the striped pattern did not appear to play a role in the novelty detection, otherwise it would have provided the animals with a stable reference and would have been expected to enable the accurate detection of change in most of the tests. This failure to use the sole visual cue has been confirmed in a further study relying on the same procedure (Thinus-Blanc, Durup & Poucet, 1992). Removing the striped pattern just before the test session with a spatial change did not alter the reaction to this change.

At a more general level, these conclusions address theoretical issues related to the processing of spatial information. More specifically, the conclusion that animals encode geometrical relationships might mean that they have extracted the overall geometrical structure defined by the four objects. An alternative explanation, however, would be that hamsters have stored some "local views" in which relative distances and angles between the objects as seen from one or several vantage point or points are precisely encoded. This would not imply the use of an overall representation of the object arrangement. Modifications involving changes in these relative distances (and, thereby, geometric changes) could be detected on the basis of the comparison between a stored local view of the initial arrangement and the current local view. This might provide an explanation as to why animals react to any small change affecting the relative metric relationships between the objects (see Fig. 3.6), whereas they do not re-explore when a displacement of the four objects leaves their relative distances (and therefore the shape they define) unaffected.

However, other studies with different rodent species and procedures show animals' ability to react to change using allocentred strategies (i.e., representations which are independent of the animal's position). For example, Tomlinson and Johnston (1991) conducted an experiment with hamsters using two visually identical objects differing only on the basis of their olfactory characteristics. After habituation, the two objects were both associated with the olfactory cue previously used for one of them. Thus, the novelty can be described as "a familiar odour at an unusual place". When introduced into the field at a point 180 degrees from the point of entry for the habituation phase, hamsters reinvestigated the object whose odour had been changed.

Thinus-Blanc and Ingle (1985) demonstrated that gerbils are also able to use allocentred strategies. After habituation had occurred with an object located in a field containing a sole striped pattern, another object similar to the first (including the olfactory scents deposited on it) was located symetrically with respect to the first in relation to the striped pattern (Fig. 3.8). While gerbils had access to the apparatus through door D1 during the exploratory phases, they entered the field through D2 for the test. Despite being provided with a different scene at the very beginning of the test session, they accurately re-explored the object which was new on the basis of its location alone (Fig. 3.9).

These data are convergent with those obtained in rats by Xavier, Porto Saito, and Stein (1991) and Poucet and Buhot (1989). In both experiments, the animals displayed reactions to a change even if they approached the locus of change from opposite directions.

Finally, the dishabituation paradigm allows us to manipulate to some extent the nature of the information which is used during exploration. We have suggested (see the section, Exploration and problem solving) that visual continuity while exploring may be crucial for constructing an overall spatial representation. Using the dishabituation paradigm, Save, Granon, Buhot and Thinus-Blanc (1996) have recently provided experimental arguments in support of this hypothesis. Rats were allowed to explore separately and successively the two halves of a circular open field containing objects. The partition separating the two halves was either transparent or opaque allowing or preventing animals from visualizing the whole situation. In addition, locomotor access between halves could be permitted or denied via two openings at the extremity of the partition (Fig. 3.10).

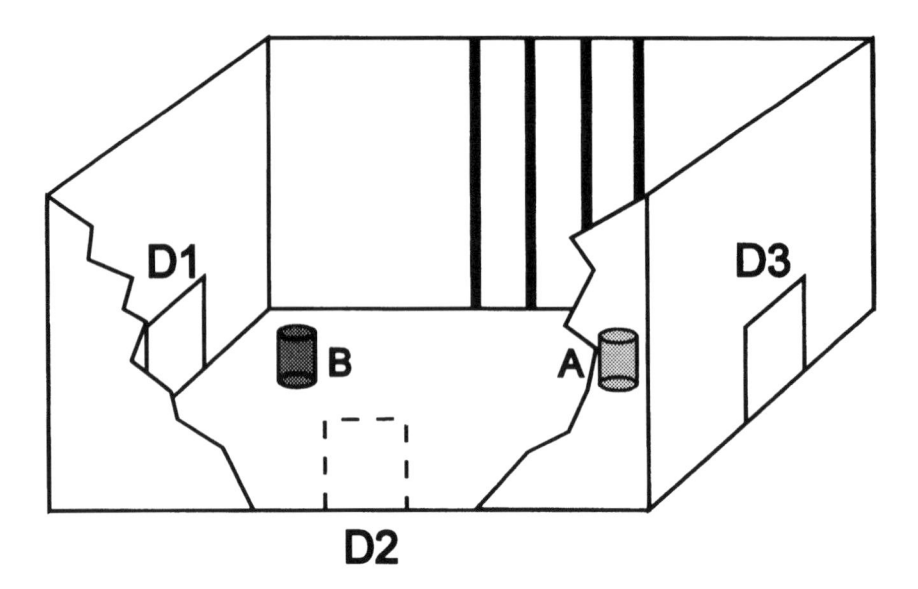

FIG. 3.8 Diagram of the apparatus with the familiar object in position B during the five trials of habituation. The "new" object identical to the familiar one is in position A in the sixth trial. Animals had access to the apparatus through door D1 in the habituation phase and through door D2 in the test phase. (Adapted from Thinus-Blanc & Ingle, 1985.)

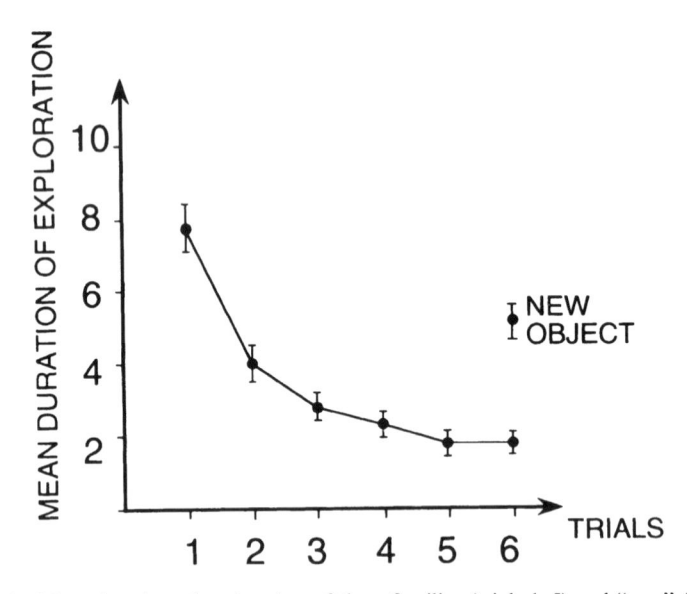

FIG. 3.9 Mean duration of exploration of three familiar (trials 1–5) and "new" (trial 6) objects. (Adapted from Thinus-Blanc & Ingle, 1985.)

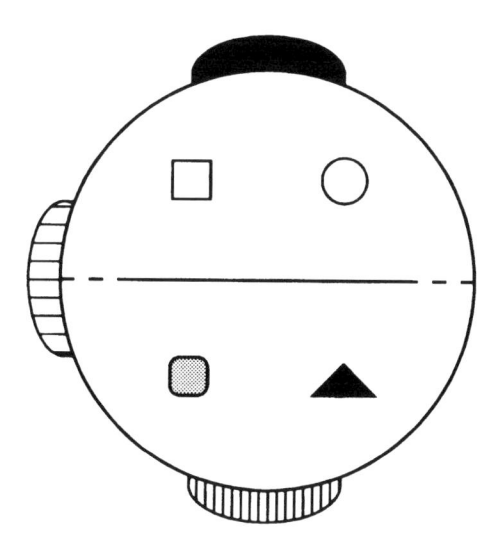

FIG. 3.10 Schematic representation of the open-field containing four objects and an opaque or transparent partition which could be opened at each of its extremities. (Adapted from Save et al., 1996.)

Therefore, four experimental conditions were defined: transparent and opened partition (group V+/L+), transparent and closed partition (group V+/L−), opaque and opened partition (group V−/L+) and opaque and closed partition (group V−/L−). Once exploration was complete, rats from these groups were exposed during the test session to the whole field without the partition. For four control groups (CNT) which had explored in the same situations as those presented above, the partition remained during the test session. The degree to which a unitary spatial organization had been built up was examined via a measure of reaction to novelty (contacts made with the objects). As in the other dishabituation experiments, the differences between the scores recorded during the test session and the last session before the test were computed, because it may happen that animals have different levels of overall exploratory activity but with the same variations (habituation and reaction to the change) during the course of the experiment. Whereas none of the CNT groups increased their exploratory activity, rats which had had visually discontinuous experience (i.e., with the opaque partition: V−/L+/EXP and V−/L−/EXP) whatever their locomotor experience of the two halves of the apparatus (whether continuous or discontinuous), displayed a renewal of exploratory activity (Fig. 3.11, A). Conversely, rats having explored with the transparent partition (V+/L+/EXP and V+/L− EXP) did not react significantly to its removal (Fig. 3.11, B). Although a tendency was observed

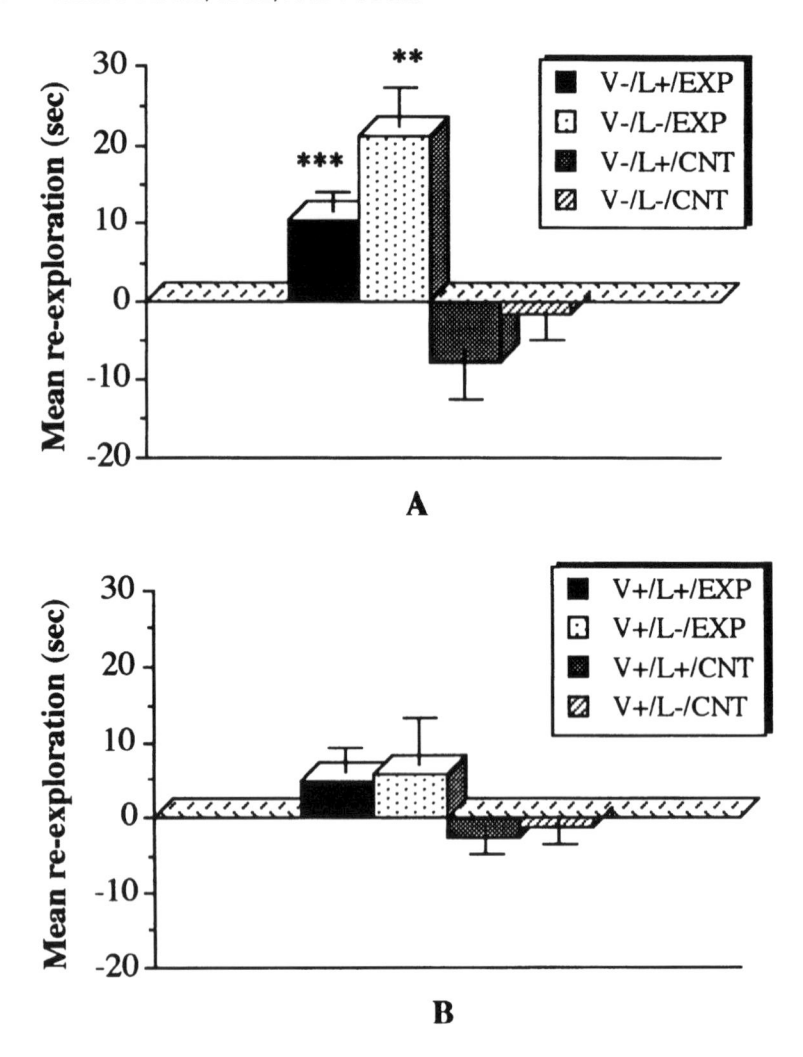

FIG. 3.11 Mean duration of re-exploration of the objects from the last session of habituation to the test session for the groups with the opaque barrier (A) and the transparent barrier (B) (see text for details). (Adapted from Save et al., 1996.)

in the group without continuous visual and locomotor experience (V−/L−/EXP) to react to the change more than the group with the discontinuous visual experience and continuous locomotor activity (V−/L+/EXP), the difference did not reach statistical significance. The lack of reactivity of rats that had been provided with a continuous visual perception of the whole situation, whatever their locomotor experience, strongly suggests that spatial relationships between the four objects had been already established on the basis of visual information since removing the partition did not induce an updating of the representation already set up.

A more parsimonious hypothesis is that the amount of re-exploration during the test reflects the process of updating the spatial representation, which varies as a function of the importance of the change. Indeed, re-exploration level was highest with the opaque partition without doors, next highest with the opaque partition with doors, next for the transparent partition without doors and least for the transparent partition with doors. However, the fact that the renewal of exploration during the test in the group V−/L−/EXP with neither visual nor locomotor continuous experience reached the same level as during the first phase of exploration (first exploration phase: m = 33sec, test: m = 32sec) supports the idea that at least this group had stored the subregions as independent representations. Accordingly, the exploratory activity of this group during the test appears to correspond to the formation of a new representation.

However one chooses to interpret these data, they demonstrate unambiguously that continuous and integrated visual experience is important for exploration.

Reaction to a missing stimulus

Other situations in the same vein have been devised which rely on habituation and dishabituation following a spatial change. The distinctive feature of these experiments is that whereas the initial environment that the animals explore is heterogeneous, the change consists of making the field uniform by removing a stimulus contained in it. Therefore, the reaction to novelty takes the form of a search pattern that can be more or less focused on the locus of the missing stimulus. For example, Corman and Shafer (1968) used a field with a black floor and a white square in the centre. After a phase of familiarization, the white square was replaced by a black one. Following this stimulus change, which made the floor visually homogeneous, an increase in approaches to the locus of the change was recorded.

Another dishabituation test of spatial memory has been devised (Thinus-Blanc, Save, Buhot & Poucet, 1991a; Thinus-Blanc, Save, Poucet & Buhot, 1991b; Save, Buhot, Foreman & Thinus-Blanc, 1992). The general procedure was the same as in the experiments in the circular open field discussed above but, instead of objects being contained in the field, a conspecific animal in a small cage was located under the glass floor. This type of "stimulus" elicits strong investigative reactions that can be readily quantified by measuring the time spent in its vicinity. After habituation had occurred, the stimulus was removed. During the test session

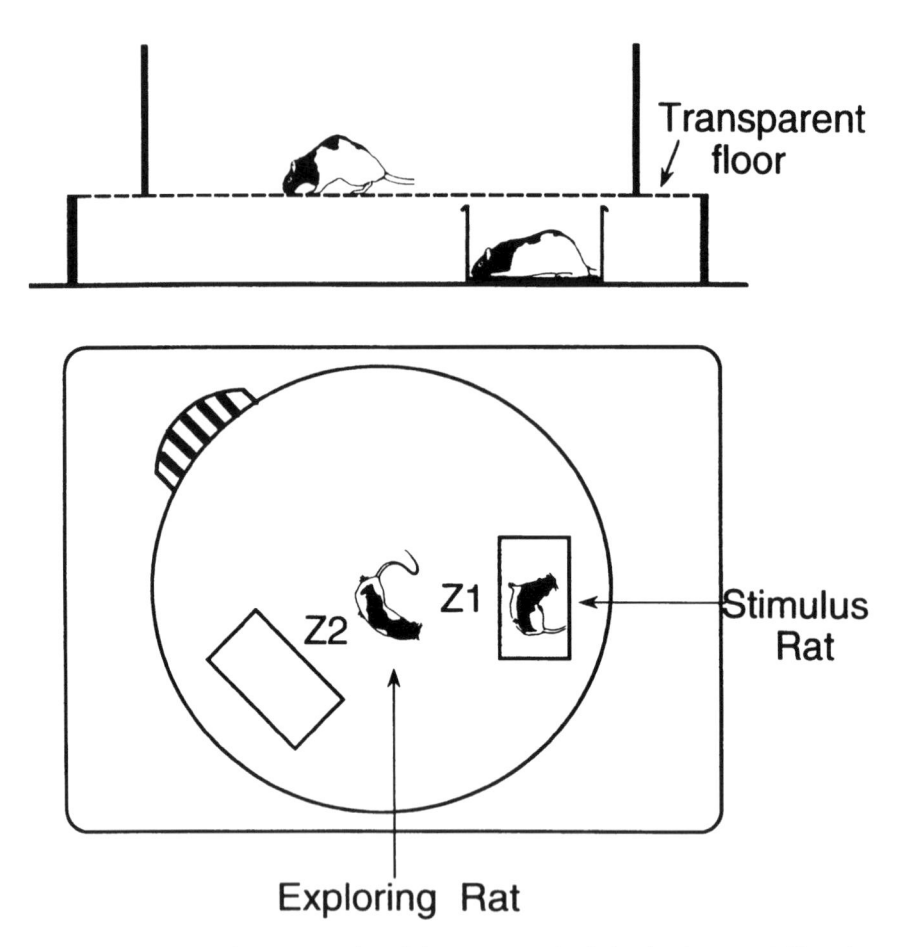

FIG. 3.12 A schematic representation of the apparatus: (top) the circular open-field used during sessions 2 and 3. During the test session (session 4), the stimulus rat was removed. The lower figure represents the corresponding video image. (Reproduced from Thinus-Blanc et al., 1991b, with kind permission of Oxford University Press, Oxford.)

the time spent at the locus corresponding to the previous location of the conspecific (Z1) was recorded and compared to the time spent above a neutral zone arbitrarily defined by the experimenter on the video screen (Fig. 3.12). Throughout the experiment, olfactory cues were carefully neutralized. After habituation had occurred, removing the stimulus rat elicited an increase in the time spent above place Z1 in comparison with the neutral zone Z2 (Fig. 3.13). This procedure has proven to be a useful tool for testing the effects of lesions or temporary inactivation of brain structures such as the hippocampus (see below).

NO OBJECT CONDITION

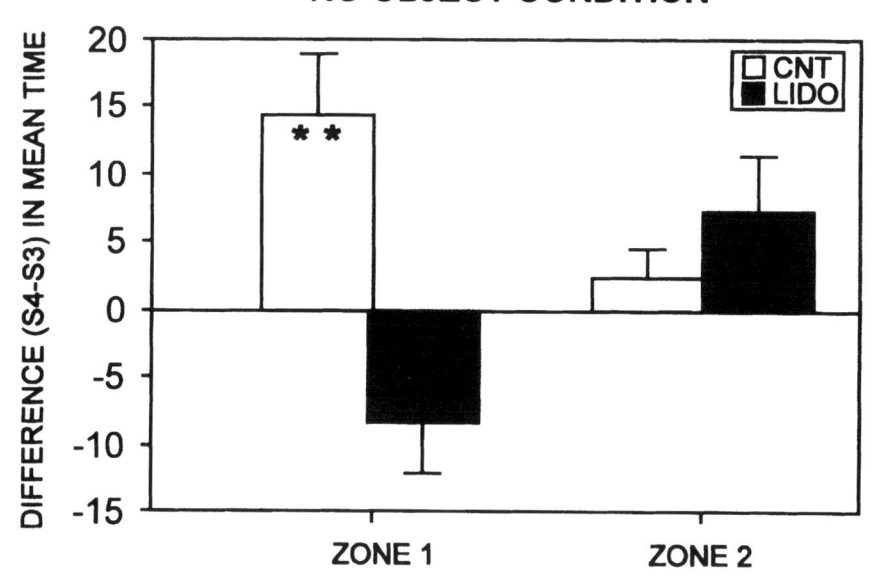

FIG. 3.13 Behaviour in response to the change. The histograms represent the difference between the time (mean ± s.e.m.) spent investigating the two zones Z1 and Z2 during the test session with the stimulus removed and the corresponding time during the preceding session with the stimulus under the floor of the apparatus, for the two groups CNT and LIDO. The stars indicate a statistically significant difference between the two sessions. (Adapted from Thinus-Blanc et al., 1991b.)

The dishabituation paradigm for studying hippocampal functions

Experiments based on the time course of exploratory activity address a short-term process and allow dissociation of acquisition and retrieval phases. In this respect, the dishabituation paradigm is useful for studying spatial memory and hippocampal function. These studies complement those described by Hodges (see Chapter 6) to examine brain damage in animals.

Lesions of the hippocampal formation in rats induce hyperactivity in the open field but this is not an expression of exploratory activity *per se*, since inter- and intra-session habituation is delayed or absent (Foreman, 1983; Poucet, 1989). Hippocampal lesions also abolish or decrease orienting responses which are components of exploratory activity (Raphelson, Isaacson & Douglas, 1985). Although sensory discriminative abilities are left intact, hippocampal rats are also impaired in terms of their reactions to spatial novelty (Markowska & Lukaszewska, 1981; Poucet, 1989).

However, it is hard to define the stage or stages of spatial information processing which is or are specifically hippocampus driven. Usually, lesions are made before the experiment which commences after a few days of postoperative recovery (Ammassari-Teule et al., 1995; Save et al., 1992). If lesions are made between the acquisition phase and the retrieval or test phases, the necessary period of postsurgical recovery inordinately lengthens the time interval between the two phases, in addition to which, surgical anaesthesia may have effects on brain functioning which could interfere with the evolution of the memory process. Therefore, with respect to the study of the functions of the hippocampus, it is necessary to design experiments in which not only the various phases of information processing can be dissociated at the behavioural level, but also in which the hippocampus function can be neutralized at any time in the course of the experiment. Of course, classical lesion techniques do not permit such an approach. In contrast, local injection of short-term tricaine anaesthetics (e.g., lidocaine or tetracaine) in animals equipped with intra-cerebral cannulae before the experiment allows us to inactivate the hippocampus. It becomes possible to test immediately, after *in situ* injections, the behavioural effects of such inactivations at any time in the course of a short-term experiment.

The following experiment in rats illustrates the advantage of combining the dishabituation paradigm with reversible inactivation of the hippocampus for studying the role of this structure in the various phases of spatial information processing (Thinus-Blanc et al., 1991a,b)

Before the experiment, all animals were surgically equipped with guide cannulae in which the injection needle could be inserted. The guide cannula was aimed at the ventral hippocampus. An injection of 1.5 microlitres at this location is completely confined to the hippocampus and blocks all of the hippocampal formation at this level (Sandkuhler, Maisch & Zimmermann, 1987). Half of the animals were injected with lidocaine while the other half were sham-injected. The injections were performed five minutes before the test session. The condition used was that described above, with a conspecific located under the glass floor. For the test, the stimulus rat was removed and time spent on the zone where the rat had been located during initial exploration was recorded.

The time spent above the neutral zone (Z2) did not vary during the course of the experiment. On the other hand, the stimulus elicited strong investigative reactions as indicated by the increased time spent by the animals above zone Z1 during Session 2 (with the stimulus rat) as compared to Session 1 (without the stimulus rat). This investigative behaviour decreased during the following Session 3 in both groups. Sham-injected rats reacted strongly to the removal of the stimulus during Session 4, by spending more time at place Z1 (where the stimulus was previously located) than at the neutral zone Z2. In contrast, lidocaine-injected rats did not display any sign of having noticed the change (Fig. 3.13). To sum up, reversible inactivation of the hippocampus did prevent the animals from reacting to the spatial change, in spite of a normal phase of exploration. This result parallels the effects of classical hippocampal lesions when they are made after spatial learning has occurred over a larger time scale.

In a second experiment, we attempted to make localization of the change easier to detect by adding some cues into the field. In fact, the "task" in the first experiment was very difficult because, except for the pattern on the wall, there were no cues inside the field. Several earlier studies have shown that hippocampal lesions have no harmful influence on performance scores when the animal is guided by cues which are closely associated with the goal. In contrast, hippocampal lesions have drastic effects when cues are far from a hidden goal, and thus when its location must be inferred from their configuration. This suggests that the hippocampus is not involved in "simple associative processing" (such as that involved in sensory guided displacements), but in "configural associative processing", such as that which underlies orientation relying on spatial cognitive maps (Sutherland & Rudy, 1989). In this case, therefore, cues closely associated with the stimulus in the exploration paradigm could facilitate the localization and search for the disappeared stimulus in lidocaine-injected animals.

The principle in this experiment was the same as for the previous experiment except that two objects were placed under the glass floor. For half of the subjects these objects were located close to the stimulus whereas, for the other half, they were spatially dissociated from the stimulus location. During the test session, the box containing the rat was removed but the objects remained at their respective locations in the field. The overall time course of exploratory activity was the same as in the first experiment. After removal of the stimulus, sham-injected rats displayed a consistent, although weak, reaction to the change, both when the objects were close to the stimulus (Fig. 3.14) and far from the stimulus (Fig. 3.15). However, the lidocaine-injected animals did not react at all to the change and continued to habituate in both conditions. The lack of reaction in the lidocaine-injected rats to the "associated object" (Fig. 3.14) condition may be due to several factors: the strength of the association between the stimulus and the cues may have been too weak either because of the absence of a strong reinforcer or because the procedure involved only a short period of acquisition time as compared to classical spatial learning procedures.

Both these experiments revealed that lidocaine injections into the hippocampus are effective in blocking the response-to-change behaviour displayed by normal animals although the initial phase of exploration occurred under normal conditions for all the animals. There are a number of explanations for the observed absence of reactions to the change in the

ASSOCIATED OBJECT CONDITION

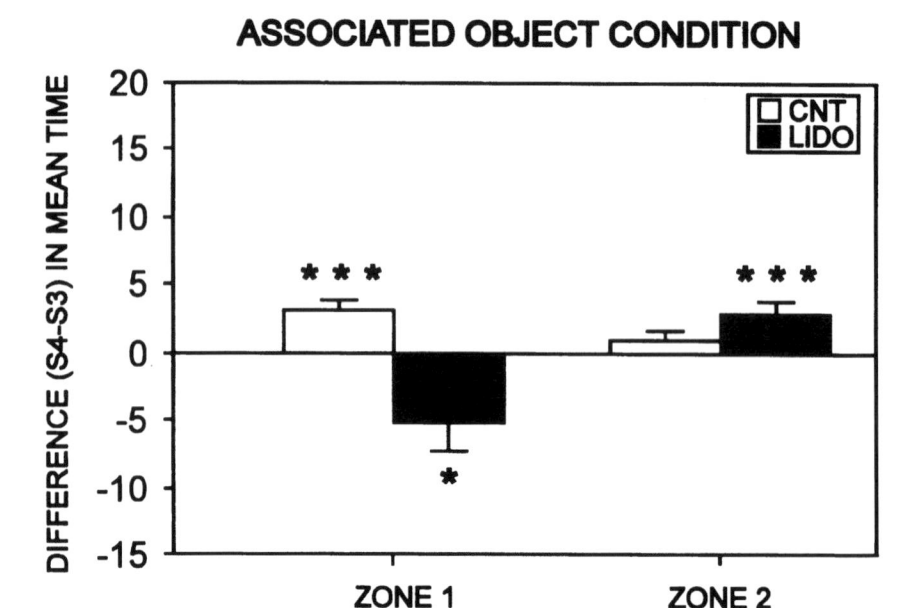

FIG. 3.14 Behaviour in response to the change in the condition with the objects close to the stimulus location. The histograms represent the difference between the time (mean ± s.e.m.) spent investigating the two zones Z1 and Z2 during the test session with the stimulus removed but the objects still in the field and the corresponding time during the preceding session with the stimulus under the floor of the apparatus for the two groups CNT and LIDO. The stars indicate a statistically significant difference between the two sessions. (Adapted from Thinus-Blanc et al., 1991b.)

injected animals, namely: (1) a failure to notice that a change has occurred; (2) an impaired spatial memory of the location of the disappeared stimulus; and, (3) a more general memory deficit. These different possibilities are discussed elsewhere (Save et al., 1992).

These experiments hint at the usefulness of the technique of reversible inactivation associated with the dishabituation paradigm in the search for the neural bases of spatial knowledge.

ADVANTAGES AND PITFALLS OF PARADIGMS BASED ON EXPLORATION

Exploratory activity is of considerable methodological interest in the study of spatial memory. The renewal of investigative activity after habituation implies that a change bearing exclusively upon the spatial features of the situation has been detected. Such a detection is possible only by referring to the initial situation, no longer present as a whole, of which the subject

DISSOCIATED OBJECT CONDITION

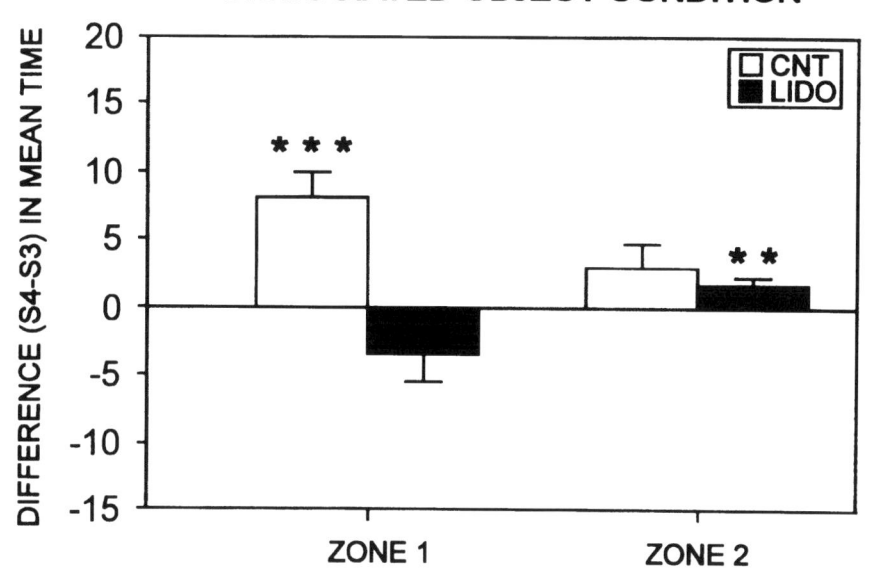

FIG. 3.15 Behaviour in response to the change in the condition with the objects dissociated from the stimulus location. The histograms represent the difference between the time (mean ± s.e.m.) spent investigating the two zones Z1 and Z2 during the test session with the stimulus removed but the objects still in the field and the corresponding time during the preceding session with the stimulus under the floor of the apparatus for the two groups CNT and LIDO. The stars indicate a statistically significant difference between the two sessions. (Adapted from Thinus-Blanc et al., 1991b.)

has stored a memory or representation that needs to be compared to the present situation. The fact that exploration that is incomplete, either in terms of its quantity or its quality, precludes solving spatial problems which demand the use of accurate representations provides complementary arguments for the idea that exploration has an important spatial function. Whereas problem-solving tasks allow us to glimpse the functional properties of spatial representations constructed during the exploratory phase, the dishabituation paradigm enables us to investigate the spatial parameters which are spontaneously encoded by animals.

In addition, concerning the study of the function of brain structures involved in spatial processing, such a method can be combined with temporary pharmacological inactivation of restricted areas. This combination enables the testing of new concepts, because it becomes possible to intervene at any given time during the experiment, and the effects can be tested immediately. This possibility of testing the role of a given brain structure at any time is of the greatest importance in the study of a process in which space and time are related dimensions. This allows us to study short- and intermediate-term spatial memory and perhaps to focus on the particular moment(s) when the hippocampus is likely to play a crucial role in spatial processes. The fact that it is possible to record hippocampal place-cell activity in freely moving animals while they are exploring strengthens the interest of this method.

This approach has other advantages. First, spontaneous behaviour is observed in animals that are not food-deprived. That simplifies the practical preparation of the experiment but, more importantly, it eliminates the risk that animals fail in a given task because they do not understand the rules, when they would have been capable of succeeding in other conditions. In addition, it is possible to determine what information about the environment is primarily important for the animal because, by comparison with classical learning tasks, these experiments are conducted within a relatively short period of time. The latter advantage may also be of interest for developmental studies since the duration of the experiment does not interfere with maturation.

It is also possible to manipulate a variety of factors, which confers upon this paradigm a potential for wide application, and interest for a large number of experiments and experimental questions. It enables us to investigate further the temporal course of memory, with different treatments during the inter-session intervals, which can themselves be varied, and to dissociate the various phases (acquisition, consolidation, retrieval) of the memory process. In studies aimed specifically at investigating spatial processing, it is possible to manipulate the amount and conditions of exploration, for example, controlling the parts of the apparatus that can be explored, fractionation of exploration, removing vision or other sensory modalities, and so on.

Nevertheless, studying exploration also has some drawbacks, though these can be easily minimized. The first one is related to neophobia. It may happen that instead of investigating, animals display freezing reactions. Such reactions are often observed in animals which have not been handled before the experiment. In addition to intensive handling, a familiarization phase in the empty test apparatus may be useful in reducing neophobia. In contrast, the fact that animals of the same species display different levels of exploratory activity is not a drawback provided that each individual displays habituation. If important within-group heterogeneity of this type distorts the results of statistical analyses, it is possible to quantify habituation (and dishabituation) in terms of percentages by reference to the maximal level of activity (number and/or duration of the contacts with objects) observed during the first session (cf., the example shown in Fig. 3.9).

A second disadvantage is that the general level of exploratory activity and the rate of habituation may vary depending on the species under study, in the same experimental situation (Poucet, Durup, & Thinus-Blanc, 1988). In addition, the general activity level interacts with the complexity of the situation, for instance the number and nature of the objects placed in the field. Therefore, if an original spatial arrangement is designed or if a species whose exploratory activity has not been extensively studied so far is used, it is necessary before undertaking the experiment proper to carefully adjust the design (vary the complexity of the situation, session duration, intersession interval, and so on) and to determine the optimal conditions for observing habituation.

It should be added that despite the increasing and justified popularity of automatic video-tracking systems, it is still necessary to observe animals' behaviour when one wants to study investigative activity. Automatic trackers have the disadvantage that they focus on the proximity of the subject to some specific location. Of course, the system is not able to determine which type of behavioural activity takes place in that location (e.g., grooming can take place close to an object, and might be mis-recorded as object investigation). This type of study requires the experimenter to interpret the nature of behaviours displayed by the animal. Finally, some elementary precautions have to be taken to control for olfactory cues. When displacing an object in the reaction-to-change situation, the nondisplaced objects also have to be touched by the experimenter in order to "equalize" the olfactory scents deposited on the objects. Moreover, it is recommended that animals of the same sex are run in an apparatus, which should be carefully cleaned before starting to test each new subject.

CONCLUSIONS

Studying exploration is a simple and reliable method of making progress toward understanding spatial processing. In addition, by approaching behaviour from another perspective, using original procedures like those described above, it is possible to design tasks that are complementary to classic learning tasks. Because of their short duration, and the fact that they do not depend upon reward being delivered by the experimenter, such studies can be conducted systematically before conditioning training and do not interfere with other paradigms.

Although the link between exploratory behaviour and spatial representations is unquestionable, there is still the pending and crucial problem of how information that is gathered during exploration is transformed so as to lead to the construction of representations, including spatial representations. Indeed, the link between exploration and representation is not obvious at first glance. Exploration is a sensorimotor activity, organized along a body-centred referent, entailing the position of the sensory receptors, the direction of the displacement, gravitational forces, and so on. In contrast, by their very nature, spatial representations, such as cognitive maps, must be independent of the subject's current position at a given time. Information is said to be allocentrically organized. It can be reasonably hypothesized that during exploration, feedback arising from the trajectory itself (vestibular, muscular, and other information) is matched with the ever-changing visual scenes of the environment. This matching would inform the subject that it is moving in a stable environment whose properties are invariant whatever their perceptual appearance. However, if it appears legitimate to conceive exploration as an interface between the physical world and its representation (whatever the complexity and level of abstraction of the latter), many further experiments are required. They will be aimed at determining how and where in the brain this transformation of information and the shift of reference frames take place.

REFERENCES

Ammassari-Teule, M., Tozzi, A., Rossi-Arnaud, C., Save, E., & Thinus-Blanc, C. (1995). Reactions to spatial and nonspatial change in two inbred strains of mice: further evidence supporting the hippocampal-dysfunction hypothesis in the DBA/2 strain. *Psychobiology, 23*, 284–289.

Berlyne, D.E. (1950). Novelty and curiosity as determinants of exploratory behaviour. *British Journal of Psychology, 41*, 68–80.

Berlyne, D.E. (1960). *Conflict, arousal, and curiosity.* New York: McGraw-Hill.

Berlyne, D.E. (1966). Curiosity and exploration. *Science, 153*, 25–33.

Buhot, M.-C., Rage, P., & Ségu, L. (1989). Changes in exploratory behaviour of hamsters following treatment with 8-hydroxy-2-(di-*n*-propilamino)tetralin (8-OH-DPAT). *Behavioural Brain Research, 35*, 163–179.

Buhot, M.-C., Soffié, M., & Poucet, B. (1989). Scopolamine affects more the cognitive processes involved in selective object exploration than locomotor activity. *Psychobiology, 17*, 409–417.

Chapuis, N., Durup, M., & Thinus-Blanc, C. (1987). The role of exploratory experience in a shorcut task by golden hamsters. *Animal Learning and Behavior, 15*, 174–178.

Cheal, M.-L. (1978). Stimulus-elicited investigation in the Mongolian gerbil (*Meriones unguiculatus*). *Journal of Biological Psychology, 20*, 26–32.

Cheng, K., & Gallistel, C.R. (1984). Testing the geometric power of an animal's spatial representation. In H.L. Roitblat, T.G. Bever & H.S. Terrace (Eds.), *Animal cognition* (pp.409–423). Hillsdale, NJ: Lawrence Erlbaum Associates Inc.

Corman, C.D., & Shafer, J.N. (1968). Open-field activity and exploration. *Psychonomic Science, 13*, 55–66.

Cowan, P.E. (1976). The new object reaction of *Rattus rattus L.*: the relative importance of various cues. *Behavioral Biology, 16*, 31–44.

Ellen, P., Parko, E.M., Wages, C., Doherty, D., & Herrmann, T. (1982). Spatial problem solving by rats: exploration and cognitive maps. *Learning and Motivation, 13*, 81–94.

Ellen, P., Soteres, B.J., & Wages, C. (1984). Problem solving in the rat: piecemeal acquisition of cognitive maps. *Animal Learning and Behavior, 12*, 232–237.

Foreman, N. (1983). Head-dipping in rats with superior collicular, medial frontal cortical and hippocampal lesions. *Physiology and Behavior, 30*, 711–717.

Glanzer, M. (1953). Stimulus satiation: an explanation of spontaneous alternation and related phenomena. *Psychological Review, 60*, 257–268.

Gray, J.A. (1982). *The neuropsychology of anxiety: an enquiry into the functions of the septo-hippocampal system*. Oxford: Oxford University Press.

Hall, C.S. (1934). Emotional behavior in the rat. I: Defecation and urine as measures of individual differences in emotionality. *Journal of Comparative Psychology, 18*, 385–403.

Joubert, A., & Vauclair, J. (1986). Reaction to novel objects in a troop of Guinea baboons: approach and manipulation. *Behaviour, 96*, 92–104.

Lukaszewska, I. (1978). The effects of exposure and retention interval on response to environmental change in rats. *Acta Neurobiologica Experimentalis, 38*, 323–331.

Maier, N.R.F. (1932). A study of orientation in the rat. *Journal of Comparative Psychology, 14*, 387–399.

Markowska, A., & Lukaszewska, I. (1981). Response to stimulus change following observation or exploration by the rat: Differential effects of hippocampal damage. *Acta Neurobiologica Experimentalis, 41*, 325–338.

Menzel, E.W., & Menzel, C.R. (1979). Cognitive developmental and social aspects of responsiveness to novel objects in a family group of marmosets (*Saguinus fuscicollis*). *Behaviour, 70*, 251–278.

Misslin, R. (1983). *Contribution neuroéthologique à l'étude des conduites néotiques chez la souris*. Thèse de Doctorat d'Etat, Université de Strasbourg.

O'Keefe, J., & Nadel, L. (1978). *The hippocampus as a cognitive map*. London: Oxford University Press.

Piaget, J. (1937). *La construction du réel chez l'enfant*. Neuchâtel, Switzerland: Delachaux & Niestlé.

Poucet, B. (1989). Object exploration, habituation and response to a spatial change following septal or medial frontal cortical damage. *Behavioral Neuroscience, 103*, 1009–1016.

Poucet, B., Bolson, B., & Herrmann, T. (1990). Spatial behaviour of normal and septal rats on alternate route maze problems. *Quarterly Journal of Experimental Psychology, 42B*, 369–384.

Poucet, B., & Buhot, M.-C. (1989). Scopolamine impairs response-to-change based on distal cues in the rat. *Physiology and Behavior, 46*, 355–359.

Poucet, B., Chapuis, N., Durup, M., & Thinus-Blanc, C. (1986). Exploratory behavior as an index of spatial knowledge in hamsters. *Animal Learning and Behavior, 14*, 93–100.

Poucet, B., Durup, M., & Thinus-Blanc, C. (1988). Short-term and long-term habituation of exploration in rats, hamsters, and gerbils. *Behavourial Processes, 16*, 203–211.

Raphelson, A.C., Isaacson, R.L., & Douglas, R.J. (1985). The effects of distracting stimuli on the runway performance of limbic-damaged rats. *Psychonomic Science, 3*, 483–484.

Renner, M.J. (1988). Learning during exploration: the role of behavioral topography during exploration in determining subsequent adaptive behavior. *International Journal of Comparative Psychology, 2*, 43–56.

Sandkuhler, J., Maisch, B., & Zimmerman, M. (1987). The use of local anesthetic microinjections to identify central pathways: A quantitative evaluation of the time course and extent of the neural block. *Experimental Brain Research, 68*, 168–178.

Save, E., Buhot, M.-C., Foreman, N., & Thinus-Blanc, C. (1992). Exploratory activity and response to a spatial change in rats with hippocampal or posterior parietal cortical lesions. *Behavioural Brain Research, 47*, 113–127.

Save, E., Granon, S., Buhot, M.C., & Thinus-Blanc, C. (1996). Effects of limitations on the use of some visual and kinesthetic information in spatial mapping during exploration in the rat. *Quarterly Journal of Experimental Psychology, 49B*, 134–147.

Sokolov, E.N. (1960). Neuronal models and the orienting reflex. In M.A.B. Brazier (Ed.), *The central nervous system* (pp.187–276). New York: Josiah Macy Jr. Foundation.

Sokolov, E.N. (1963a). *Perception and the conditioned reflex* (S.W. Waydenfield, Trans.) Oxford: Pergamon.

Sokolov, E.N. (1963b). Higher nervous functions: the orienting reflex. *Annual Review of Physiology, 25*, 545–580.

Sutherland, N.S. (1957). Spontaneous alternation and stimulus avoidance. *Journal of Comparative and Physiological Psychology, 50*, 358–362.

Sutherland, R.J., & Rudy, J.W. (1989). Configurational association theory: the role of the hippocampal formation in learning, memory and amnesia. *Psychobiology, 17*, 129–144.

Thinus-Blanc, C. (1988). Animal spatial cognition. In L. Weiskrantz (Ed.), *Thought without language* (pp.371–395). Oxford: Oxford University Press.

Thinus-Blanc, C., Bouzouba, L., Chaix, K., Chapuis, N., Durup, M., & Poucet, B. (1987). A study of spatial parameters encoded during exploration in hamsters. *Journal of Experimental Psychology: Animal Behavior Processes, 13*, 418–427.

Thinus-Blanc, C., Durup, M., & Poucet, B. (1992). The spatial parameters encoded by hamsters during exploration: a further study. *Behavioural Processes, 26*, 43–57.

Thinus-Blanc, C., & Ingle, D. (1985). Spatial behaviour in gerbils. *Journal of Comparative Psychology, 99*, 311–315.

Thinus-Blanc, C., Save, E., Buhot, & M.-C., Poucet, B. (1991a). The hippocampus, exploratory activity and spatial memory. In J. Paillard (Ed.), *Brain and space* (pp.334–352). Oxford: Oxford University Press.

Thinus-Blanc, C., Save, E., Poucet, B., & Buhot, M.-C. (1991b). The effects of reversible inactivation of hippocampus on exploratory activity and spatial memory. *Hippocampus, 1*, 363–369.

Thorpe, W.H. (1963). *Learning and instinct in animals*. London: Methuen (1st ed. 1956).

Tomlinson, W.T., & Johnston, T.D. (1991). Hamsters remember spatial information derived from olfactory cues. *Animal Learning and Behavior, 19*, 185–190

Wilz, K.J., & Bolton, R.L. (1971). Exploratory behaviour in response to the spatial rearrangement of familiar stimuli. *Psychonomic Science, 24*, 117–118.

Xavier, G.F., Porto Saito, M.I., & Stein, C.S. (1991). Habituation of exploratory activity to new stimuli, to the absence of a previously presented stimulus and to new contexts, in rats. *Quarterly Journal of Experimental Psychology, 43B*, 157–173.

4 The Radial Arm Maze: Twenty Years On

Nigel Foreman
University of Leicester, UK
Irina Ermakova
Institute of Higher Nervous Activity and Neurophysiology, Moscow, Russia

INTRODUCTION

The radial arm maze (RAM) is arguably one of the most prominent and influential experimental paradigms to have been used in spatial research. Since its introduction by the late David Olton (Olton, 1977; Olton & Samuelson, 1976) it has been used in its original form, and in modified forms, with several species, to investigate a broad range of issues. These include natural foraging behaviour, short- and long-term memory, spatial and nonspatial memory, working and reference memory, drug effects on behaviour, ageing effects, strain and species differences in spatial competence, and strategic choice behaviour, among others. It remains a popular paradigm, and its importance is widely recognised (e.g., see Brown & Huggins, 1993). Surprisingly, with the exception of brief reviews in literature concerned more generally with spatial cognition in animals (e.g., Thinus-Blanc, 1996), no comprehensive review has hitherto been published that charts its evolution and status. It is the purpose of this chapter to review the history of the paradigm over the twenty years since its introduction, and indicate the ways in which several theoretical problems have been been highlighted and/or resolved using the RAM in that period of time. Current practice in the use of the apparatus will be reviewed. Past and current uses of alternative test paradigms will be illustrated. It is hoped that this will be of use to scientists beginning to use the apparatus *de novo*, and who will want to be aware of the progress and pitfalls that have characterised the use of the RAM in the past.

The apparatus in its original form is built of wood or plastic, consisting of a central hub, raised off the ground (sometimes within centimetres of the ground, sometimes raised a metre above floor level), from which a number of alley ways radiate like the spokes of a wheel (Fig. 4.1, later in the chapter, illustrates the configuration; see also Hodges, Chapter 6 in this volume). The subject is placed at the centre of the maze and in the original form of the task, is typically allowed to investigate freely, taking small pieces of food placed in wells at the ends of each maze arm.

EVOLUTION OF THE RADIAL ARM MAZE

This task is the most recent in a series of what have been called Multiple-Goal Tasks (Horner, 1984), employed to investigate the ways in which animals learn about and remember places and, arguably, their spatial interrelationships. It combines aspects of many spatial tasks in which locations in the environment are connected by tunnels or runways, and animals must learn to approach particular places, take short cuts or detours, use familiar or unfamiliar routes, and/or discriminate places already visited from those unvisited within a trial.

Laboratory testing of animals' spatial skills via exploratory tasks and spatially distributed cues is covered elsewhere in this volume (Thinus-Blanc, Poucet & Save, Chapter 3) although it is worth reviewing here the features of test apparatuses that culminated in the emergence of the RAM *per se*. One example of an earlier form of spatial test was the Maier (1932) three-table reasoning task in which an animal that has explored three tables connected by runways had to return to one on which it has been fed when released from either of the other two. In order for the animal to perform effectively, it needed to discriminate locations in the environment and label one as "visited", others "unvisited". Perhaps the most direct antecedent of the RAM is the sunburst maze, used by Tolman, Ritchie and Kalish (1946). In this apparatus a rat had to run from a platform to a goal, G, taking a single available route incorporating a series of left and right turns. When they were running at asymptotic speed, the trained route was blocked off and animals had a choice among 17 straight radiating alleyways, spanning 180 degrees, leading from the central hub like rays of the sun. Animals demonstrated that they had discriminated the location of goal G by choosing, in most cases, alleys that pointed in the general direction of G. Unfortunately, the goal, G, was labelled by the presence of an overhead bulb, which may have encouraged simple body-orienting strategies. The original results of Tolman et al. (1946) have been challenged on both empirical (Gentry, Brown & Kaplan, 1947; Gentry, Brown & Lee, 1948) and theoretical grounds (Olton, 1978), although Harley (1979), using a 7-arm radial version of the task was able to confirm the original finding. Evidence from other types of study (error patterns in the Dashiell maze, e.g., Dashiell, 1930) also lend support to Tolman et al.'s (1946) conclusion, that rats show a "... tendency to approach the general location of the goal even when, at a distance, the goal-box itself must have been invisible" (see Thinus Blanc, 1996, pp.4–14, also Mackintosh, 1983, p.262). As Olton (1978) concluded, Tolman et al.'s (1946) experiment was itself not definitive, but the conclusions drawn from it are correct.

Although the above tasks all require spatial judgments and an appreciation of spatial relationships among widely distributed locations, they differ from the RAM insofar as they each demand that the animal *visits* a familiar place in order to obtain a reward. To trace the history of the RAM it is necessary to consider another type of task, requiring that subjects make multiple choices to distributed locations, *avoiding* locations previously visited within that particular trial. These can be traced back to multiple path elimination problems (Crannell, 1942; Lachman, 1965, 1966, 1969, 1971; Lachman & Brown, 1957). In these, the animal must make successive choices among an array of elevated radiating alleys visiting each just once; to do this, the animal must encode and remember which it has previously visited (eliminated) and avoid returning to those for the rest of that test session. Lachman used between 3 and 5 alleys (Lachman, 1965–1971; Lachman & Brown, 1957). Interestingly, instead of making sequential choices (e.g., to alleys 1-2-3-4-5 in that order) rats usually made choices that were maximally divergent one from the next, giving sequences such as 4-1-5-2-3. Buhot and Poucet (1986) have used more elaborate branching versions of the original test, also finding a general preference in rats for highly divergent choices.

The RAM is thus essentially a path elimination task in which, in order to perform the task using a spatial strategy, animals must identify and remember (and distribute choices nonredundantly among) several widely distributed spatial locations.

The RAM configuration was first employed in its conventional form by Potegal (1969), although in his study, animals made only one choice per trial. Released into one of 12 arms, rats were required to emerge from the starting arm and choose another by making a specific angle of turn consistently in a particular direction (e.g., 30 degrees clockwise, or 60 degrees anticlockwise). Rats with lesions in the frontal subcortex (striatum) were poor at this "egocentric" localising task. However, Potegal made an interesting incidental observation, that when animals failed to find the correct arm, their exploration of the remaining arms was remarkably well organised. In particular they would move from one arm to another, avoiding return visits to previously sampled arms. Thus, with minimal training, the animals seemed to be sampling efficiently and systematically, as Lachman and Brown (1957) had observed in their own discrete trial version of a similar task. In particular, rats ran preferentially to arms that they had not previously visited on that day, reminiscent of the well-known spontaneous alternation that animals display in two-choice T- and Y-mazes (Dember, 1956; Dennis, 1939; Douglas, 1966), a tendency that is so strong that perseverative responding to a consistent location can be very difficult to train (Petrinovich & Bolles, 1954; Foreman, Toates & Donohoe, 1990). In single trial alternation tasks, despite the fact that animals are in unnatural laboratory conditions (being manually handled and replaced at a start position following each spatial response) their high degree of organisation and strong dispositions suggests that alternation behaviour reflects an innate tendency, perhaps related to natural foraging behaviour when searching for food. The RAM took maximum advantage of this tendency, allowing animals complete freedom to explore and make a sequence of self-paced choices.

Olton and Samuelson (1976) and Olton (1977) first introduced the "continuous" version of the task, in which every arm of the maze is baited with a reward, and in which, therefore, the optimal strategy is for the animal to run from arm to arm, visiting each arm once within a trial. Note that a *trial* in this context refers to the whole sequence of arm choices prior to the subject's removal from the maze.

PROCEDURES: PRACTICAL CONSIDERATIONS FOR THE STANDARD CONTINUOUS VERSION OF THE TASK

Although inconvenient features have been pointed out (Masuda et al., 1994), the continuing popularity of the RAM must derive in part from the ease with which it can be administered and scored, the simplicity of the apparatus (particularly in its original form), and the speed with which a high level of performance is acquired by most species that have been tested (Mundy & Iwamoto, 1988; Rawlins & Deacon, 1993). Practical difficulties are few. Helpful hints on the practical running of the task are provided by Rawlins and Deacon (1993). However, the apparatus has often been modified or used in different ways across laboratories (see Roullet & Lassalle, 1992). Although the apparatus can be enclosed, it is usually located in a cue-rich laboratory room, containing items of furniture and pieces of apparatus. It is usually illuminated via a single light source positioned 2 metres above the central platform. Extraneous sounds can be masked via a radio, or the noise of ventilation equipment. The experimenter is usually seated in the test room, although a curtained environment can be used with an overhead mirror enabling observation to be made from outside the curtained environment (e.g., Packard & White, 1991) or closed-circuit video enabling observation from an adjacent room (e.g., Toumane, Durkin,

Marighetto, Galey & Jaffard, 1988). The apparatus can be automated (e.g., Peele & Baron, 1988; Pallage, Orenstein & Will, 1992).

Animals have been tested using various rewards, including liquids (plain water, for water-deprived animals, or milk), solid foods such as cereal grains or seeds, or sweet foods such as chocolate where treatments induce a dry mouth and/or reduce food palatability (e.g., Watson, Hewitt, Fone, Dickinson & Bennett, 1994). Food reward in the form of 45mg saccharin pellets is most usual, in which case animals are typically tested 23 hours hungry, being allowed access to food for 1 hour per day following each test session. Pretraining can utilise just a central platform with a single arm attached, which is useful if a pretraining (food retrieval latency) criterion is required to equate groups prior to treatments (Gallagher, Bostock & King, 1985). More usually pretraining involves the entire apparatus, the surface having been liberally scattered with food pellets, on which animals are allowed to eat freely. Food is progressively restricted to the ends of the maze arms over 2–3 days, at which point animals must run the full length of each maze arm to retrieve the reward (usually 1–2 saccharin pellets per arm, contained in a shallow food well a few centimetres from the distal end of the arm.) Experimenters may place pellets at the entrance to a maze arm, at the midpoint and in a food well at the end of the arm on day 1, middle and food well on day 2, and food well only on day 3 (e.g., see Caprioli, Ghirardi, Giuliani, Ramacci & Angelucci, 1991). The availability of food during pretraining is not a trivial issue, since lesion effects on performance can depend upon reinforcer exposure during pretraining (Packard, Winocur & White, 1992).

DATA COLLECTION

It is important to set an appropriate criterion for an arm visit, particularly since animals sometimes engage in vicarious sampling from the centre platform prior to making arm choices (Einon, 1980; Brown, 1992). For the rat, a subject is often said to have visited an arm if all four paws have been placed on the arm (Olton & Werz, 1978; Roberts & Dale, 1981; Simon, Taghzouti & Le Moal, 1986) although some authors only record a visit when an animal has passed an arbitrary landmark along the arm (Pico & Davis, 1984; Poucet, Lucchessi & Thinus-Blanc, 1991) or progressed to the distal end of the arm (Suzuki, Augerinos & Black, 1980). Although in recent reports, experimenters frequently omit to specify such parameters, this is not a trivial issue, since information signifying choice may variously be acquired at the point where the animal enters the maze arm, while it traverses it, or from the point where it is rewarded (see Poucet et al., 1991). It is likely that most experimenters adopt a compromise criterion such as the animal's "having to pass the half-way point on the maze arm with both paws" (cf. Kobayashi, Kametani, Ugawa & Osanai, 1988), which is indeed probably the most suitable. The majority of responses made by food-deprived rats involve running to the distal ends of maze arms even when the animals are not reinforced on the maze; only nondeprived rats often fail to do this (Timberlake & White, 1990).

Animals occasionally fail to take food on a maze arm, or jump off an arm, particularly if the maze is close to floor level, though these phenomena occur only on days 1–3 of testing and almost never thereafter. Data collection usually begins (day 1) on the day when the animal first makes at least N choices on an N-arm maze, this being the first day on which a conventional accuracy score can be calculated (the number of separate arms visited in choices 1–N). A normal rat will typically obtain a high score on a 4-arm maze in 10 training sessions (Olton, 1978), on an 8-arm maze (7.5–7.8 separate arms visited in choices 1–8) in 20 sessions (Olton, 1977; Olton & Samuelson, 1976) and on a 17-arm maze (15 separate arms visited in choices 1–17) after about 50 trials (Olton, Collison & Werz, 1977). Thus, the RAM can be employed, and performance quantified for groups of subjects, with considerable confidence. Its widespread use means that an inexperienced experimenter has an accurate impression of how a normal subject ought to behave, and thus an accurate baseline against which to compare treatment effects.

MEASUREMENT OF PERFORMANCE: THE STANDARD CONTINUOUS VERSION OF THE TASK

Animals are usually deemed to have reached a performance criterion when they first achieve what is roughly the typical asymptotic number of correct choices in choices 1–N for an N-arm maze [e.g., 3 out of 4 correct in a 4-arm maze (Mundy & Iwamoto, 1988), 7.5 or 7.8 correct in choices 1–8 on an 8-arm maze (Olton, 1977), 10.75 correct on a 12-arm maze (Buresova, 1980), and 14.5 on a 17-arm maze (Olton et al., 1977)]. Alternatively, a criterion can be set across days, such that on an 8-arm maze animals may be required to make 7 correct responses in the first 8 choices and obtain all 8 rewards within the first 10 choices, for 5 consecutive trials (Deyo & Conner, 1989). Of course, "N" represents the minimum number of choices in which the subject can visit all arms once if performing perfectly. However, variability on this score is low, particularly on the 8-arm maze where it usually ranges between 5 or 6 [scores which approximate chance; see Eckerman (1980), Lanke, Mansson, Bjerkemo & Kjellstrand (1993)] and 8 (perfect performance) on an 8-arm maze, and between 8 and 12 on a 12-arm maze (Buresova, 1980). Clearly, it does not represent an interval scale of measurement and therefore ought to be analysed using distribution-free, nonparametric statistical procedures although parametric statistics are most often applied to group means. This measure has the disadvantage that it utilises data from the first N choices only. In some studies, criterion has been set only marginally above chance, for example, where animals must make "no more than 2 errors per trial for 3 consecutive trials" in an 8-arm maze (Bostock, Gallagher & King, 1988).

Alternative measures of choice accuracy that take into account all of the choices permitted are: (a) the number of choices made prior to the first error, (b) the total number of visits required by a subject to visit all N maze arms, (c) the total number of errors made before all N have been visited, or (d) the number of errors made in the course of finding the last k visited arms (see Bruto & Anisman, 1983; Dachir, Kadar, Robinson & Levy, 1993; Fagioli, Rossi-Arnaud, & Ammassari-Teule, 1991; Simon et al., 1986). A subject may make several error revisits in the course of finding one or more "missed" maze arm(s) late in the trial, and so these latter measures can be sensitive discriminators between groups. Nevertheless, such measures also have drawbacks, in that to allow subjects an unlimited number of choices would mean that incompetent subjects get a great deal more exploratory experience than those more competent. It is difficult to say whether such "extra" maze experience is beneficial (perhaps enhancing the integration of spatial information), or detrimental (perhaps creating interference between choices, both correct and incorrect choices.) In either case, there is a confounding effect. Some authors do continue testing until all arms have been visited, however long that takes (Godding, Rush & Beatty, 1982; Roberts & Dale, 1981; Simon et al., 1986), but in most cases where trials involve errors, the total number of choices permitted in any one trial is subject to a limit of some sort. Authors vary in the protocol adopted; most remove animals after a particular number of choices *or* an elapsed time, whichever occurs first. For example, for the 8-arm maze, until all 8 have been visited or 10 minutes have elapsed (Olton & Samuelson, 1976; Pearson, Raskin, Shaywitz, Anderson & Cohen, 1984; Winocur, 1982); 16 choices or 10 minutes (Yoerg & Kamil, 1982); 16 choices or 7 minutes (Goodale & Dale, 1981); 17 choices or 5 minutes (Foreman, 1985; Foreman & Stevens, 1982; Watts, Stevens & Robinson, 1981); and 16 choices or 20 minutes (Einon, 1980). Bond, Cook and Lamb (1981) allowed up to 30 minutes per trial when testing pigeons, although it must be doubtful whether responses made at the start and end of such a long interval can always be *assumed* to form components of a unitary choice sequence, despite the reported longevity of spatial working memory, discussed in a later section ("Persistence of memory").

Although it is important to avoid overexposure to the maze in certain subjects, it is equally important to ensure that poor performance does not prevent a subject from experiencing particular maze arms, for example when animals are removed after having made N choices on an N-arm maze (Buresova, 1980). On such terminated trials, it is usual to manually transport a subject to any arms still remaining unvisited and allow them to take the unretrieved rewards. These practical issues are important, to minimise variability among laboratories. Since exploratory experience is important to the establishment of cognitive representations (Thinus-Blanc, Save & Poucet, Chapter 3, this volume), the degree of exploration, time on the maze, and locations visited are potentially significant variables. Handling practice is also potentially important. Grobety and Schenk (1992) allowed 1 minute of exploration or 3 extra choices following the last correct arm choice, presumably to ensure that a final correct choice was not differentially rewarded or punished by handling and removal from the maze.

Time taken to complete the task (visit all maze arms) is often measured and used as an index of competence (e.g., Lichtman, Dimen & Martin, 1995), though there are many influences on this variable, which is at best an indirect measure of spatial competence. Activity rate (running speed, or exploratory persistence) could be important for learning; more active strains of mice make more correct arm choices than slower ones (Ammassari-Teule & Caprioli, 1985), although activity level has been reported by others to be independent of choice accuracy (e.g., Cocchetto, Miller, Miller & Bjornsson, 1985). Clearly, hyperkinetic activity must be distinguished from exploratory activity; hippocampally damaged animals that show excessive activity are hypoexploratory, and they perform the radial maze particularly badly (see Foreman & Stevens, 1987). In conjunction with other measures, "time to complete" might distinguish deficits due to exploratory changes from memorial deficits (Watson et al., 1994).

It is interesting to note that most measures of choice accuracy are more robust for larger mazes, which are preferable if a large apparatus can be accommodated. Practically, the average laboratory room may have difficulty accommodating a 17-arm configuration, or 24-arm hierarchical configuration (Roberts, 1979). It is interesting that Olton (1977) originally described the 8-arm maze, which rapidly became the laboratory standard. The 8-arm RAM is quick to run, and animals learn rapidly. However, in terms of performance measurement, as we shall see later, other configurations (e.g., 12 arms) might be preferable for several reasons.

The probability of making a correct response (i.e., an initial visit to a maze arm) by chance alone diminishes with successive correct choices, as more locations become eliminated from the to-be-visited set. Thus, statistical models have been developed for various response contingencies, both on path elimination tasks (Lachman & Brown, 1957) and for the RAM (Buresova, 1980; Eckerman, 1980; Foreman, 1985; Leis, Pallage, Toniolo & Will, 1984; Spetch & Wilkie, 1980). The P(cor) measure (Kessler, Markowitsch & Otto, 1982), which adjusts for the likelihood of a correct response being made by chance, at each choice point, was devised when the radial maze was first introduced (Olton et al., 1977; Olton & Samuelson, 1976). P(cor) is in fact not a probability but a probability ratio, reflecting the degree to which performance at any point in the choice sequence exceeds or falls below what would be expected by chance. A score of zero represents random choice, although a perseverative subject can obtain a negative value. Shors and Dryver (1992) have provided an alternative within-trial measure that reflects the degree of success or perseverative error behaviour exhibited at successive choice points; animals acquire increasing numbers of positive points as the trial progresses and choices become more difficult, from which are deducted error points. The points weightings seem arbitrary (since changes in chance probabilities are nonlinear), yet the formula does provide a useful and simple measure of performance.

Classic occupancy models can be used to calculate the size of the memory buffer being used by a subject in making choices across a trial (Gillett, 1994, 1996; Glassman, Garvey, Elkins, Kasal & Couillard, 1994), to estimate the number of arms that would be visited by chance in m choices, and hence the degree to which subjects improve upon this score at any point in the choice sequence (8–20 choices: Buresova, Bures, Oitzl & Zahalka, 1985; Leis et al., 1984), the degree to which performance deviates from chance according to informational analysis, using first- and second-order informational terms (Suzuki et al., 1980), and to correct for obviously nonrandom behaviours. For example, unless animals have suffered vestibular damage (Ossenkopp & Hargreaves, 1993) or stressful stimulation immediately prior to testing (Shors & Dryver, 1992), it is almost unknown for them to make successive responses to the same maze arm. Incorporating this constraint into a Monte Carlo simulation, Eckerman (1980) was able to increase the accuracy of probability estimates of correct scores against which treatment effects may be compared. Spetch and Wilkie (1980) have used simulation programmes to generate many thousands of random choice sequences; when compared with occupancy calculations (Eckerman, 1980; Olton, 1978) they are in close agreement, as Spetch and Wilkie (1980) point out. Chance performance on an 8-arm maze (that generated by the hypothetical memoryless animal) is usually taken as 5.25, but a score of 5.5–5.6 is said to be more appropriate by Lanke et al. (1993), who describe a model which partitions scores according to the contributions made by spatial choice and by response patterning.

CONTAINMENT BETWEEN CHOICES

In Olton and Samuelson's (1976) and Olton's (1977) original reports, it was implied that rats performing the task chose randomly among the arms available, although at an early stage in the development of the apparatus, these authors began to use a containment procedure in order to reduce the influence of nonrandom choice strategies, such as choosing the next adjacent arm in a consistently clockwise or anticlockwise direction, on each successive choice. These algorithmic strategies (which will be considered in detail in a later section) only provide optimal performance in the version of the maze in which all arms are baited; although they appear in versions of the task in which only some maze arms contain reward they do not promote a high level of performance, and diminish gradually (Okaichi & Oshima, 1990). Containment (or "post-selection confinement" [Peele & Baron, 1988]) procedures involve the equipping of the RAM with guillotine doors surrounding the central platform, each giving access to one maze arm. The doors can be raised and lowered individually or simultaneously. A trial begins with the doors open, thus giving the animal a relatively unrestricted view of the test environment. When the animal runs through a door, all other doors are closed. The animal may now return from the chosen arm via the open door on to the central platform, whereupon the final door is closed and the animal is contained for a fixed period of time [usually around 5 seconds (e.g., Molina-Holgado, Gonzalez & Leret, 1994), or 8 seconds (e.g., Van Gool, Mirmiran & Van Haaren, 1985)], before all doors are simultaneously opened for the next choice to be made. Confinement periods of 20 seconds are sometimes used. Duration is probably not important, in view of the longevity of spatial working memory, which is at least 4 hours for several arm choices (Beatty & Shavalia, 1980a,b). The open door is only closed when the animal has made a full run to the distal end of the maze arm and returned to the centre, thus avoiding the problem of setting "maze visit" criteria. The same scoring methods are applied as for the continuous version, although "time to complete a trial" will vary according to confinement intervals (e.g., by delaying the initiation of responses on occasions) and is thus a less valid measure on the containment version.

Containment greatly reduces algorithmic responses made to adjacent maze arms (Olton & Werz, 1978; Stevens, 1981), but it has the drawback that it is unnatural, being experimenter- rather than subject-paced; it also requires more training than the free-choice version, and the movement of guillotine doors is never silent and animals can be seriously distracted, especially in early training (see Pearson et al., 1984). Occasionally a rat will sit by the "next" door in algorithmic sequence between choices (R. Stevens, personal communication). Peele and Baron (1988) used an automated version of the maze with programmed postselection confinement intervals of 0.5sec. or 100sec. Interval length had no effect on choice accuracy but did affect the patterning of arm selection. Roullet and Lassalle (1992) developed a RAM for mice which gives good control over confinement via a drawbridge system, which the authors argue avoids stressing animals in automated equipment where occasionally they might be caught by falling doors.

For tasks in which response sequences are interrupted via long delays interpolated at the half-way point in the choice sequence, either by removal from the maze or containment, Cook, Brown and Riley (1985; p.456) provide a "deviation from control" performance measure based on the expected probability of revisiting arms chosen prior to the delay, against which postdelay choice behaviour can be evaluated. This complements the usual range of measures (applied to the noncontainment version) such as the number of choices required to visit all previously unvisited maze arms.

PERFORMANCE STRATEGIES: HOW DO ANIMALS COMPLETE THE TASK?

It is likely that the ways in which choices are made throughout a trial undergoes change, as familiarity with the apparatus improves, and/or strategic choice behaviour changes. It may also depend on the performance level of individual animals, whether they are contained on the centre platform between choices, the degree of algorithmic constraint on choices, and cue availability. Moreover, the way in which choices occur in relation to spatial cues is a matter of debate (Thinus-Blanc, 1996). The cognitive mapping theory of O'Keefe and Nadel (1978) is frequently used as the framework within which choice behaviour is analysed. According to this model, each maze arm represents a separate and discriminable place within the subject's internal map of the experimental environment. Following a series of responses, animals may be predisposed to approach locations ("places") which remain within the yet-to-be-visited subset. Every response is assumed to be made within the context of the overall spatial environment, and the spatial interrelationships between arms guide responses. Other authors regard maze arms as being treated independently of one another in the form of a "list", and remembered (chosen or rejected) without reference to their relative positions within a spatial map (Brown, 1992; Olton, 1978). Gaffan and Davies (1981) argued that alternation between maze arms is the basis of choice, but that to implement alternation, when approaching a particular maze arm, the animal needs to make a go-no go choice as to whether the arm is perceived as familiar (i.e., recently visited; therefore no-go and choose another arm) or unfamiliar (novel, recently unvisited, therefore go and visit). More recently, studies by Brown (Brown, 1990, 1992; Brown, Rish, Von Culin & Edberg, 1993) have examined the behaviour of rats on the central platform as they make exploratory movements toward maze arms (termed "microchoices"), and the relationship between microchoices and actual maze visits ("macrochoices"). Analyses of videotapes of rats' choice behaviour showed that, in the standard version of the maze especially, microchoices were not targeted selectively toward baited maze arms. Toward the end of a trial, many arms that were the target of microchoices were correctly rejected (were not visited) prior to the selection of a correct (previously unvisited) arm. A simulation model based in part on the behaviour of actual rats (microchoice distributions and acceptance/rejection rates), but which made no assumptions about relative maze arm positions in a cognitive map, generated choice accuracy scores similar to those achieved by actual rats.

However, not all macrochoices are preceded by microchoices, and probably different rats perform the task in different ways; there is considerable variation among animals in hesitancy on the centre platform and individuals vary in the range and variability of algorithmic reponse sequences exhibited at different points in training (cf., Foreman, 1985). Of course, the assumption that an animal makes a discrimination at each arm entry on the basis of familiarity (Brown, 1992; Gaffan & Davies, 1981) is not entirely inconsistent with cognitive mapping theory, if it is assumed that spatially distributed cues, rather than a single label, are used to discriminate an arm as familiar (visited) or unfamiliar (unvisited). Indeed, as Brown recognises (Brown, 1992), there are data showing that animals' arm choices in mazes are more accurate the wider the view that they have of the surrounding environment (Mazmanian & Roberts, 1983), and rats have been shown to use cue arrays rather than specific arm labels via cue rotation studies using arm-specific labels (Suzuki et al., 1980). They acquire information both when embarking on an arm visit and in the course of the subsequent arm traverse, and they remember and use such multicue information when making subsequent choices (Roberts, 1984; Poucet et al., 1991). Moreover, they can track individual maze arms successfully, as though using a "cognitive map" (Wathen & Roberts, 1994). In a hierarchical RAM, consisting of 8 primary arms radiating from a central platform, with 3 arms leading from the end of each, rats appear to integrate the spatial features of the maze, apparently holding separate "maps" for primary and secondary arm configurations (Roberts, 1979). It is unlikely that any one model provides a universal explanation of radial maze choice, although it is clear from Brown's data (1992) that there are at least some choice processes involved in RAM performance which cannot be assumed to be survey map based (Thinus-Blanc, 1996). A feature of the RAM is that the entire environment is visible from the central platform, and thus the use of distributed visuospatial cues and local views of cue configurations may represent a form of "mapping" that is different from cue use when moving and navigating among visually disconnected spaces. Note that if "snapshot" memory (Collett, Cartwright & Smith, 1986), or local view information (Thinus-Blanc, 1996) is used to identify maze arms, small movements of the RAM from one part of a test room to another should be sufficient to disrupt performance. To the authors' knowledge this has not been tested. Hodges (Chapter 6, this volume) provides evidence for the combined use of spatial and associative learning in RAM performance, which she relates to neural mechanisms.

Staddon (1983) suggested that performance can be based upon three levels of strategy: random choice, algorithmic patterning, or true spatial mapping, to which Dallal and Meck (1990) added a fourth, arguing that hierarchical spatial labelling or "chunking" represents the highest, most sophisticated, use of spatial information.

Extramaze vs. intramaze spatial cues

It might have been expected that animals could be tested on a cued version of the radial maze simply by leaving salient food rewards at the distal end of each maze arm that are clearly visible from the centre platform; however, this does not prove successful (rat: authors' own observations; pigeon: Dale, 1988). Similarly, it ought to be a simple matter to use cue subtraction and substitution to identify the environmental stimuli that are used in the radial maze, but spatial cues are subject to substantial redundancy. Most radial maze studies have been carried out in cue-rich environments; the standard cluttered laboratory typically provides an ideal environment, although in order to manipulate cues, a more restricted situation with curtaining and a small number of salient cues is preferable. For example, the removal of one or two out of four salient distal cues leaves performance unaffected on the RAM, but the removal of three has a significant effect (Pico, Gerbrandt, Pondel & Ivy, 1985), suggesting that relative cue positions are used to specify arm locations. However, since cues are subject to considerable redundancy, gross manipulations of multiple objects scattered about the test room usually have remarkably little effect on performance, as shown in examples cited below.

Animals may be biologically predisposed to learn about complex arrays of distributed cues. Suzuki et al. (1980) have shown via cue manipulations that rats on the radial maze rely upon overall cue configurations rather than individual cues or arm "labels" (see Roitblat, 1982). This is perhaps the more surprising since in the Suzuki et al. study, specific cues were used to "label" maze arms. Studies in which cues have been relocated within test environments have shown that maze rotations rendering extramaze cues ambiguous adversely affect performance (Foreman, 1985), as does a shift in room geometry (Williams, Barnett & Meck, 1990; cf. Gallistel, 1990). Intramaze cues can also be used as the basis of performance (Brown et al., 1993; Kraemer, Gilbert & Innis, 1983; Woodruff, Baisden & Cannon, 1993, although see Olton & Collinson, 1979), particularly in animals lacking vision (Dale & Innis, 1986), and can perhaps be used more effectively than extramaze cues in terms of acquistion speed, yet when intramaze and extramaze cues are compounded, extramaze cues take precedence (Diez-Chamizo, Sterio & Mackintosh, 1985). Byatt and Dalrymple-Alford (1996) found, using a differentially baited maze (see below), that removal of distal extramaze cues reduced both working and reference memory accuracy, but that removal of proximal intramaze cues produced a further drop in performance. Bostock et al. (1988) identified animals using intramaze strategies on the conventional RAM by conducting trials with rotated maze arms when animals reached criterion; those returning to previously visited places in order to traverse a *maze arm* that had not been previously traversed were deemed to be using an intramaze cue strategy. Approximately 70% of animals were found to be using extramaze cues. When animals are specifically forced to use intramaze visual cues in a win-stay design, such as when rewarded maze arms (which vary from day to day) are marked with lights, animals find this task difficult. Packard and White (1991) required rats to make two rewarded visits per day to each of the 4 out of 8 maze arms marked with a light—intuitively an easy task. Nevertheless, after 10 days of training (more than 80 choices), control animals were obtaining scores 60–70% correct, (i.e., making only approximately 5/8 "stay" responses within the illuminated set). In that case, the use of specific labels appears to conflict with the more natural tendency to alternate among locations discriminated via extramaze cues.

Cue use has been shown to change with time (Ellen, 1980) and vary among subjects (Bostock et al., 1988), and therefore the effects of manipulations or cue rotations on performance may vary depending on the point in training at which they are carried out. Gross rearrangement of room cues has remarkably little effect on performance in well-trained animals, unless *all* such cues are rendered ambiguous with respect to maze arm positions (Foreman, 1985; Williams et al., 1990.) Wilkie and Slobin (1983) found that when gerbils trained in one room were switched to testing in a different room altogether, performance fell, but only slightly. This suggests that animals' performance on the maze is in part due to an alternation tendency that is independent of room cues, plus a learned component relating to the particular test environment, reflected by the constant, repeated, rapid acquisition seen when rats are tested in successive new rooms compared to the slower acquisition seen in initial training (Gallagher et al., 1985). The systematic ambiguous reorganisation of familiar room cues (rat: Foreman, 1985) may create interference in memory, perhaps explaining why this disturbs performance more than the replacement of cues by a totally new set (gerbil: Wilkie & Slobin, 1983). The rearrangement of landmarks from day-to-day delays learning in pigeons (Spetch & Honig, 1988).

The spatial nature of the radial maze, and the wide spatial separation of maze arms, does seem to be an important factor in determining the ease with which it is learned. Rats generally prefer to take maximally divergent routes when performing path elimination tasks (Buhot & Poucet, 1986; Lachman, 1965–71; Lachman & Brown, 1957). It is intuitively easier to maintain in memory a distinct separation between alternative routes (and the places to which they lead) when these are widely separated in space, or to operate a more rigorous criterion when alternating among them (see Brown, 1992). Of course, more divergent choices are also consequently more widely separated in time, which might be a crucial factor (Buhot & Poucet, 1986). The more overlap that exists between cues associated with adjacent response locations, the greater would be the ambiguity in their spatial memory representations. Indeed, spontaneous alternation in a 2-choice runway is equally affected by alley separation—alternation rate falls dramatically when the angle between choice alleys is made acute (Douglas, Mitchell & Del Valle, 1974) and rats find difficulty in other tasks if different paths lead to the same goal (Horner, 1984; Sutherland, 1957). Thus, it is not surprising that in a parallel arm equivalent of the radial maze, animals learn comparatively slowly (Buresova & Bures, 1981; Dale, 1982; Horner, 1984), although overtrained rats can perform quite well (Buresova, 1980). This phenomenon applies to a variety of animals: cattle also perform less well in a parallel than a radial maze (Bailey, Rittenhouse, Hart & Richards, 1989.) When rats are given a choice between radiating and parallel arms within the same apparatus, the parallel arms tend to be neglected (Schenk & Grobety, 1992.) Buresova (1980) has shown that when maze alleys are dismantled and spread over a two-square metre area, rats will still make visits to each with an accuracy approaching that exhibited when the alleys are arranged radially, thus the radial arrangement of choice arms is not crucial to the task, though in general, spatial separation appears to be. Grobety and Schenk (1992) arranged for subsets of arms to be either vertically ramped or horizontally separated by small or large angles. Ramping of some maze arms had little effect on performance, perhaps slightly improving it by comparison with performance in the standard version of the task. With maze arms separated by differing angles, animals preferred maximally horizontally diverging arms, making algorithmic patterned responses among them, and they relied preferentially on horizontal (extramaze) cues rather than (intramaze) ramping to guide choices when the maze was rotated.

Poucet et al. (1991) used versions of the same 8-arm maze, with 4/8 arms baited (see the section on differential baiting, below). Mazes had either straight arms, or angled arms that required animals to make a sharp turn at the end to reach the goal. This enables the direction of trajectory to be separated from the place at which reward is consumed. In transfer tests, animals were more affected by having to learn new reward locations than when forced to learn new response trajectories. The authors conclude that cues sampled at both the place of reward and during locomotion on maze arms are used in a complementary fashion to guide choices, though the former are the most crucial.

Choice criteria

Brown (1990, 1992) has considered the choice criteria which animals apply, particularly in larger mazes and with maze arms of different lengths. The fact that performance can be improved by making it more difficult for animals to gain access to maze arms suggests the operation of a choice criterion (Brown, 1990; Brown & Huggins, 1993). Twelve- and 16-arm mazes, particularly with short arms (30cm as against the usual 70cm or 80cm) produce a poorer performance, not because the arms cannot be easily differentiated, but apparently because animals adopt a lax response criterion when choosing. Where mazes have long and short arms, the short arms are visited first (even when made comparatively less attractive by painting them white, against black long arms); they are visited most often repeatedly in error, and they are most likely to be visited in the course of algorithmically patterned sequences. Free choices following forced choices are more likely to be subject to errors if short arms are offered rather than long ones.

Olfaction

It might be expected that species having highly developed olfaction would use odours and the decay of their own recent odour trails to identify visited from unvisited arms within a trial. Most experimenters do take the precaution of rotating the maze arms between days, and wiping urine from the woodwork (e.g., with a dilute acetic acid solution) between subjects. Nevertheless, Olton (1977) and Olton and Collison (1979) took the trouble to examine whether rats were using odour cues by rotating the woodwork between choices. Rats continued to choose places they had not visited despite having to traverse runway arms that had been previously visited in the trial. The nonuse of odours is perhaps surprising in such highly olfactory animals (cf. Durup, 1964; Means, Hardy, Gabriel & Uphold, 1971). That olfaction is unimportant in the radial maze is illustrated by the fact that rats, particularly in early training, will revisit an arm in which their own faeces have previously been deposited on an earlier visit in the trial (own observations). Of course, this is not to say that animals are incapable of using alternative modalities to vision, and Buresova and Bures (1981) have found that odours can contribute to maze performance in an enclosed maze. A variety of sensory cues can be used by animals deprived of some sensory (particularly visual) input (Dale & Innis, 1986; Suzuki et al., 1980; Zoladek & Roberts, 1978). Lavenex and Schenk (1989) have found that supplementary olfactory cues were used explicitly only in darkness, and that the spatial configuration of olfactory cues required linkage to the external spatial frame of reference.

LEARNING AND MEMORY

Spatial vs. operant learning

A number of authors have been concerned with the nature of spatial learning, and whether the learning that is required to visit or avoid previously visited places involves different mechanisms from nonspatial discriminations. Most authors agree that it does (Mackintosh, 1983, p.264; Morris, 1982; Mundy & Iwamoto, 1988; Nadel & MacDonald, 1980; Roberts & Dale, 1981; Sutherland & Mackintosh, 1971). In particular, spatial learning appears to involve the simultaneous, rapid processing of much contextual, distributed, relevant-redundant information in a seemingly unlimited spatial memory store. It appears to have greater ecological validity than conventional operant learning of fixed stimulus-response associations (Olton, 1978). Indeed, in the initial stages of learning an operant discrimination "spatial solutions" (place strategies) are adopted and then rejected by the animal before it gravitates to the appropriate stimulus dimension (O'Keefe & Nadel, 1978). Olton (1978, p.341) has commented that:

> ... the vast majority of current experimental procedures have made spatial learning impossible. Either spatial cues are minimized as in an operant box, or spatial cues are explicitly made irrelevant as in a visual discrimination task in a T maze. This situation is paradoxical. On the one hand, rats consistently demonstrate a preference for solving discrimination tasks on the basis of spatial cues; on the other hand, experimenters just as consistently prevent the rats from effectively exploiting this preference.

Thus, compared with spatial learning, the learning of operant, N-choice discriminations is slow (Sutherland & Mackintosh, 1971, p.279). It seems unlikely, as Mackintosh (1983, p.264) points out, that: "... a rat would be able to remember which of eight, let alone 17 auditory stimuli had occurred as a CS on a given conditioning trial". Across a diverse range of species, from nectar-feeding birds such as the Hawaiian Honeycreeper, Amakihi (Kamil, 1978) to children searching hundreds of locations (Cornell & Heth, 1986), many spatial choices can be made before repetition occurs—many times the number of cue labels that could be remembered individually (or could conceivably be discriminated). Spatial learning is clearly a particularly fast and accurate form of learning about multiple cues, and complex arrangements of cues. Conventional reinforcement principles appear to apply to revisit frequency when the arms of a RAM are baited according to a fixed interval schedule (Elsmore & McBride, 1994). Various models of associative, conditional and configural memory have been discussed in connection with spatial behaviours generally (see Hirsch, 1980; Maki, 1987; Sutherland & Rudy, 1989; Thinus-Blanc, 1996) and to the selective disruption of memory functions by brain damage (see Hodges, Chapter 6 in this volume).

Memory: Parameters of performance

Whatever form of representation (Roberts, 1984; Roitblat, 1982) is used to discriminate and choose among maze arms, the N arms can be regarded as a set of to-be-remembered locations. Behaviour on the maze is frequently considered with reference to human and primate memory models, including examination of serial position effects, interference, and the effects of interchoice delays. The major findings, mainly from rodents, have been discussed previously in various forms (Mizumori, Rosenzweig & Kermish, 1982; Olton, 1978; Roitblat, 1982).

Proactive interference

The task is subject to proactive interference (PI), particularly when trials are massed within a day and when long retention intervals are used (Maki, Beatty & Clouse, 1984a; Roberts & Dale, 1981; Wilkie & Slobin, 1983). Memory for arms visited survives interruption of a choice sequence via removal of the animal from the maze (Beatty & Shavalia, 1980a,b) for long periods of time. Olton (1977, 1987) argued that when a trial has been fully completed (i.e., all arms have been visited) the short-term spatial working memory store is almost fully reset. Nevertheless, such "flushing" of memory (Roitblat, 1982, p.354) is at best an imperfect process (perhaps similar to the "directed forgetting" phenomenon in humans), since accuracy as a function of choices declines less sharply on the first trial than on subsequent trials within a day (Roberts & Dale, 1981). Memory may be fully reset if stimuli are relocated between trials (Suzuki et al., 1980). PI decays (Maki et al., 1984a) and is greatest when interfering tasks involve repetition of arm choices (Hoffman & Maki, 1986).

Retroactive interference

Several authors have used interpolated tasks to investigate retroactive effects in rats, either finding no retroactive interference (RI) (Maki, Brokofski & Berg, 1979), or a little in exceptional circumstances: where interpolated experiences consisted of forced choices on 1, 2 or 3 mazes in separate rooms from the test room, RI was only evident with 3 interpolated mazes. Thus rats can keep at least three trials separate in memory, perhaps using a separate store for each (Pico & Davis, 1984).

Capacity

Rats can remember as many as 32 separate places on a hierarchical radial maze (Roberts, 1979), and thus spatial memory in rats is said to clearly exceed Miller's (1956) 7 ± 2 bits, applicable to human short-term memory (Pico & Davis, 1984; Magni, Krekule & Bures, 1979). However, Glassman et al. (1994) argue that rat and human memory capacities both approximate Miller's magic number.

Serial position

A clear recency effect occurs, since revisits are more commonly made to arms visited in the early part of the choice sequence than later choices (Kesner & Novak, 1982).

Persistence of memory

Spatial memory can last for up to 6 hours, and perhaps even for 24 hours in rats (Beatty & Shavalia, 1980a,b) and in some birds (see Balda & Kamil, 1988 for a discussion). Markowska, Buresova and Bures (1983) reported a figure of only 40 minutes in rats. They argue that for foraging/patrolling purposes, a more durable memory for places recently visited would not be ecologically advantageous to a rodent. However, Maki, Beatty, Hoffman, Bierley and Clouse (1984b) criticised the latter conclusion, believing that Markowska et al.'s (1983) results may have been affected by their use of a small-diameter radial maze apparatus. On both 8- or 12-arm radial mazes, administration of amphetamine immediately after animals have chosen half of the arms reportedly improves recall over an 11-hour retention interval (Strupp, Bunsey, Levitsky & Kesler, 1991).

Prospective vs. retrospective memory

Cook et al. (1985) have argued for a dual-code model, in which already-visited arms are stored in memory until the mid-point in the reponse sequence when working memory suddenly converts to holding prospective information about still-to-be-visited locations. Of course, as the authors recognise, there is a functional equivalence between these two forms, inasmuch as one can be deduced from the other. They found that interpolated interruptions had greatest disruptive effects when introducd at the mid-point of the choice sequence. Brown, Wheeler and Riley (1989) argued that the data are better explained by assuming that animals adopt a progressively tighter choice criterion as a trial progresses.

Memory for patterns of reinforcement among maze arms

Researchers have attempted to identify the limits of performance on the task, and the ability of animals to remember rules, by baiting patterned subsets of arms from trial to trial. Wathen and Roberts (1994) used a maze in which 4 out of 8 arms were baited with 2 pellets (b) or no pellets (n) in a regular fashion (such as b-n-b-n, or b-b-n-n-b-b-n-n). Animals demonstrated tracking by entering baited arms early in the choice sequence, and were capable of tracking up to four different patterns. Animals were most successful when access to maze arms was made difficult, perhaps because this encourages greater attention to environmental cues, or tightens animals' choice criteria. Increasing stereotypy in arm choice across trials (which is an uncommon phenomenon in radial maze studies) suggested to the authors that in their study, animals used a trial-number hypothesis (discriminating the ordinal position of successive trials) rather than applying a patterning rule to individual maze arms. Clearly, a wide range of "chunking" strategies may be available according to the particular task demands and reward configurations.

DIFFERENTIAL BAITING: WORKING VS. REFERENCE MEMORY

An alternative version of the radial maze, which has been used by some experimenters in tandem with the continuous and/or cued versions in both rats and mice (Ammassari-Teule & Hoffmann, 1993; Ikegami, Nihonmatsu, Hatanaka, Takei & Kawamura, 1989; Mundy & Iwamoto, 1988; Ward, Mason & Abraham, 1990), can be used to distinguish between two components of memory that are arguably conflated within the continuous, free-choice version, namely, working memory (WM) and reference memory (RM) (Olton, Becker & Handelmann, 1979). The neural bases of this dissociation are discussed by Hodges (Chapter 6, this volume).Working memory (WM) is defined as within-trial memory for information that must be retained during a single trial and which enables the animal to remember, at any point in the trial, which maze arms have been already visited. It is essential that WM is erased between trials, to avoid proactive interference from earlier choice sequences, since a choice sequence is uninformative once completed. In contrast, reference memory (RM) is defined as memory for invariant information that must crucially be retained from day to day, and trial to trial, allowing performance to improve across days/trials. This distinction was first emphasised in the present context by Honig (1978). When only a subset of radial maze arms is baited consistently over days, animals must restrict their choices to the baited subset (i.e., using RM) and in addition avoid revisits to arms within the baited subset which have already been visited within that trial (i.e., using WM). Control rats take about 40 trials to learn which 8 arms out of 17 are baited, irrespective of the configurations of the baited arms (Olton et al., 1979).

Differential baiting paradigms have become popular in studies using rats and mice, numerous studies using 4 arms baited out of 8 (e.g., Jarrard, 1980; Stokes & Best, 1988; Whishaw & Tomie, 1996), others 3/8 (Gage, 1985; Toumane et al., 1988) or 5/8 (Davis, Baranowski, Pulsinelli & Volpe, 1987), also 6/12 (Byatt and Dalrymple-Alford, 1996), 8/12 (Brown et al., 1993; Williams et al., 1990), 9/16 (Murray & Fibiger, 1986), 8/17 (Olton et al., 1979; Olton & Papas, 1979), 4/7 with hamsters (Jones, McGhee & Wilkie, 1990) and 4/8 with lemurs (Picq, 1993). Performance criteria on the 8-arm maze (with 4/8 arms baited) have been set at: 3 correct choices in the first 4 (Mundy & Iwamoto, 1988) or stabilisation of response accuracy at >= 87% over 4 days (implying that 4 out of 4 unrepeated baited arm entries occur in the first 4 choices on at least 2 of the 4 days), taking about 30 days to achieve (Wirsching, Beninger, Jhamandas & Boegman, 1989). Both rats and mice successfully reverse a 4/8 configuration, in terms of both errors and re-entries, in about 4 days (Whishaw & Tomie, 1996; see below). Young rats appear to acquire reference memory in advance of working memory (Green & Stanton, 1989).

Note that a form of spatial reversal learning can be tested since the subsets of baited and unbaited arms can be reversed (Keseberg & Schmidt, 1995; Whishaw & Tomie, 1996) or modified following an initial period of training (Gage, 1985). Indeed, authors have used the differential version of the task to obtain multiple measures of acquisition, reversal and extinction within a single study, generating a large number of data (mice: Toumane et al., 1988). Okaichi and Oshima (1990) have used both cued and place versions. In the place version, 4 out of 8 maze arms (i.e., places) were always baited. In the cued version, 8 intramaze patterned inserts were placed on the maze arms, 4 of which were always baited (though their locations varied from day to day). Rats were capable of learning these two tasks in parallel.

Some have questioned whether working memory and reference memory can really be separated as independent entities; they may be needlessly juxtaposed (Pico & Davis, 1984). Certainly, there would appear to be differences in brain cholinergic activity following training in RM or WM, the former giving rise to prolonged elevated activity in frontal cerebral cortex, the latter short duration activity in cortex and hippocampus (Durkin & Toumane, 1992). Pharmacological treatments that destroy "spatial" neural structures can induce both RM and WM errors (Murray & Fibiger, 1983). Morris (1982) has pointed out that most tasks require a combination of both; animals must use reference information about the task and apparatus in order to respond at all, even in using WM. Indeed, in natural environments, where only a subset of all plausible food locations will be visited according to the past likelihood of success, many locations will need to be sampled and explored before such a subset can be established. Thus, knowledge of spatial reference cues is needed before working memory can be employed, perhaps explaining why the development of RM appears to precede WM ontogenetically (Green & Stanton, 1989). Also consistent is the fact that while prenatal treatments with drugs enhancing cholinergic function affect both WM and RM (Meck, Smith & Williams, 1989), RM performance can improve independently of WM (Murray & Fibiger, 1986). For further discussion of treatment effects on the RAM, see Hodges (Chapter 6, this volume).

Procedural factors may explain some of the variability among studies. Olton et al. (1979) reported hippocampally damaged rats to be selectively deficient in WM, yet Leis et al. (1984) and Nadel and MacDonald (1980) found them to make both RM amd WM errors. It is worth noting that the test period was short in the Leis et al. (1984) study, and hippocampals may have been extraordinarily deficient having been tested on a cued version of the task involving textured floor inserts before transferring to the uncued condition. This may have selectively disrupted hippocampals' subsequent testing, since they are notoriously prone to proactive interference (see Foreman and Stevens, 1987 for a review). Similarly, in the Nadel and MacDonald (1980) study, animals were preoperatively trained on both cue and place conditions in separate rooms, again inviting interference between the two tasks. Colombo, Davis and Volpe (1989) reported that rats with lesions in the caudate nucleus were selectively impaired in reference memory, showing a normal frequency of arm revisits. Many studies have found that damage within the hippocampal system or anticholinergic treatments (Higashida & Ogawa, 1987; Lydon & Nakajima, 1992; Robinson, Hambrecht & Lyeth, 1988; Wirsching et al., 1989), N-methyl-D-aspartate (NMDA) antagonists (Ward et al., 1990), or electroconvulsive shock (ECS) (Beatty, Bierley & Rush, 1985) produce selective impairments in WM, although the anticholinergic, scopolamine, has been reported to affect both RM and WM (Okaichi, Oshima & Jarrard, 1989). Nevertheless, at high doses this treatment also substantially increases time to complete the task, and thus reduced WM and RM performance may be a by-product of a general, nonspecific performance deficit, perhaps general lethargy (see Watson et al., 1994). These issues reinforce the need to take multiple measures of performance when using the RAM.

A difficulty in interpreting errors in the differential version of the task occurs when animals make repeat visits to previously sampled arms within the unbaited subset (Lydon & Nakajima, 1992; Marighetto, Micheau & Jaffard, 1993.) As Thinus-Blanc (1996) points out, these may be regarded as both WM and RM errors; indeed, where they have been analysed, they have been referred to as RWM errors (Lydon & Nakajima, 1992). A degree of random responding is likely to occur on the RAM. Ironically, to achieve accurate performance in a differentially baited maze effectively requires animals to suppress exploratory tendencies. In real environments, occasional visits to recently unrewarded locations are advisable; the no-reward status of a radial maze arm might reasonably be checked from time to time. A reference memory "error" in the laboratory is only an error according to the experimenter's criterion.

It is probably advisable that WM scores are calculated both including all repeat responses and just those repeat responses occurring within the baited subset. WM errors are more rare than RM errors (cf., Mundy & Iwamoto, 1988). Moreover, measures of WM and RM are not independent, inasmuch as improvement in RM (greater restriction of choices within the baited subset) places more demands on WM to avoid repeat visits among the baited arms. Improvements in RM can be seen independently of WM performance, if re-entries into both baited and baited subsets are combined for the WM measure (Furukawa & Iwasaki, 1989). Brown et al. (1993) and Keseberg and Schmidt (1995) in rats, also Toumane et al. (1988) in mice, found that in control subjects WM errors fell rapidly over the first 20 trials, and more slowly thereafter, while RM errors showed a linear decline over 50–80 trials. The earlier drop in repeat errors is counterintuitive. Performance might be expected to pass through phases in which (a) RM improves, followed by (b) a progressive improvement in WM. However, this does not happen in practice in humans (Foreman, Gillett & Jones, 1994) or in rodents (Brown et al., 1993; Keseberg & Schmidt, 1995; Toumane et al., 1988), suggesting that subjects learn *simultaneously* to respond within the rewarded subset and to avoid WM error revisits within a trial. However, this applies to the acquisition of RM and WM, yet were performance to deteriorate, the interpretation of data could be ambiguous. Were RM to improve, coupled with a fall in WM performance, the latter would be uninterpretable as being due to the treatment applied or to a greater pressure on WM (see Foreman et al., 1994). Were WM to deteriorate, resulting in proportionately greater numbers of repeat responses within the baited subset, this would automatically lead to improved RM scores, whether or not RM had in fact "improved".

In summary, the differentially baited version of the task generates a great number of data, but it needs cautious interpretation. In mice, the standard 8-choice version of the radial arm maze has been found to discriminate spatially competent from incompetent strains more effectively than cued or differentially baited versions of the task (Ammassari-Teule & Hoffmann, 1993), and large posterior cortical lesions, while giving rise to deficits in the 8-arm RAM (Foreman & Stevens, 1982; Goodale & Dale, 1981), fail to do so when the first 4 choices per day are considered in a 4/8 differentially baited maze (Kesner, DiMattia & Crutcher, 1987), and thus results from the two types of paradigm may differ in important ways.

INFLUENCES ON RAM PERFORMANCE

One of the main reasons for the development of, and continuing interest in, the radial arm maze is its sensitivity to manipulations such as brain damage (particularly hippocampus and related structures; see Olton et al., 1979) and pharmacological treatments (particularly related to cholinergic neurotransmission; see reviews in Levin, 1988; and Olton, 1987). It is sensitive to memorial and nonmemorial changes in behaviour, enabling differentiation among specific and nonspecific drug and lesion effects (e.g., Ennaceur, 1994). The effects of neural damage and pharmacological treatments are addressed elsewhere in this volume (Hodges, Chapter 6) and will not be covered here. Several individual and environmental variables have been examined with respect to animals' predisposition to perform well or poorly. In view of the inter- and intralaboratory variability in algorithm use and other factors, caution should be applied when relying upon single, unreplicated studies.

Early experience

Early experience appears to have a substantial influence on later ability to perform the RAM. Rats reared in complex environments (of either sex) make fewer errors in a 17-arm maze than those reared in standard laboratory cages (Seymoure, Dou & Juraska, 1996). Cramer, Pfister and Haig (1988) found that rat pups given many distributed food sources (nipples available) in the suckling period reached radial maze criterion in one-third of the number of trials required by pups having a few nipples available. Gerbils reared in enviornments in which the food and water source varied between days learned a 17-arm maze more quickly than those exposed to a constant source (Takai & Wilkie, 1985).

Sex differences

Male rats tend to make more errors on the standard version of the task than females, although the effect is usually small (Beatty, 1984; Einon, 1980; Luine & Rodriguez, 1994), not observed in all studies (cf., Juraska, Henderson & Muller, 1984), and where such is observed, it may be in part related to differences in exploratory activity between the sexes (Van Haaren, Wouters & Van de Poll, 1987). Seymoure et al. (1996) investigated the effects of rearing environment, room cues and baiting regime using a 17-arm RAM and found reliably superior performance in male rats across all conditions. In dark-reared animals tested on a 17-arm maze, male superiority is reportedly particularly evident (Tees, Midgley & Nesbit, 1981).

Age

Very young rats have been tested on the RAM, showing that extramaze spatial cues begin to be used at around 25 days of age (Rauch & Raskin, 1984). In mice, Chapillon, Roullet and Lassalle (1995) found that 23-day-old mice could perform the task, but apparently only when able to use algorithmic response strategies, using distal cues effectively for orientation only after 37 days of age. In studies examining age-span development, it has usually been found that ageing animals make more errors than juveniles, though the data are variable (see Luine & Rodriguez, 1994). Older females perform worse than older males (Kobayashi et al., 1988). Gallagher et al. (1985) found 4–6 month rats to make fewer errors than 24–26 month animals, particularly when shifted to a novel test environment. Caprioli et al. (1991) tested male Sprague-Dawley rats at 4, 13 and 25 months (unusually, in an environment devoid of extramaze cues). Those tested at each age performed well at 25 months, but animals first tested at 25 months made more errors than naive controls. Algorithmic response patterning has been observed to change systematically over time (Kobayashi et al., 1988). Juvenile rats tend to adopt more adjacent-arm responding than older ones, in both Long-Evans and Sprague-Dawley strains (Hall & Berman, 1995). Where algorithms are prevented by using a differentially rewarded maze, the ageing deficit seems to be particularly in reference rather than working memory (Furukawa & Iwasaki, 1989), although work with aged lemurs suggests a predominantly working memory deficit (Picq, 1993). Strain may interact with age. Age-related deterioration in performance is evident as early as 12 months in Wistar rats

(Kadar, Silbermann, Brandeis & Levy, 1990). Performance is poor, but still above chance in Brown Norway rats as old as 30–33 months (Van Gool et al., 1985).

Several pharmacological treatments have been said to improve radial search performance in senescent rats. For example, Caprioli, Ghirardi, Ramacci and Angelucci (1990) have shown that chronic treatment with acetyl-L-carnitine delays age-related deterioration in performance, treatment with the calcium channel blocker nimodipine enhanced performance in aged rats in a dose-dependent manner (Levere & Walker, 1992) and nicotine treatment improves both working and reference memory in aged rats tested in a 17-arm maze (Arendash, Sanberg & Sengstock, 1995). Free radicals have been implicated in age-related deficits in mice, which can be reversed pharmacologically (Fredriksson and Archer, 1996).

Stress

Performance on the RAM following aversive stimulation is poorer than in nonstressed animals, arguably because stress has the effect of enhancing the learning of simple associations at the expense of the learning of spatial associations (Shors & Dryver, 1992). The effect of stress may be indirect, via reduction of animals' exploration of the maze (Shors & Dryver, 1992), which would be expected to retard the formation of spatial representations of the maze (see Thinus-Blanc, Save & Poucet, Chapter 3, this volume). Chronic restraint stress produces a temporary and reversible impairment (Luine, Villegas, Martinez & McEwen, 1994). Regarding the stress which might arise in the course of its use as a laboratory tool, the radial maze is clearly less stressful than forcing animals to escape by swimming to a platform in water, despite the mild stress associated with daily food restriction (Shors & Dryver, 1992).

Species and strain variation in performance

Among the behavioural studies conducted using the radial maze in the period 1985–97, the majority were conducted using rats (85.4%), but others used mice (9.9%), birds (2%) and other species (2.7%) including monkey, cattle, dog, hamster, gerbil, opossum, lemur, and tree shrew. Although only mammals, birds and fish have been tested extensively on the radial maze, the garter snake is also said to perform well (Kunka, Bernstein & Kubie, 1979). Bees may be tested on a simplified version of the task (Burmeister, Couvillon & Bitterman, 1995).

Species other than the rat (which is the laboratory standard) have often been suggested to have poorer radial maze performance, although this has aroused controversy. Levy, Kluge and Elsmore (1983) have pointed out several variables (deprivation factors, apparatus variables, animals' running speed) that might influence performance idiosyncratically in particular species, though it has to be said that the radial maze compares favourably with other spatial paradigms, particularly the Water maze, in making fewer discriminatory demands on rodent species (Whishaw & Tomie, 1996). Limited data are available for obscure species. As indicated earlier, it is unwise to draw firm conclusions on the basis of single studies, particularly given that it may be necessary to tailor a task to the ecology of the particular species in order to obtain valid estimates of their memory capacity. Hoarding species, such as hamster, collect sunflower seeds in cheek pouches rather than eating the reward after each choice (Jones et al., 1990) and lemurs have been tested by allowing them reinforcing access to a homecage, following the final choice in a 4/8 differential baiting paradigm (Picq, 1993). Such modifications limit comparability with studies conducted conventionally.

Humans have been tested on 8, 13 and 17 choice versions of the RAM with success. Performance improves over years 2–5 (Aadland, Beatty & Maki, 1985; Foreman, Arber & Savage, 1984), and RM emerges before WM (Foreman et al., 1994) when 6/12 room locations were differentially rewarded. On 13- and 17-choice versions, human memory span is arguably similar to that of the rat (Glassman et al., 1994). The use of analagous search tasks with children has been discussed elsewhere in this Handbook (Deloache and Marzolf, Volume 1, Chapter 5). Menzel (1978) showed that chimps were capable of strategic gathering in an open field version of the RAM. Monkey species for which food is naturally abundant show poorer performance than those having to search for sparse food sources in the wild (Andrews, 1988). Tree shrews (Tomonaga, Kinoshita & Ohta, 1990) and opossum (Kimble & Whishaw, 1994) learn the task but may accomplish it by adopting nonspatial strategies (see below). Note that where emotional species and vertical climbers, such as tree shrews, need to be tested in a covered apparatus, the covers mask extramaze cues, and the resulting sparseness of cues may tend to promote algorithmic responding, as in the rat (e.g., Einon, 1980; Suzuki et al., 1980). In the case of tree-dwelling animals that may not forage horizontally, a vertical radial structure might be more appropriate, although none has been used to date. Grazing cattle perform as competently as rats, and also show better performance when tested on a radial configuration than on a parallel version (Bailey et al., 1989).

The humming bird, which is highly selective in choosing among response sites—a fact that stimulated interest in the radial maze in the first place (see Olton, 1977)—performs well when tested on a laboratory version of the task (Healy & Hurly, 1995), and storing species of tit reportedly exhibit superior performance on an analogue of the radial maze task than nonstorers

(Hilton & Krebs, 1990). Pigeons were once though to perform poorly (Bond et al., 1981), but both doves (Wilkie, Spetch & Chew, 1981) and pigeons (Honig, 1986; Roberts & Van Veldhuisen, 1985) can perform well with extensive training, and particularly when tested using analogues of the RAM (e.g., an array of feeding boxes arranged in a radial fashion at floor level). Indeed, pigeon spatial memory exhibits many of the features of rat memory, including prioritisation of locations according to reward size and both WM and RM components of memory (Roberts & Van Veldhuizen, 1985). Nevertheless, the durability of pigeon spatial memory may be less than the rat (Dale, 1988), and perhaps less than other avian species, such as Clark's nutcrackers (Balda & Kamil, 1988). Fish spatial memory appears to be somewhat limited (Roitblat, Tham & Golub, 1982).

Among rodents, the rat is the laboratory standard. However, mice have been used in a number of studies, often concerned with strain differences in performance and their correlation with anatomical measures, particularly within "spatial" neural structures such as the hippocampus. Roullet and Lassalle (1992) are critical of the unsystematic way in which the RAM has been used with mice. First reports tended to favour the use of a 6-arm maze with mice, arguing that the 8-arm maze was too difficult for them (Levy et al., 1983; Mizumori et al., 1982; both using the CD-1 strain) although reports soon appeared rebutting this (Bruto & Anisman, 1983; Bernstein et al., 1985: C57BL/6J; Pico & Davis, 1984: CD-1). Reinstein, DeBoissiere, Robinson and Wurtman (1993) obtained good performance in C57Br/cdj and Balb/cj strains but poor performance in the C57Bl/6j strain. On the other hand, Ammassari-Teule and Caprioli (1985) obtained good performance from the C57BL/6 strain in a 6–arm maze (which used a variety of strategies to solve the task), but poor performance in DBA/2, which they attributed to the inactivity of the latter strain. Activity level might influence performance, although some disagree (Cocchetto et al., 1985), and Higashida and Ogawa (1987) could find no evidence for activity effects on performance when comparing Fischer 344, Sprague Dawley and Wistar strains of rat. Pico and Davis (1984) attributed the deficient performance of Mizumori et al.'s (1982) mice to apparatus variables. Certainly there is evidence of wide differences among mouse strains in choice accuracy and algorithm use (Roullet & Lassalle, 1992). Gerbils perform well, as rats, and are often the species of choice in studies of ischemia and anaesthesia effects (e.g., Wishart, Ijaz & Shuaib, 1994).

WIN-SHIFT AND NATURAL FORAGING

In early studies of the RAM, it was assumed that performance is underpinned by a "win-shift" strategy, that ensures efficiency and optimisation in natural foraging (Andrews, 1988; Bond et al., 1981; Elsmore & McBride,

1994; Haig, Rawlins, Olton, Mead & Taylor, 1983; Olton, 1977; Olton et al., 1979; Phelps & Roberts, 1989; Seamans & Phillips, 1994; Yoerg & Kamil, 1982). According to this view, being reinforced by finding and eating food at the distal end of a maze arm causes an animal to shift subsequent responses away from that location, to others that still contain food. This contrasts with operant learning tasks, in which reinforcement has the effect of increasing the likelihood of a behaviour/response that immediately preceded it ("win-stay") (Yoerg & Kamil, 1982). Assuming that the maze is not continuously rebaited, to return to an arm that has just been emptied of food would clearly not be efficient or productive; indeed, intact rats almost never make that response (Eckerman, 1980). Win-stay behaviour is notoriously difficult to achieve in spatial tasks compared with win-shift in many species (Roitblat et al., 1982), although win-stay behaviour can be induced in rats using appropriate paradigms (Evenden & Robbins, 1984).

Most damaging for a win-shift interpretation of RAM behaviour is that alternation between alternative spatial choices is more likely to occur after nonreward than reward, and thus it would not seem to be the "winning" that is important (Gaffan & Davies, 1981). Gaffan, Hansel and Smith (1983) caution against equating laboratory "foraging" with natural foraging. Haig et al. (1983, p.347) emphasise the range of factors that influence laboratory behaviour, including natural predispositions, but also conclude that primarily, animals are: ". . . strongly predisposed to remember the relative familiarity of different places and to follow a shift strategy based on this memory". Versions of the radial maze have been exploited which require no deprivation of subjects *per se*, and where no rewards are "won", yet animals still distribute their choices (Masuda et al., 1994; Wilkie, Mumby, Needham & Smeele, 1992). Using the conventional version of the RAM, Timberlake and White (1990) found that food-deprived rats that were rewarded (or nonrewarded) on the maze performed equally well, although nondeprived rats chose maze arms at chance levels and failed to traverse maze arms to the distal ends. When transferred to the conventional version of the task from versions of the maze in which arms were (a) always unbaited, or (b) continuously rebaited with food following visits, rats showed a better performance following unbaited training, suggesting that in this instance, repeated indiscriminate reward (winning) had induced a degree of win-stay perseveration (Ammassari-Teule & Maho, 1992).

Nevertheless, it would seem strange to conclude that, for food-deprived animals in a conventional RAM, what they find at the end of a maze arm is irrelevant. Certainly, reinforcing arm choices has the effect of speeding testing, and reward does influence choice behaviours. Pigeons and rats tend to order their arm visits beginning with those containing the greatest amount of food (Roberts, 1981) and rats rapidly change their exploratory strategies when the volumes of food in maze arms are altered (Ryabinskaya & Ashikhmina, 1988). Batson, Best, Phillips, Patel and Gilleland (1986) found that rats modify preferences according to whether they like what they find at the end of an arm. The trade-off between food accessibility and quantity (Ilersich, Mazmanian & Roberts, 1988), and the exercise of preferences and response biases, modified by maze size, hub size and arm length, are consistent with optimal foraging principles (Brown, 1990; Grobety & Schenk, 1992). In conclusion, although winning "isn't everything" (Timberlake & White, 1990), it does seem to be influential on choice behaviour, which is strategically organised on the RAM according to rules that would enhance the efficiency of food-gathering in foraging situations. Yet it is not *primarily* the winning of food which promotes shifting to new maze arms, rather a tendency to alternate ("shift") between spatial locations, which may be more related (in the wild and in the laboratory) to the acquisition of environmental information than consumption of rewards *per se*. It takes only a few days for rats to achieve high scores on the 8-arm version of the task, but some 30 days to achieve competence in finding just 4/8 arms that are baited in a differential baiting task (Wirsching et al., 1989). Were animals concerned only with food locations, the latter might be expected to be the simpler task. Of course, the differential task requires the suppression of responses to nonrewarded arms as well as learning which are baited. However, the difference between these paradigms might be taken as evidence that finding food in a maze arm is secondary to having visited and sampled that arm.

ALGORITHMIC RESPONSE PATTERNING

Early reports of radial maze performance (Olton, 1977; Olton & Samuelson, 1976) implied that rats make random choices among the maze arms available. However, Potegal (1969) had previously detected preferences in rats for particular angles of turn (120 degrees, 150 degrees, etc.) when exploring a 12-arm maze. Several authors were somewhat dismissive of algorithmic response patterning as a factor influencing performance. Maki et al. (1984a) indicated that it was unusual to observe algorithmic patterns, and Mizumori et al. (1982, p.38) reported that, overall, choices made by their rats and mice ". . . seemed random". This is surprising since Roberts and Dale (1981), for example, reported that all of their animals in one study used a consecutive arm choice strategy (e.g., "take the next arm on the left"), the incidence of this pattern eventually rising to 90% of all responses (see also Dale & Innis, 1986). Other similar examples are given below. Magni et al. (1979) reanalysed Olton and Samuelson's (1976) data and found the response distribution to be distributed nonrandomly in favour of 45° turns. Olton (1978) accepted this argument, additionally pointing out that direction of turn was also nonrandom. The discovery that animals do use such strategies is important because, as Eckerman (1980) pointed out, the use of a simple algorithm (e.g., the adjacent arm algorithm), to determine every choice could be made entirely independently of spatial encoding and memory. The algorithm is clearly an "optimal" choice strategy (Ammassari-Teule, D'Amato, Sansone & Oliverio, 1988), but it is inappropriate to regard it as a "mapping" strategy (Ammassari-Teule & Maho, 1985; Leis et al., 1984), in the sense in which "mapping" is used elsewhere (O'Keefe & Nadel, 1978; Thinus-Blanc, 1996). The consistent use of a consecutive arm algorithm cannot be assumed to have any spatial component in so far as the subject needs to know nothing about the spatial arrangement of maze arms in order obtain a 100% score on a maze of any size. Theoretically, the animal needs only to retain a memory of the last arm visited on each choice, as several authors have pointed out (Kraemer et al., 1983; Markowska et al., 1983; Olton, 1978; Roberts & Dale, 1981; Roitblat, 1982). Of course, not all algorithms require a constant angle of turn; most discussion centres on the consecutive arm algorithm, yet this is only one of several algorithmic response strategies that can be used. As discussed in the following section, systematic algorithmic turning throughout a trial may or may not generate a perfect score; spatial judgments may still be required, depending on the number of arms in the maze.

Scoring of algorithmic response patterns

Olton (1978, p.344) and Roitblat (1982, p.360) draw a distinction between an algorithm and a response chain. The former is a "... relational code that specifies the next arm to be visited, based on its relationship to the current arm", whereas a response chain would involve the choice of arms according to a chain sequence that may or may not be specified in terms of absolute spatial positions. Were an animal to use the same sequence of choices on successive days, or a fixed sequence of turns, this would be difficult to detect from simple observation, yet it ought to become clear during the analysis of turn angles although this has not been reported, nor observed by the present authors. At most, a small tendency may exist for animals to begin choosing at a particular maze arm (Bierley, Rixen, Troster & Beatty, 1986). Indeed, memory for a fixed sequence of successive turns would be a more difficult solution (akin to a discrimination task) than a spatial solution, particularly on larger mazes, although rats are capable of tracking arm patterns (Wathen & Roberts, 1994). The present review is concerned with algorithmic response strategies, on mazes of various sizes, though the use of software to compare animals' response chains on successive days might be advisable in versions of the radial maze with small numbers of maze arms.

As many authors have pointed out (e.g., Einon, 1980; Foreman, 1985; Watts et al., 1981), algorithmic aspects of performance can be categorised. Some have labelled algorithms according to whether their use logically requires a spatial decision at any choice point (Watts et al., 1981) or whether overall performance suggests to several independent raters that an animal is making spatial decisions (Leis et al., 1984). For data presentation purposes, however, individual subjects' data are often not described. For example, the percentage of adjacent arm entries may be computed (Roberts & Dale, 1981), or distributions of the longest sequences of successively selected adjacent arms can be displayed, along with corresponding random simulations (Magni et al., 1979), although these measures ignore all the other angles of algorithmic turn that can occur. Okaichi and Oshima (1990) reported the percentage of subjects using a single turn angle throughout an entire trial, which misses many aspects of algorithmic behaviour. Note that the direction of turn (left or right) is important, since some animals will switch between using clockwise and anticlockwise algorithms variously across successive days. (Whether direction of turn is related to the direction in which animals turn at the distal end of maze arms has not been investigated). Direction of turn needs to be taken into account when analysing data, particularly within a test session. Rats do tend to show intrasession consistency in their preferred direction of turn (Einon, 1980), although data are often grouped independently of direction of turn across test sessions. Of course, responses are directionally ambiguous, since on an 8-arm maze, a response from arm 1 to arm 4 could be classified as +3 or −5, and if direction of turn is a significant factor in an experiment, there is no alternative to recording the direction of movement and number of arms passed over between successive responses. Authors may report the transitional angles of successive responses as means or frequencies for individual subjects (e.g., Watts et al., 1981) or groups (Caprioli et al., 1991a,b; Olton & Werz, 1978), plot the relative incidence of the various angles of turn available (Foreman, 1985; Winocur, 1982), calculate the percentage of string repetition (Pallage et al., 1992), or show diagramatically the mean angle of turn, on successive choices with respect to the position of the first choice (Buresova, 1980, her figure 4), or illustrate representative response patterns with respect to the just-exited arm position (Stokes & Best, 1988, their figure 4). Total turn angle, or mean turn angle (cf., Lachman & Brown, 1957), used by Ammassari-Teule and Maho (1985), degree of divergence (Maho, Dutrieux & Ammassari-Teule, 1988), delta arm score (Peele & Baron, 1988), mean transition size (Dale & Roberts, 1986; Leis et al., 1984) or mean angle frequency (Loh & Beck, 1989) is of limited descriptive value in RAM studies except as a very gross measure of algorithmic tendencies. It is unhelpful to label choice sequences as partially or totally sequential, in view of the range of possible algorithms that can be employed (cf., Higashida & Ogawa, 1987). Measures of response variability (Loh & Beck, 1989) quantify repetition within choice sequences but not algorithms *per se*. No single measure provides a full description of algorithmic behaviour. It would seem essential to identify the precise relationship between algorithmic behaviour and the main variables under examination, in order to determine the most suitable form of analysis. In many cases, classification of individual animals' algorithm use and its possible influence on choice accuracy is critical. For groups of animals, a combination of measures is necessary to summarise effectively algorithm use, except where turning behaviour is very systematic and homogeneous within a group, in which case a single measure could be adequate.

Figure 4.1 shows the algorithms that are possible on an 8-arm radial maze. These are usually labelled according to either the angle of turn made between choices (45°, 120°, etc.) or the number of 45-degree segments between successively chosen arms (± 1, ± 2, ± 3, for example). Some algorithms cannot be classified this way.

The "1" strategy

This is often referred to as the consecutive arm or adjacent arm strategy (e.g., Innis & Macgillivray, 1987; Roberts & Dale, 1981). Note that to term this choice simply "clockwise-anticlockwise" (Ammassari-Teule & Maho, 1985; Toumane et al., 1988) is in error since other algorithms (e.g., the "3" type) are also directional. In the earliest reports of RAM performance, Olton et al. (1977) and Olton and Werz (1978) reported a 30–40% use of "1" turns, using a 17-arm maze, though others have since reported much higher rates of "1" responding, particularly on larger mazes (see Foreman, 1985). The adjacent turn pattern is said to occur more frequently in animals tested with water rewards on the maze than food rewards (Innis & Macgillivray, 1987; Dale & Roberts, 1986) and in younger more than older rats (Hall & Berman, 1995).

The "2" strategy

This involves 90° turns between choices on an 8-arm maze. This may seem a nonoptimal solution, since the algorithm, when strictly applied, would cause errors to be committed following choice 4. In order to use this algorithm, an animal must switch between orthogonal subsets of maze arms making a "spatial discrimination" (Watts et al., 1981) at choice 5. As is shown in Table 4.1, similar strategies on larger mazes can be grouped according to the number of such spatial decisions that are needed to

120 FOREMAN AND ERMAKOVA

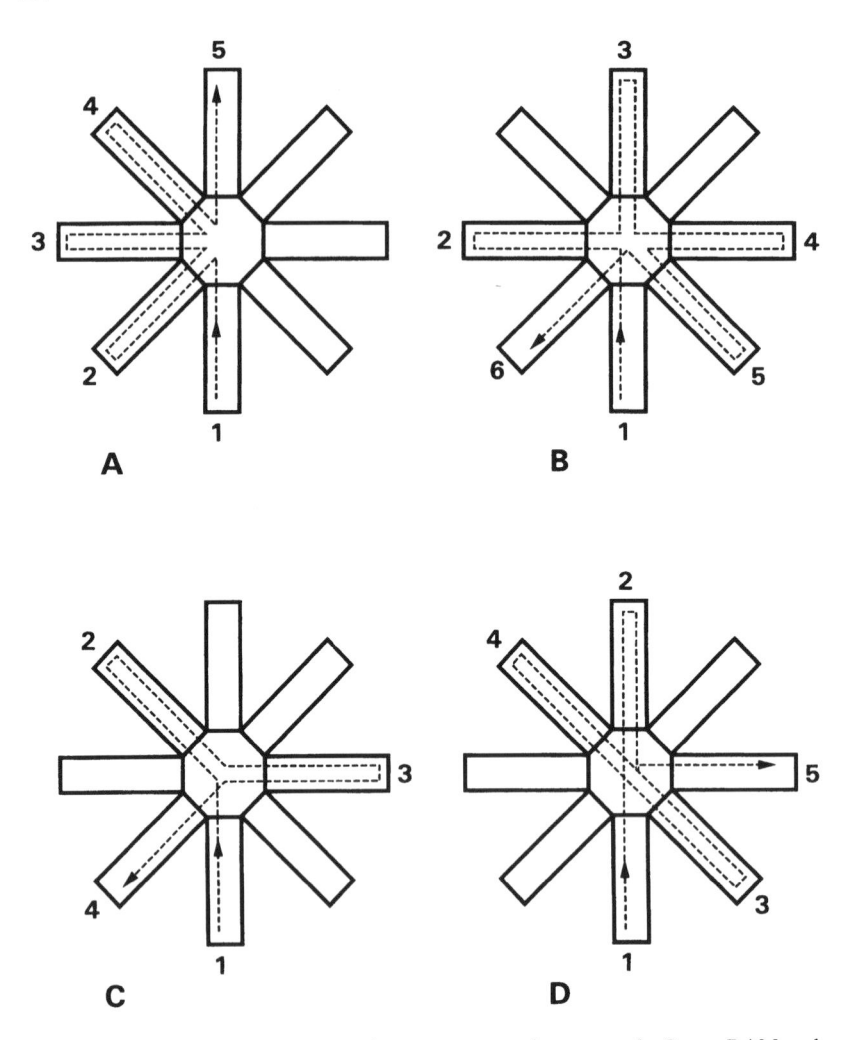

FIG. 4.1 Algorithmic response strategies most commonly seen on the 8-arm RAM and described in the text (A, "1", B, "2", C, "3", D, "4–3").

achieve 100% performance, over and above the score generated by the algorithm alone, thus placing all algorithmic strategies on a single continuum. Caprioli et al. (1991a) have suggested that the "2" algorithm becomes more popular as rats age.

The "3" strategy

This involves successive 135° turns, to left, or right, consistently and generates 100% performance. The 135° turn is a popular one in rats, whose mean angle of turn is usually about 100° when moving about on platforms

(Leis et al., 1984), although clearly this preference is subject to modification. There is some disagreement in the literature over rats' preferred turning angles. The frequency with which some groups of animals adopt a 45° turning preference (cf. Dale & Innis, 1986; Brown, 1992) indicates that they will rapidly adopt acute angles in preference to obtuse angles, despite the extra effort involved in stopping and turning sharply on reentering the central platform. It is interesting to note that animals' spontaneous and learned alternation responses are usually tested using T-mazes, requiring 90° turns (Douglas, 1966; Petrinovich & Bolles, 1954), yet given a choice animals will actually prefer non-90° turns, sometimes preferring more obtuse (Leis et al., 1984), but sometimes more acute angles (O'Keefe, 1982). They rarely prefer 90° *per se*. In other words, rats possess a turning vector, although not a 90° turning vector (O'Keefe, 1982). From the commencement of testing in the radial maze, some rats will make consistent obtuse angle responses (see Stokes & Best, 1988), making 90° or 135° turns, a phenomenon that is also evident when animals are making swimming responses in an aquatic version of the task (Buresova et al., 1985; Ermakova, Loseva, Valouskova & Bures, 1989). The importance of vestibular input in T-maze alternation was demonstrated by Douglas et al. (1974) and it is interesting to note that vestibular input appears to be an important factor in the development of algorithmic behaviours, since Ossenkopp and Hargreaves (1993) has shown that algorithms are disrupted by labrynthectomy, using an enclosed maze.

The "4-3" strategy

This is sometimes referred to as the 180° algorithm, though it actually involves 180° responses alternating with 135° responses (the latter in a consistent clockwise or anticlockwise direction.) Animals will occasionally use this algorithm (Foreman, 1985; Okaichi & Oshima, 1990). Note that "3-4" does not work, leaving one arm unentered at the end of the choice sequence. Presenting frequency-of-turn data, including 180° responses (cf., Loh & Beck, 1989), without reference to subsequent choices, is insufficient to show whether this turn pattern is being used algorithmically or not.

Algorithms may occur by chance from time to time, and Foreman (1985) has calculated the chance of obtaining 4-response algorithmic strings by generating all possible choice sequences and scanning these for instances of 4 successive responses of any of the above algorithmic types. Of the 8! (40,320) possible choice sequences on an 8-arm maze, 11.9% of these contain at least one 4-choice ARM, but the rate at which algorithms are observed typically exceeds this figure (Foreman, 1985). Longer sequences can be scanned for above-chance rates of patterned responding (Pearson et al., 1984).

TABLE 4.1
Algorithmic turn: Arm separation of successive choices

No. of maze arms	Interarm angle (°)	"1"	"2"	"3"	"4"	"5"	"6"	"7"	"8"	"9"	Successful algorithms	Turns available (%)
3	120	0									1	100
4	90	0	0								1	100
5	72	0	1	0							2	100
6	60	0	0	0							1	50
7	51.4	0	1	0	0						3	100
8	45	0	0	0	1						2	67
9	40	0	1	2	0						3	75
10	36	0	0	0	0	0					2	50
11	32.7	0	0	0	0	0					5	100
12	30	0	1	2	3	0					2	40
13	27.7	0	0	0	0	0	0				6	100
14	25.7	0	1	0	1	0	1				3	50
15	24	0	0	2	0	4	2	0			4	57
16	22.5	0	1	0	3	0	1	0			4	57
17	21.2	0	0	0	0	0	0	0	0		8	100
18	20	0	1	2	1	0	5	0	1		3	38
19	19	0	0	0	0	0	0	0	0	0	9	100
20	18	0	1	0	3	4	1	0	3	0	4	44

The tabulated value ascribed to each algorithm represents the number of "spatial" choices that would be required in each case to complete an otherwise algorithmic sequence and obtain a 100% performance. Thus, the higher the value, the less successful is the algorithm in generating a high score alone. A value of 0 indicates that an algorithmic turn can be adopted on every choice to give a 100% performance, a value of 1 indicates that a "spatial" choice is required at one choice point. As an example, an animal using a "2" algorithm on an 8-arm maze, having made orthogonal choices to arms 1, 3, 5, 7, has to make one "spatial" response to avoid an error (return to arm 1, which would be the next in the algorithmic sequence). The animal can then continue with "2" responses to complete the sequence, making their remaining responses within the unvisited orthogonal subset (e.g., arms 8, 2, 4, then 6).

Multiples of the above algorithms (e.g., a "9" response) would amount practically to a "1" response and would be as successful, although such circuitous running would surely have been noticed but has never been reported. Other patterns (such as +1, −2, +3, −4 . . .) would be successful but would also have been obvious and have never been seen. There are other possible response chains or patterns such as omitting 1, then 2, then 3 . . . between successive choices, or +2, +3, +2, +3; these have not been seen, and in any case they are unsuccessful in generating 100% performance in virtually all cases. These strings are bound to occur by chance from time to time, yet to include these as genuine strategies would imply that rats can count the number of arms between choices. Undoubtedly, the only basis for an algorithmic strategy is likely to be a preferred (acute) angle of turn. It would be absurd to regard a 45° turn to the right as being a 225° turn to the left; no algorithmic turn can realistically exceed 180° (see above).

The present review has concerned only those algorithms that are regularly seen and reported, and which potentially allow subjects to perform perfectly without the use of any "spatial" response that would indicate the use of spatial encoding or memory. It must be emphasised, however, that other patterns of choice or partial algorithms could be useful to the subject, and may promote better performance than would be possible for that subject without their use. However, the algorithm in that case is not a potential mask for deficient performance (since spatial decisions *are* still required to obtain a high score), and is thus of less theoretical interest in the context of the present review.

On smaller or larger mazes than the 8-arm version, several of the above algorithms can promote perfect performance, and some maze configurations lend themselves to algorithmic solutions more than others. Table 4.1 shows the number of "spatial" choices (discrimination that an arm has been previously visited) that must be made with each algorithm on each size of RAM. The "1" algorithm will be an option on any maze; the "3" algorithm works on a 10-arm maze; the "2" (next-but-one) algorithm can be used continuously to obtain 100% only where a maze has an odd number of maze arms. On mazes with even numbers of maze arms, the switch between subsets of arms is not a 90° switch between "orthogonal" subsets of arms (as on the 8-arm maze) but occurs after N/2 choices where N represents the total number of maze arms. A 90° turn on a 16-arm maze would be a "4" algorithm, for example. Understandably, most authors have used the 8-arm version of the maze, for which the most detailed empirical, comparative and statistical data are available, and since it was the first to be widely reported (Olton, 1977), yet it is clear from Table 4.1 that 2 out of 3 consistent angles of turn (67%) promote 100% performance on the 8-arm version, whereas on the 10- and 12-arm versions, only 50% and 40% of available turn angles respectively promote perfect performance, and thus on algorithmic grounds these maze configurations are to be preferred.

Predominant algorithms: Influencing factors

There is substantial interlaboratory and interstudy variation in animals' use of algorithmic strategies. Apparatus differences may account for some of the variability, although the predominant type of algorithm cannot be predicted from the type or size of apparatus, neural damage or treatment, or laboratory. Some possible determining factors are, *inter alia*: (a) species or strain, (b) apparatus dimensions and design, (c) reward type, (d) test period length, (e) the cue richness of the test environment, (f) cue visibility, (g) subject variables such as age, sex and experience, or a combination of these. A few of these factors have been formally examined.

Yoerg and Kamil (1982) found that increasing the size of the hub of the maze increased the incidence of "1" type responses. Indeed, using a platform 88 cm in diameter, their animals almost exclusively used this type of algorithm, but when a smaller platform (34cm) was used, this was not the case. Unfortunately, other authors using platform sizes similar to the smaller of Yoerg and Kamil's platforms have also obtained predominantly "1" responses (e.g., Dale, 1986; Einon, 1980). Olton and Werz (1978) and Foreman (1985) found that increasing the number of maze arms from 8 to 10 or more has a dramatic effect in causing exclusive use of the "1" algorithm. Yoerg and Kamil (1982) invite us to imagine a radial maze 1km in diameter, yet this cannot realistically be viewed as a radial maze. Some nidifugous animals (i.e., central-place foragers; see Hamilton & Watt, 1970) may forage from a nesting site and make frequent returns to a central base, although the likelihood is that most animals evolve foraging routes having constructions which are *related to* but not *centred upon* the location of the nest site. Decreasing the hub-food distance also induced the consecutive arm strategy (see Maki et al., 1984b), perhaps because visited places that are close together are easily confused. Brown (1992) has argued that a lax choice criterion is applied to short maze arms, giving rise to a high proportion of indiscriminate visits, and increased patterning of reponses. It is not uncommon to find that apparatus dimensions and configuration affect general exploration in rodents, and in predictable ways (Uster, Bättig & Nageli, 1976), and they may also affect foraging among radial locations. Magni et al. (1979) found that when animals are required to reverse out of a chosen alley (in an enclosed maze), they were most likely to choose a consecutive arm strategy. These authors recommend the use of a version of the radial maze in which animals return to the centre between choices via routes beneath the maze arms; this situation proved to be the least likely to induce algorithmic responses. A replication of this study would reinforce the conclusion. Stokes and Best (1988) found that rats with lesions in the thalamic nucleus medialis dorsalis were especially likely to make obtuse angle ("3") responses, as are animals given high doses of diazepam (Loh & Beck, 1989).

Suzuki et al. (1980) and Einon (1980) found that algorithms occurred most often when testing took place in a poorly cued environment or enclosed maze, which may be interpreted as reflecting the increased difficulty associated with cue sparseness (Stokes & Best, 1988). Einon (1980) found less algorithm use in socially reared than isolated rats. Dale and Roberts (1986) and Innis and Macgillivray (1987) report that water-deprived rats tend to make many more "1" responses than food-deprived, which might be taken as support for the suggestion (Petrinovich & Bolles, 1954) that water deprivation induces perseverative spatial responding in rats, at least in so far as a "1" response represents a perseverative "stay" response. However, in a study by Foreman et al. (1980), water-deprived rats failed to show more algorithmic turning than food-deprived, and these authors concluded that the *water-perseveration, food-alternation* hypothesis does not bear close scrutiny. Kobayashi et al. (1988) reported young rats to use more adjacent arm responses than older ones, although older rats will continue to use a "1" algorithm predominantly if they were formerly tested as juveniles (Hall & Berman, 1995). Caprioli et al. (1991a,b) found that the distribution shifted from predominantly "1" responses in rats aged 13 months or less in favour of "2" responses in 25-month-olds.

Grobety and Schenk (1992) found algorithmic choices to occur among widely separated maze arms, while animals avoided those separated by small angles. However, they point out that sharp interarm angles do not appear to deter adjacent choices since in Olton et al.'s (1977) 17-arm maze, with 21° interarm angles, "1" algorithms were very common.

Breaking algorithmic response patterns (strategies)

Several authors have introduced protocols and apparatus adaptations that prevent algorithmic responding, including the use of a guillotine door arrangement, to restrain the animal between choices (see "Containment between choices" above), or alternatively the addition of wire mesh covers over the proximal portion of maze arms (e.g., Stevens, 1981) or the addition of perspex guards that may line the maze arms, enclose them, or enclose the entire apparatus (Caprioli et al., 1991a,b); these do not reduce the visibility of room cues, yet they do prevent the animal from stepping between adjacent alleys (which they will sometimes do before reaching the centre platform).

Algorithmic responses are not as problematic when a version of the task is used in which only a subset of maze arms are baited, although the "1" algorithm was used persistently by control rats in one such study, in both cue and place versions of the task (Okaichi & Oshima, 1990), and for 70–80% of turns by mice in a 3/8 differentially baited RAM (Toumane et al., 1988). Where some choices are forced prior to free choice, algorithms will not promote good performance, or where an animal is removed from the maze between choices N/2 and N/2+1 (see the following section), although in the latter case algorithms may still contaminate the data, since an animal could in theory just remember the last-visited maze arm prior to removal and recommence the algorithm on replacement.

Treatment with amphetamine can apparently eliminate "1" responding (Bruto & Anisman, 1983) although it is unclear why this should be. Buresova et al. (1985) found that animals tested after anorexigenic doses of amphetamine do not perform on the standard radial maze at all and are better tested using an aquatic, aversively motivated version.

Do algorithms mask poor performance?

Algorithms might be thought to enhance performance in animals that are spatially impaired; the adoption of a simple turning preference might be seen, on the free choice version of the task, to enable artificially high scores to be achieved without recourse to any form of spatial encoding (Pico & Davis, 1984). Hall and Berman (1995, p.195) report that "juvenile rats use an adjacent arm strategy to solve the radial eight-arm maze". Suzuki et al. (1980, p.6) suggested that some use of algorithms seemed necessary for high performance, particularly when the test room is sparsely cued, and others believe that the algorithm occurs when the task is more difficult, as when maze arms are short (Brown, 1992; Maki et al., 1984b). Pearson et al. (1984) attributed the success of 6-OHDA-treated animals to the excessive use of algorithms, and other treatment effects have been arguably masked in this way (Leis et al., 1984). Stevens (1981) reported that the use of containment on the central platform standardised the adverse effects of scopolamine on performance, perhaps because it eliminated the availability of nonspatial strategies reportedly used by scopolamine treated animals when no containment is used (Watts et al., 1981). Similarly, medial septally lesioned animals appear to learn the RAM, but only when allowed to use algorithmic strategies (Beatty & Carbone, 1980; Janis, Bishop & Dunbar, 1994). As Pallage et al. (1992) comment, the use of an algorithm by the septal lesioned animal may represent "the best of a bad job". However, other studies show that algorithms do not "protect" deficient performance. Hippocampally damaged animals are reliably deficient on all versions of the radial arm maze (Olton et al., 1979). They notably fail to adopt the "1" algorithm when intact animals do so, and use a variety of other response types but with equal lack of success; this does not reflect a general learning deficit since they can use cueing with some success to complete versions of the task (Okaichi & Oshima, 1990; Winocur, 1982). It might be intuitively expected that the adoption of strong algorithmic responding would be an easier compensation. Some types of deficiency might therefore be compensatable via algorithm use, others not.

Roberts and Dale (1981) argued that algorithms protect performance against proactive interference. They found that animals' use of algorithms built up as successive trials were administered during a day, on successive days, and after long delay intervals. A similar argument was made for gerbils (Wilkie & Slobin, 1983). Dale and Innis (1986, p.23) found that sighted, naive blinded and experienced blinded animals all used algorithms, but all used memorial strategies as well, extramaze cues in the case of sighted animals and intramaze for the blinded animals, indicating that rats can "... select from several search strategies" and "... use several strategies simultaneously".

An algorithm could be regarded as form of response "perseveration". However, while treatment with scopolamine produces response perseveration (consistent direction of turn) in a T-maze (Douglas & Truncer, 1977) it does not produce algorithmic responding on the RAM. Nor is algorithmic behaviour explicable in terms of simple learning paradigms. Algorithms are variable from day to day, between and within animals; they may be dropped, swapped, and resumed days later on the 8-arm radial maze (Foreman 1985; Leis et al., 1984; Okaichi & Oshima, 1990), appearing in partial form or several rearranged forms prior to being adopted throughout an entire trial. Thus, although Suzuki et al. (1980) and Pico and Davies (1984) refer to algorithms as "proprioceptive" or "kinaesthetic" strategies, they are clearly not simple muscular programmes, in which a "45° turn to the right" is consistently reinforced, for example. They appear only when subjects have become familiar with the environment (Dale & Innis, 1986; Foreman, 1985; Peele & Baron, 1988). The issue is to what extent responses remain "spatial" when most or all are made algorithmically. Roullet and Lassalle (1992) found that three out of five mouse strains that were using a radial strategy when tested without confinement were nevertheless reactive to spatial changes in the environment. Bierley et al. (1986), using rats, alternated free choice trials with trials on which algorithms were broken by an initial forced four choices. Animals using algorithms performed equally well on trials when they were prevented from using them. Einon (1980) noted that rats would sometimes complement an algorithmic string with a "spatial" choice to an arm omitted from the algorithmic sequence (e.g., 3, 4, 5, 7, 8,1, 2, . . . 6). This is evident in the data appended by Yoerg and Kamil (1982) to their paper, and has been seen by the present authors many times. Rats using algorithms consistently are not always immune to the effects of spatially ambiguous maze rotations, such as a 22.5° rotation of an 8-arm maze (Foreman, 1985). Dale (1986) and Dale and Innis (1986) found that rats often stop after obtaining all 8 rewards in an 8-arm maze, despite having obtained all rewards via algorithmic choices, as though "aware" that all available baited locations have been exhausted prior to making a first unrewarded error visit.

It would clearly be wrong to attribute the adoption of an algorithmic pattern to sophisticated goal-directedness on the part of the animal. Whether an algorithm is being "employed strategically" by an animals is perhaps unanswerable, begging definitions and raising anthropomorphic issues (cf. the "purposivism" of Krechevsky, 1932, discussed by Sutherland & Mackintosh, 1971, p.171), and experience-related metacognition (see Caprioli et al., 1991a,b). Clearly, an animal may adopt a "strategy" of some sort that does not involve algorithmic turning, though it is hard to know how such a strategy might be identified (cf., Peele & Baron, 1988). It is perhaps most likely that the algorithm begins as a simple turning preference—one that minimises errors, increases efficiency and optimises reinforcement density as its stereotypy increases with time, and as it becomes employed increasingly systematically throughout a trial. Toumane et al. (1988) have argued that a "1" algorithm has the effect of distributing responses and decreasing RM errors in a differentially baited RAM, thus promoting choice accuracy in the early stages of learning. Ammassari-Teule et al. (1988) regard algorithms as "optimising" exploratory sequences, and evidence of "enhanced cognitive development".

In short, the algorithm is probably a convenient rather than primarily strategic way of completing the task, minimising load upon working memory and in turn, cognitive effort. It is idiosyncratic, and non-essential for accurate performance (Bierley et al., 1986; Foreman, 1985). It is probably not appropriate to regard algorithm use as the engagement of reference memory (Bierley et al., 1986), in view of the changes in predominant algorithms across days. The algorithm may perform a similar function to mnemonic devices in human memory (Cornell & Heth, 1983; Foreman, 1985).

AN AQUATIC VERSION: THE RADIAL WATER MAZE

At the time of its introduction, the Morris water maze (see Schenk, Chapter 5, this volume) apparently provided a purer method of testing animals' spatial memory than the RAM, since it eliminated odour trails and other intramaze cues, also providing a mildly aversive situation in which rats were well motivated to succeed, in order to escape from a bath of opaque water on to a dry platform. Consequently, some laboratories which had used the radial maze developed an aquatic radial maze, the RWM, which retained the benefits of a multiple-choice task in which working and reference memory could be studied, yet incorporated the benefits of the water maze. Ermakova et al. (1989) used a circular pool, 120cm in diameter and 60cm high, filled 30cm deep with water (20°C), made opaque by the addition of nontoxic white paint. Eight 12cm wide channels formed by 40cm × 40cm plastic walls projected radially from a central arena 30cm in diameter. A circular platform (15cm in diameter) could be raised or lowered via a pneumatic device between a position 1cm above water level to the floor of the pool. At the distal end of each channel was located a horizontal platform (10cm × 11cm). The eight platforms could also be raised 1cm out of the water, or dropped to the pool floor. Rats were placed for 10sec on the centre platform at the start of a trial, after which the plaftorm was lowered and the rat had to swim to a channel to reach an escape platform. After 20sec, the platform in the chosen channel was collapsed, and the rat had to return to the central platform which had now been raised out of the water. After 10sec it was dropped, forcing the rat to swim to another channel. The channel benches remained collapsed once visited, so that animals had to choose nonvisited channels in order to reach an elevated platform. An error visit was recorded when an animal made a channel choice (traversing three-quarters of the length of the channel) but the animal was left to continue swimming until a correct choice was made. Note that the central platform was not raised after an erroneous choice. Animals received one trial per day. The trial terminated when 8 channels had been visited, after 16 choices had been made, or 10 minutes had elapsed, after which animals were dried in a waiting cage with a heating lamp before being returned to their cages. Measures were taken of the number of errors in the first 8 choices, the total choices required to visit all 8 channels, and time taken to complete the trial. Control animals' correct choices in visits 1–8 averaged 6.0 at the start of testing but increased to 7.5 within 15 trials. Rats achieve 7–8–8.0 after 30 trials (Bolhuis, Buresova & Bures 1985). Pool water must be agitated between trials. Animals do defecate in the water, but it would seem impossible for any form of odour trail to be used in this version of the maze, compared with the dry version.

Bures and co-workers (Bolhuis et al., 1985; Ermakova et al., 1989) rarely observed the algorithmic strategies that are seen frequently on the dry version of the RAM. Certainly there was no evidence of consistent responding to adjacent channels (the "+1" algorithm) in control animals, which usually made choices +2 or +3 from the channel just visited in the study by Ermakova et al. (1989). A detailed analysis was carried out of 360 trials from 60 control animals, representing the final 6 trials of an 18-trial learning period (Ermakova, unpublished data). This showed that 43/60 animals used algorithms to a lesser or greater degree. The predominant algorithm (based on cases where 3 or more transitions were of this type within a trial) was the "2" variety (11%), then the "1" algorithm (8%), "3"

(5.3%) and others 0.3%. In only five trials were algorithms used consistently for all 7 interchannel transitions (2 of the "1" type, 3 of the "3" type). The aquatic form of the maze may indeed, therefore, provide a measure of spatial memory that is less contaminated by algorithms than the dry version. The aquatic version requires no food restriction, thus avoiding deprivation stress, and it might also be especially useful where animals are poorly motivated to eat food (Buresova et al., 1985). Water mazes may be especially difficult for mice, compared with rats (Whishaw & Tomie, 1996), although no species comparisons have so far been conducted to compare aquatic and dry versions of the RAM.

ALTERNATIVE PARADIGMS USING A RADIAL STRUCTURE

The above review concerns versions of the RAM that allow free choice among multiple locations. However, tasks that involve a radial arrangement of choice locations have been developed that require animals to make specific predetermined responses to target locations. Kessler et al. (1982) used a RAM which consisted of six goal locations connected by a complex but symmetrical array of alleyways. Animals are manually transported back to the centre platform after correct choices. This apparatus is essentially comparable to the RAM but data interpretation is complicated by the manual handling and the multiple paths that animals can take to goals. Kessler et al. (1982) argue that the task is more spatially demanding, yet it is not clear that the use of such a configuration is beneficial in view of the doubts over data comparability with the standard version.

The Barnes platform task

Barnes (1979, pp.76–77) introduced a radial task which is arguably more suited to the testing of senescent subjects than the traditional RAM, capitalising upon rats' ". . . tendency . . . to avoid brightly lit, unenclosed surfaces and to seek darkened enclosed shelter". It consisted of a platform rather than radiating arms, having 18 equidistant holes at the perimeter. Animals could escape from a selected hole into a darkened recess. Changing the position of the hole between trial blocks required the animal to discriminate the location of the target hole with reference to surrounding visuospatial cues and to respond only to the experimenter-chosen target. The platform surface can be rotated between choices, preventing the use of odour trails to find the target tunnel (although this does not eliminate the distracting effects of odours, which might lead animals on occasions to make errors). Barnes (1979) found that 28 to 34-month-old rats made more errors than 10 to 16 month animals. This task imposes some similar requirements to the RAM, although, despite its radial nature, the data yielded may not be directly comparable with RAM measures. Elsewhere, WM measures using a head-dipping holeboard (with 4/16 holes baited) failed to correlate with measures simultaneously generated from a RAM in which 4/8 arms were baited (Van Luijtelaar, Van der Staay & Kerbusch, 1989).

Matching and nonmatching to sample

Experimenters wishing to employ the radial structure of the RAM to examine specific aspects of memory, have used tasks that more closely approximate the visual-spatial tasks used with primates and human subjects with amnesic disorders, such as matching and nonmatching to sample, either to spatial sample (i.e., place), or cue sample via stimuli inserted in maze arms, and using delayed versions of these tasks. Such versions can be automated (e.g., Lebrun, Durkin, Marighetto & Jaffard, 1990), and have the advantage that task difficulty can be titrated by varying the interval over which animals must retain memories of chosen arms. Like the standard version of the RAM, spatial versions of the task require the use of spatial memory to discriminate the room locations of specific maze arms, but scoring is on a choice-by-choice basis. These paradigms require more training than the RAM and diverge from the seminatural behaviour required on the conventional RAM, resembling more closely the more standard laboratory operant tasks. The basic protocols have been summarised by Rawlins and Deacon (1993) and discussed in more detail by Hodges in Chapter 6 of this volume.

CONCLUDING REMARKS

The radial arm maze continues to be an important piece of apparatus, indeed a laboratory standard, for measuring various aspects of spatial behaviour. In its original free-choice form, the maze suffers interpretative problems resulting from the use by many animals of algorithmic choice strategies, though this version of the task is of interest because it may best elicit nonredundant choice distribution of the sort used in natural foraging situations. Algorithmic choice patterns are nonessential for high performance to be achieved, and secondary to the use of spatial encoding and memory. Nevertheless, the possibility that they might become strategic during long test periods suggests that when comparing treatments, it is preferable to use a prechoice confinement interval via the use of doors that restrict arm access. The aquatic version of the task, while more difficult to use than the dry version, appears to suffer less from confounding effects of algorithms. Cue manipulation studies, involving movement of the apparatus between test environments, intramaze versus extramaze cue substitution, and further examination of choice criteria would further illuminate the bases of subjects' spatial choices. Individual differences in performance should always be considered when group data are reported. The diversity of applications, the richness of data generated, and the usefulness of the RAM are likely to maintain the popularity of the apparatus in future research.

ACKNOWLEDGEMENTS

The authors would like to thank Dr C Thinus-Blanc for helpful comments on the chapter.

REFERENCES

Aadland, J., Beatty, W.W., & Maki, R.H. (1985). Spatial memory of children and adults assessed in the radial maze. *Developmental Psychobiology, 18*, 163–172.

Ammassari-Teule, M., & Caprioli, A. (1985). Spatial learning and memory, maze running strategies and cholinergic mechanisms in two inbred strains of mice. *Behavioral Brain Research, 17*, 9–16.

Ammassari-Teule, M., D'Amato, F.R., Sansone, M., & Oliverio, A. (1988). Enhancement of radial maze performances in CD1 mice after prenatal exposure to oxiracetam: Possible role of sustained investigative responses developed during ontogeny. *Physiology and Behavior, 42*, 281–285.

Ammassari-Teule, M., & Hoffmann, H.J. (1993). Learning in inbred mice: Strain-specific abilities across three radial maze problems. *Behavior Genetics, 23*, 405–412.

Ammassari-Teule, M., & Maho, C. (1985). Properties of mapping induced by fornix damages: Learning and memorizing the radial maze task. *Physiological Psychology, 13*, 230–234.

Ammassari-Teule, M., & Maho, C. (1992). Choice behavior of fornix-damaged rats in radial maze error-free situations and subsequent learning. *Physiology and Behavior, 51*, 563–567.

Andrews, M.W. (1988). Selection of food sites by Callicebus moloch and Saimiri sciureus under spatially and temporally varying food distribution. *Learning and Motivation, 19*, 254–268.

Arendash, G.W., Sanberg, P.R., & Sengstock, G.J. (1995). Nicotine enhances the learning and memory of aged rats. *Pharmacology, Biochemistry and Behavior, 52*, 517–523.

Bailey, D.W., Rittenhouse, L.R., Hart, R.H., & Richards, R.W. (1989). Characteristics of spatial memory in cattle. *Applied Animal Behavior Science, 23*, 331–340.

Balda, R.P., & Kamil, A.C. (1988). The spatial memory of Clark's nutcrackers (Nucifraga columbiana) in an analogue of the radial arm maze. *Animal Learning and Behavior, 16*, 116–122.

Barnes, C.A. (1979). Memory deficits associated with senescence: A neurophysiological and behavioral study in the rat. *Journal of Comparative and Physiological Psychology, 93*, 74–104.

Batson, J.D., Best, M.R., Phillips, D.L., Patel, H., & Gilleland, K.R. (1986). Foraging on the radial-arm maze: Effects of altering the reward at target location. *Animal Learning and Behavior, 14*, 241–248.

Beatty, W.W. (1984). Hormonal organization of sex differences in play fighting and spatial behavior. *Progress in Brain Research, 61*, 320–324.

Beatty, W.W., Bierley, R.A., & Rush, J.R. (1985). Spatial memory in rats: Electroconvulsive shock selectively disrupts working memory but spares reference memory. *Behavioral and Neural Biology, 44*, 403–414.

Beatty, W.W., & Carbone, C.P. (1980). Septal lesions, intramaze cues and spatial behavior in rats. *Physiology and Behavior, 24*, 675–678.

Beatty, W.W., & Shavalia, D.A. (1980a). Rat spatial memory: Resistance to retroactive interference at long-term retention intervals. *Animal Learning and Behavior, 8*, 550–552.

Beatty, W.W., & Shavalia, D.A. (1980b). Time-course of working memory and effects of anestheticss. *Behavioral and Neural Biology, 28*, 454–462.

Bernstein, D., Olton, D.S., Ingram, D.K., Waller, S.B., Reynolds, M.A., & London, E.D. (1985). Radial maze performance in young and aged mice: Neurochemical correlates. *Pharmacology Biochemistry and Behavior, 22*, 301–307.

Bierley, R.A., Rixen, G.J., Troster, A.I., & Beatty, W.W. (1986). Preserved spatial memory in old rats survives 10 months without training. *Behavioral and Neural Biology, 45*, 223–229.

Bolhuis, J.J., Buresova, O., & Bures, J. (1985). Persistence of working memory of rats in an aversively motivated radial maze task. *Behavioural Brain Research, 15*, 43–49.

Bond, A.B., Cook, R.G., & Lamb, M.R. (1981). Spatial memory and the performance of rats and pigeons in the radial-arm maze. *Animal Learning and Behavior, 9*, 575–580.

Bostock, E., Gallagher, M., & King, R.A. (1988). Effects of opioid microinjections into the medial septal area on spatial memory in rats. *Behavioral Neuroscience, 102*, 643–652.

Brown, M.F. (1990). The effects of maze-arm length on performance in the radial-arm maze. *Animal Learning and Behavior, 18*, 13–22.

Brown, M.F. (1992). Does a cognitive map guide choices in the radial arm maze? *Journal of Experimental Psychology: Animal Behavior Processes, 18*, 56–66.

Brown, M.F., & Huggins, C.K. (1993). Maze-arm length affects a choice criterion in the radial arm maze. *Animal Learning and Behavior, 21*, 68–72.

Brown, M.F., Rish, P.A., VonCulin, J.E., & Edberg, J. (1993). Spatial guidance of choice behavior in the radial arm maze. *Journal of Experimental Psychology: Animal Behavior Processes, 19*, 195–214.

Brown, M.F., Wheeler, E.A., & Riley, D.A. (1989). Evidence for a shift in the choice criterion of rats in a 12-arm radial maze. *Animal Learning and Behavior, 17*, 12–20.

Bruto, V., & Anisman, H. (1983). Acute and chronic amphetamine treatment: Differential modification of exploratory behavior in a radial maze. *Pharmacology Biochemistry and Behavior, 19*, 487–496.

Buhot, M-C., & Poucet, B. (1986). Role of the spatial structure in multiple choice problem-solving by golden hamsters. In P. Ellen & C. Thinus-Blanc (Eds.), *Cognitive processes and spatial orientation in animals and man*. Dordrecht: Martinus Nijhof.

Buresova, O. (1980). Spatial memory and instrumental conditioning. *Acta Neurobiologiae Experimentalis, 40*, 51–65.

Buresova, O., & Bures, J. (1981). Role of olfactory cues in the radial maze performance of rats. *Behavioral Brain Research, 3*, 405–409.

Buresova, O., Bures, J., Oitzl, M.S., & Zahalka, A. (1985). Radial maze in the water tank: An aversively motivated spatial working memory task. *Physiology and Behavior, 34*, 1003–1005.

Burmeister, S., Couvillon, P.A., & Bitterman, M.E. (1995). Performance of honeybees in analogues of the rodent radial maze. *Animal Learning and Behavior, 23*, 369–375.

Byatt, G., & Dalrymple-Alford, J.C. (1996). Both anteromedial and anteroventral thalamic lesions impair radial maze learning in rats. *Behavioral Neuroscience, 110*, 1335–1348.

Caprioli, A., Ghirardi, O., Giuliani, A., Ramacci, M.T., & Angelucci, L. (1991a). Spatial learning and memory in the radial maze: A longitudinal study in rats from 4 to 25 months of age. *Neurobiology of Aging, 12*, 605–607.

Caprioli, A., Ghirardi, O., Giuliani, A., Ramacci, M.T., & Angelucci, L. (1991b). Age-dependent deficits in radial maze performance in the rat: Effect of chronic treatment with acetyl-L-carnitine. *Progress in Neuropharmacology and Biological Psychiatry, 14*, 359–369.

Chapillon, P., Roullet, P., & Lassalle, J.M. (1995). Ontogeny of orientation and spatial learning on the radial maze in mice. *Developmental Psychobiology, 28*, 429–442.

Cocchetto, D.M., Miller, D.B., Miller, L.L., & Bjornsson, T.D. (1985). Behavioral perturbations in the vitamin K-deficient rat. *Physiology and Behavior, 34*, 727–734.

Collett, T.S., Cartwright, B.A., & Smith, B.A. (1986). Landmark learning and visuo-spatial memories in gerbils. *Journal of Comparative Physiology, 158*, 835–851.

Colombo, P.J., Davis, H.P., & Volpe, B.T. (1989). Allocentric spatial and tactile memory impairments in rats with dorsal caudate lesions are affected by preoperative behavioral training. *Behavioral Neuroscience, 103*, 1242–1250.

Cook, R.G., Brown, M.F., & Riley, D.A. (1985). Flexible memory processing by rats: Use of prospective and retrospective information in the radial maze. *Journal of Experimental Psychology: Animal Behavior Processes, 11*, 453–469.

Cornell, E.H., & Heth, C.D. (1986). The spatial organization of hiding and recovery of objects by children. *Child Development, 57*, 603–615.

Cramer, C.P., Pfister, J.P., & Haig, K.A. (1988). Experience during suckling alters later spatial learning. *Developmental Psychobiology, 21*, 1–24.

Crannell, C.W. (1942). The choice point behavior of rats in a multiple path elimination problem. *Journal of Psychology, 13*, 201–222.

Dachir, S., Kadar, T., Robinzon, B., & Levy, A. (1993). Cognitive defects induced in young rats by long-term corticosterone administration. *Behavioral and Neural Biology, 60*, 103–109.

Dale, R.H.I. (1982). Parallel-arm maze behavior in blind and sighted rats: Spatial memory. *Behaviour Analysis Letters, 2*, 127–139.

Dale, R.H. (1986). Spatial and temporal response patterns on the eight arm radial maze. *Physiology and Behavior, 36*, 787–790.

Dale, R.H. (1988). Spatial memory in pigeons on a four-arm radial maze. *Canadian Journal of Psychology, 42*, 78–83.

Dale, R.H.I., & Innis, N.K. (1986). Interactions between response stereotype and memory strategies on the eight-arm radial maze. *Behavioral Brain Research, 19*, 17–25.

Dale, R.H.I., & Roberts, W.A. (1986). Variations in radial maze performance under different levels of food and water deprivation. *Animal Learning and Behavior, 14*, 60–64.

Dallal, N.L., & Meck, W.H. (1990). Hierarchical structures: Chunking by food type facilitates spatial memory. *Journal of Experimental Psychology: Animal Behavior Processes, 16*, 69–84.

Dashiell, J.F. (1930). Direction orientation in maze running by the white rat. *Comparative Psychology Monographs, 7*, 1–72.

Davis, H.P., Baranowski, J.R., Pulsinelli, W.A., & Volpe, B.T. (1987). Retention of reference memory following ischemic hippocampal damage. *Physiology and Behavior, 39*, 783–786.

Dember, W.N. (1956). Response by the rat to environmental change. *Journal of Comparative and Physiological Psychology, 49*, 93–95.

Dennis, W. (1939). Spontaneous alternation in rats as indicator of the persistence of stimulus effects. *Journal of Comparative Psychology, 28*, 305–312.

Deyo, R.A., & Conner, R.L. (1989). Microinjections of leupeptin in the frontal cortex or dorsal hippocampus block spatial learning in the rat. *Behavioral and Neural Biology, 52*, 213–221.

Diez-Chamizo, V., Sterio, D., & Mackintosh, N.J. (1985). Blocking and overshadowing between intra-maze and extra-maze cues: A test of the independence of locale and guidance learning. *Quarterly Journal of Experimental Psychology, 37B*, 235–253.

Douglas, R.J. (1966). Cues for spontaneous altervation. *Journal of Comparative and Physiological Psychology, 62*, 171–181.

Douglas, R.J., Mitchell, D., & Del Valle, R. (1974). Angle between choice alleys as a critical factor in spontaneous alternation. *Animal Learning and Behavior, 2*, 218–220.

Douglas, R.J., & Truncer, P.C. (1977). Parallel but independent effects of pentobarbital and scopolamine on hippocampus-related behavior. *Behavioral Biology, 18*, 359–367.

Durkin, T.P., & Toumane, A. (1992). Septo-hippocampal and nBM-cortical cholinergic neurones exhibit differential time-courses of activation as a function of both type and duration of spatial memory testing in mice. *Behavioral Brain Research, 50*, 43–52.

Durup, H. (1964). Influence des traces odorantes laissées par le hamster lors de son apprentissage d'un labrinthe à discrimination olfactive. *Psychologie Française, 9*, 165–180.

Eckerman, D. (1980). Monte Carlo estimation for chance performance for the radial arm maze. *Bulletin of the Psychonomic Society, 15*, 93–95.

Einon, D. (1980). Spatial memory and response strategies in rats: Age, sex and rearing differences. *Quarterly Journal of Experimental Psychology, 32*, 473–489.

Ellen, P. (1980). Cognitive maps and the hippocampus. *Physiological Psychology, 8*, 168–174.

Elsmore, T.F., & McBride, S.A. (1994). An eight-alternative concurrent schedule: Foraging in a radial maze. *Journal of the Experimental Analysis of Behavior, 61*, 331–348.

Ennaceur, A. (1994). Effects of amphetamine and medial septal lesions on acquisition and retention of radial maze learning in rats. *Brain Research, 636*, 277–285.

Ermakova, I.V., Loseva, E.V., Valouskova, V., & Bures, J. (1989). The effect of embryonal amygdala grafts on the impairment of spatial working memory elicited in rats by kainate-induced amygdaloid damage. *Physiology and Behavior, 45*, 235–241.

Evenden, J.L., & Robbins, T.W. (1984). Win-stay behaviour in the rat. *Quarterly Journal of Experimental Psychology, 36B*, 1–26.

Fagioli, S., Rossi-Arnaud, C., & Ammassari-Teule, M. (1991). Open field behaviours and spatial learning performance in C57BL/6 mice: Early stage effects of chronic GM1 ganglioside administration. *Psychopharmacology, 105*, 209–212.

Foreman, N. (1985). Algorithmic responding on the radial maze in rats does not always imply the absence of spatial encoding. *Quarterly Journal of Experimental Psychology, 37B*, 333–358.

Foreman, N., Arber, M., & Savage, J. (1984). Spatial memory in preschool infants. *Developmental Psychobiology, 17*, 129–137.

Foreman, N., Gillett, R., & Jones, S. (1994). Choice autonomy and memory for spatial locations in six-year-old children. *British Journal of Psychology, 85*, 17–27.

Foreman, N., & Stevens, R. (1982). Visual lesions and radial maze performance in rats. *Behavioral and Neural Biology, 36*, 126–136.

Foreman, N., & Stevens, R. (1987). Relationships between superior colliculus and hippocampus: Neural and behavioral considerations. *Behavioral and Brain Sciences, 10*, 101–151.

Foreman, N., Toates, F., & Donohoe, T. (1990). Spontaneous and learned turning behaviour in food- or water-restricted hooded rats. *Quarterly Journal of Experimental Psychology, 42B*, 153–173.

Fredriksson, A., & Archer, T. (1996). Alpha-phenyl-*t*-butylnitrone (PBN) reverses age-related maze learning performance and motor activity deficits in C57 BL/6 mice. *Behavioural Pharmacology, 7*, 245–253.

Furukawa, S., & Iwasaki, T. (1989). Deficits in radial-arm maze learning in aged rats. *Japanese Journal of Psychology, 60*, 192–195.

Gaffan, E.A., & Davies, J. (1981). The role of exploration in win-shift and win-stay performance on a radial maze. *Learning and Motivation, 12*, 282–299.

Gaffan, E.A., Hansel, M.C., & Smith, L.E. (1983). Does reward depletion influence spatial memory performance? *Learning and Motivation, 14*, 58–74.

Gage, P.D. (1985). Performance of hippocampectomized rats in a reference/working memory task: Effects of preoperative versus postoperative training. *Physiological Psychology, 13*, 235–242.

Gallagher, M., Bostock, E., & King, R. (1985). Effects of opiate antagonists on spatial memory in young and aged rats. *Behavioral and Neural Biology, 44*, 374–385.

Gallistel, C.R. (1990). *The organization of learning*. Cambridge, MA: Bradford Books/MIT Press.

Gentry, G., Brown, W.L., & Kaplan, S. (1947). An experimental analysis of the spatial location hypothesis in learning. *Journal of Comparative and Physiological Psychology, 40*, 309–322.

Gentry, G., Brown, W.L., & Lee, H. (1948). Spatial location in the learning of a multiple T-maze. *Journal of Comparative and Physiological Psychology, 41*, 312–318.

Gillett, R. (1994). The exact null distribution for radial maze statistics: A FORTRAN 77 program. *Behavior Research Methods, Instruments and Computers, 26*, 70–73.

Gillett, R. (1996). Testing hypotheses about working memory capacity. *British Journal of Mathematical and Statistical Psychology, 49*, 241–252.

Glassman, R.B., Garvey, K.J., Elkins, K.M., Kasal, K.L., & Couillard, N.L. (1994). Spatial working memory score of humans in a large radial maze, similar to published score of rats, implies capacity close to the magic number 7 +/− 2. *Brain Research Bulletin, 34*, 151–159.

Godding, P.R., Rush, J.R., & Beatty, W.W. (1982). Scopolamine does not disrupt spatial working memory in rats. *Pharmacology Biochemistry and Behaviour, 16*, 919–923.

Goodale, M.A., & Dale, R.H.I. (1981). Radial-maze performance in the rat following lesions of posterior neocortex. *Behavioral Brain Research, 4*, 273–288.

Green, R.J., & Stanton, M.E. (1989). Differential ontogeny of working memory and reference memory in the rat. *Behavioral Neuroscience, 103*, 98–105.

Grobety, M.C., & Schenk, F. (1992). The influence of spatial irregularity upon radial maze performance in the rat. *Animal Learning and Behavior, 20*, 393–400.

Haig, K.A., Olton, D.S., Rawlins, J.N.P., Mead, A., & Taylor, B. (1983). Food searching strategies of rats: Variables affecting the relative strength of syat and shift strategies. *Journal of Experimental Psychology: Animal Behaviour Processes, 9*, 337–348.

Hall, J., & Berman, R.F. (1995). Juvenile experience alters strategies used to solve the radial arm maze in rats. *Psychobiology, 23*, 195–198.

Hamilton, W.J., & Watt, K.E.F. (1970). Refuging. *Annual Review of Ecology and Systematics, 1*, 263–286.

Harley, C. (1979). Arm-choice in the sunburst maze: Effects of hippocampectomy in the rat. *Physiology and Behavior, 23*, 283–290.

Healy, S.D., & Hurly, T.A. (1995). Spatial memory in rufous hummingbirds (*Selasphorus rufus*): A field test. *Animal Learning and Behavior, 23*, 63–68.

Higashida, A., & Ogawa, N. (1987). Radial maze performance in three strains of rats: The role of spatial strategy. *Research Communications in Psychology, Psychiatry and Behavior, 12*, 118–128.

Hilton, S.C., & Krebs, J.K. (1990). Spatial memory of four species of *Parus*: Performance in an open-field analogue of a radial maze. *Quarterly Journal of Experimental Psychology, 42B*, 345–368.

Hirsch, R. (1980). The hippocampus, conditional operations, and cognition. *Physiological Psychology, 8*, 175–182.

Hoffman, N., & Maki, W.S. (1986). Two sources of proactive interference in spatial working memory: Multiple effects of repeated trials on radial maze performance by rats. *Animal Learning and Behavior, 14*, 65–72.

Honig, W.K. (1978). Studies of working memory in the pigeon. In S.H. Hulse, H. Fowler, & W.K. Honig (Eds.), *Cognitive processes in animal behaviour*. Hillsdale, NJ: Lawrence Erlbaum Associates.

Honig, W.K. (1986). The problem of the stimulus in the cognitive control of behaviour. In P. Ellen & C. Thinus-Blanc (Eds.), *Cognitive processes and spatial orientation in animals and man*. Dordrecht: Martinus Nijhof.

Horner, J. (1984). The effect of maze structure upon the performance of a multiple-goal task. *Animal Learning and Behavior, 12*, 55–61.

Ikegami, S., Nihonmatsu, I., Hatanaka, H., Takei, N., & Kawamura, H. (1989). Transplantation of septal cholinergic neurons to the hippocampus improves memory impairments of spatial learning in rats treated with AF64A. *Brain Research, 496*, 321–326.

Ilersich, T.J., Mazmanian, D.S., & Roberts, W.A. (1988). Foraging for covered and uncovered food on a radial maze. *Animal Learning and Behavior, 16*, 388–394.

Innis, N.K., & Macgillivray, M. (1987). Radial maze performance under food and water deprivation. *Behavioural Processes, 15*, 167–179.

Janis, L.S., Bishop, T.W., & Dunbar, G.L. (1994). Medial septal lesions in rats produce permanent deficits for strategy selection in a spatial memory task. *Behavioral Neuroscience, 108*, 892–898.

Jarrard, L.E. (1980). Selective hippocampal lesions and behaviour. *Phsyiological Psychology, 8*, 198–206.

Jones, C.H., McGhee, R., & Wilkie, D.M. (1990). Hamsters (*Mesocricetus auratus*) use spatial memory in foraging for food to hoard. *Behavioural Processes, 21*, 179–187.

Juraska, J.M., Henderson, C., & Muller, J. (1984). Differential rearing experience, gender and radial maze performance. *Developmental Psychobiology, 17*, 209–215.

Kadar, T., Silbermann, M., Brandeis, R., & Levy, A. (1990). Age-related structural changes in the rat hippocampus: Correlation with working memory deficiency. *Brain Research, 512*, 113–120.

Kamil, A.C. (1978). Systematic foraging by a nectar-feeding bird, the Amakihi (*Loxops virens*). *Journal of Comparative and Physiological Psychology, 92*, 388–396.

Keseberg, U., & Schmidt, W.J. (1995). Low-dose challenges by the NMDA receptor antagonist dizocilpine exacerbates the spatial learning deficit in entorhinal cortex-lesioned rats. *Behavioural Brain Research, 67*, 255–261.

Kesner, R.P., DiMattia, B.V., & Crutcher, K.A. (1987). Evidence for neocortical involvement in reference memory. *Behavioral and Neural Biology, 47*, 40–53.

Kesner, R.P., & Novak, J.M. (1982). Serial position curve in rats: Role of the dorsal hippocampus. *Science, 218*, 173–175.

Kessler, J., Markowitsch, H.J., & Otto, B. (1982). Subtle but distinct impairments of rats with chemical lesions in the thalamic mediodorsal nucleus, tested in a radial arm maze. *Journal of Comparative and Physiological Psychology, 96*, 712–720.

Kimble, D., & Whishaw, I.Q. (1994). Spatial behavior in the Brazilian short-tailed opossum (*Monodelphis domestica*): Comparison with the Norway rat (*Rattus norvegicus*) in the Morris water maze and radial arm maze. *Journal of Comparative Psychology, 108*, 148–155.

Kobayashi, S., Kametani, H., Ugawa, Y., & Osanai, M. (1988). Age difference of response strategy in radial maze performance of Fischer-344 rats. *Physiology and Behavior, 42*, 277–280.

Kraemer, P.J., Gilbert, M.E., & Innis, N.K. (1983). The influence of cue type and configuration upon radial maze performance in the rat. *Animal Learning and Behavior, 11*, 373–380.

Krechevsky, I. (1932). "Hypotheses" in rats. *Psychological Review, 38*, 516–532.

Kunka, M.G., Bernstein, I.G., & Kubie, J.L. (1979, April). *Short-term spatial memory in garter snakes*. Paper presented at the meeting of the Eastern Psychological Association, Philadelphia, PA.

Lachman, S.J. (1965). Behaviour in a multiple-choice elimination problem involving five paths. *Journal of Psychology, 61,* 193–202.

Lachman, S.J. (1966). Stereotypy and variability of behaviour in a complex learning situation. *Psychological Reports, 18,* 223–230.

Lachman, S.J. (1969). Behaviour in a complex learning situation involving five stimulus-differentiated paths. *Psychonomic Science, 17,* 36–37.

Lachman, S.J. (1971). Behaviour in a complex learning situation involving three levels of difficulty. *Journal of Psychology, 77,* 119–126.

Lachman, S.J., & Brown, C.R. (1957). Behaviour in a free choice multiple path elimination problem. *Journal of Psychology, 13,* 101–109.

Lanke, J., Mansson, L., Bjerkemo, M., & Kjellstrand, P. (1993). Spatial memory and stereotypic behaviour of animals in radial arm mazes. *Brain Research, 605,* 221–228.

Lavenex, P., & Schenk, F. (1989). Integration of olfactory information in a spatial representation enabling accurate arm choice in the radial arm maze. *Learning and Memory, 2,* 299–319.

Lebrun, C., Durkin, T.P., Marighetto, A., & Jaffard, R. (1990). A comparison of the working memory performance of young and aged mice combined with parallel measures of testing and drug-induced activations of septo-hippocampal and nbm-cortical cholinergic neurones. *Neurobiology of Aging, 11,* 515–521.

Leis, T., Pallage, V., Toniolo, G., & Will, B. (1984). Working memory theory of hippocampal function needs qualification. *Behavioral and Neural Biology, 42,* 140–157.

Levere, T.E., & Walker, A. (1992). Old age and cognition: Enhancement of recent memory in aged rats by the calcium channel blocker nimodipine. *Neurobiology of Aging, 13,* 63–66.

Levin, E.D. (1988). Psychopharmacological effects in the radial arm maze. *Neuroscience and Biobehavioral Reviews, 12,* 169–175.

Levy, A., Kluge, P.B., & Elsmore, T.F. (1983). Radial arm maze performance of mice: Acquisition and atropine effects. *Behavioral and Neural Biology, 39,* 229–240.

Lichtman, A.H., Dimen, K.R., & Martin, B.R. (1995). Systemic or intrahippocampal cannabinoid administration impairs spatial memory in rats. *Psychopharmacology, 119,* 282–290.

Loh, E.A., & Beck, C.H.M. (1989). Rats treated chronically with the benzodiazepine, Diazepam, or with ethanol exhibit reduced variability of behavior. *Alcohol, 6,* 311–316.

Luine, V., & Rodriguez, M. (1994). Effects of estradiol on radial arm maze performance of young and aged rats. *Behavioral and Neural Biology, 62,* 230–236.

Luine, V., Villegas, M., Martinez, C., & McEwen, B.S. (1994). Repeated stress causes reversible impairments of spatial memory performance. *Brain Research, 639,* 167–170.

Lydon, R.G., & Nakajima, S. (1992). Differential effects of scopolamine on working and reference memory depend upon level of training. *Pharmacology, Biochemistry and Behavior, 43,* 645–650.

Mackintosh, N.J. (1983). *Conditioning and associative learning.* Oxford: Clarendon.

Magni, S., Krekule, I., & Bures, J. (1979). Radial maze type as a determinant of the choice behaviour of rats. *Journal of Neuroscience Methods, 1,* 343–352.

Maho, C., Dutrieux, G., & Ammassari-Teule, M. (1988). Parallel modifications of spatial memory performances, exploration patterns, and hippocampal theta rhythms in fornix-damaged rats: Reversal of oxotremorine. *Behavioral Neuroscience, 102,* 601–604.

Maier, N.R.F. (1932). A study of orientation in the rat. *Journal of Comparative Psychology, 14,* 387–399.

Maki, W.S. (1987). On the non-associative nature of working memory. *Learning and Motivation, 18,* 99–117.

Maki, W.S., Beatty, W.W., & Clouse, B.A. (1984a). Item and order information in spatial memory. *Journal of Experimental Psychology: Animal Behavior Processes, 10,* 437–452.

Maki, W.S., Beatty, W.W., Hoffman, N., Brierley, R.A., & Clouse, B.A. (1984b). Spatial memory over long retention intervals: Nonmemorial factors are not necessary for accurate performance in the radial-arm maze by rats. *Behavioral and Neural Biology, 41,* 1–6.

Maki, W.S., Brokofsky, S., & Berg, B. (1979). Spatial memory in rats: Resistance to proactive interference. *Animal Learning and Behavior, 7,* 25–30.

Marighetto, A., Micheau, J., & Jaffard, R. (1993). Relationships between testing-induced alterations of hippocampal cholinergic activity and memory performance on two spatial tasks in mice. *Behavioural Brain Research, 56,* 133–144.

Markowska, A., Buresova, O., & Bures, J. (1983). An attempt to account for controversial estimates of working memory persistence in the radial maze. *Behavioral and Neural Biology, 38,* 97–112.

Masuda, Y., Odashima, J., Murai, S., Saito, H., Itoh, M., & Itoh, T. (1994). Radial arm maze behavior in mice when a return to the home cage serves as the reinforcer. *Physiology and Behavior, 56,* 785–788.

Mazmanian, D.S., & Roberts, W.A. (1983). Spatial memory in rats under restricted viewing conditions. *Learning and Motivation, 14,* 123–139.

Means, L.W., Hardy, W.T., Gabriel, M., & Uphold, J.D. (1971). Utilization of odour trails by rats in maze learning. *Journal of Comparative and Physiological Psychology, 76,* 160–164.

Meck, W.H., Smith, R.A., & Williams, C.L. (1989). Organizational changes in cholinergic activity and enhanced visuospatial memory as a function of choline administered prenatally or postnatally or both. *Behavioral Neuroscience, 103,* 1234–1241.

Menzel, E.W. (1978). Cognitive mapping in chimpanzees. In S.H. Hulse, H. Fowler, & W.K. Honig (Eds.), *Cognitive processes in animal behavior.* Hillsdale, NJ: Lawrence Erlbaum Associates Inc.

Miller, G.A. (1956). The magical number seven, plus or minus two: Some limits on our capacity for processing information. *Psychological Review, 63,* 81–97.

Mizumori, S.J.Y., Rosenzweig, M.R., & Kermisch, M.G. (1982). Failure of mice to demonstrate spatial memory in the radial maze. *Behavioral and Neural Biology, 35,* 33–45.

Molina-Holgado, M., Gonzalez, M.I., & Leret, M.L. (1994). Effect of delta-8-tetrahydrocannabinol on short term memory in the rat. *Physiology and Behavior, 57,* 177–179.

Morris, R.G.M. (1982). An attempt to dissociate 'spatial mapping' and 'working-memory' theories of hippocampal function. In W. Siefert (Ed.), *Neurobiology of the hippocampus.* London: Academic Press.

Mundy, W.R., & Iwamoto, E.T. (1988). Nicotine impairs acquisition of radial maze performance in rats. *Pharmacology, Biochemistry and Behavior, 30,* 119–122.

Murray, C.L., & Fibiger, H.C. (1983). Learning and memory deficits after lesions of the nucleus basalis magnocellularis: Reversal by physostigmine. *Neuroscience, 14,* 1025–1032.

Murray, C.L., & Fibiger, H.C. (1986). The effects of pramiracetam (CI-879) on the acquisition of a radial arm maze task. *Psychopharmacology, 89,* 378–381.

Nadel, L. & MacDonald, L. (1980). Hippocampus: Cognitive map or working memory? *Behavioral and Neural Biology, 29,* 405–409.

Okaichi, H., & Oshima, Y. (1990). Choice behavior of hippocampectomized rats in the radial arm maze. *Psychobiology, 18,* 416–421.

Okaichi, H., Oshima, Y., & Jarrard, L.E. (1989). Scopolamine impairs both working and reference memory in rats: A replication and extension. *Pharmacology, Biochemistry and Behavior, 34,* 599–602.

O'Keefe, J. (1982). Spatial memory within and without the hippocampal system. In W. Seifert (Ed.), *Neurobiology of the hippocampus.* London: Academic Press.

O'Keefe, J., & Nadel, L. (1978). *The hippocampus as a cognitive map.* Oxford: Oxford University Press.

Olton, D.S. (1977). Spatial memory. *Scientific American, 236,* 82–98.

Olton, D.S. (1978). Characteristics of spatial memory. In S.H. Hulse, W.K. Honig, & H. Fowler (Eds.), *Cognitive aspects of animal behavior.* Hillsdale, NJ: Lawrence Erlbaum Associates.

Olton, D.S. (1987). The radial maze as a tool in behavioral pharmacology. *Physiology and Behavior, 40,* 793–797.

Olton, D.S., Becker, J.T., & Handelmann, G.E. (1979). Hippocampus, space and memory. *Behavioral and Brain Sciences, 2,* 313–365.

Olton, D.S., & Collison, C. (1979). Intramaze cues and "odour trails" fail to direct choice behaviour on an elevated maze. *Animal Learning and Behavior, 7,* 221–223.

Olton, D.S., Collison, C., & Werz, M.A. (1977). Spatial memory and radial-arm maze performance of rats. *Learning and Motivation, 8,* 289–314.

Olton, D.S., & Papas, B.C. (1979). Spatial memory and hippocampal system function. *Neuropsychologia, 17,* 690–682.

Olton, D.S., & Samuelson, R.J. (1976). Remembrance of places passed: Spatial memory in rats. *Journal of Experimental Psychology: Animal Behavior Processes, 2,* 97–116.

Olton, D.S., & Werz, M.A. (1978). Hippocampal function and behaviour: Spatial discrimination and response inhibition. *Physiology and Behavior, 20,* 597–605.

Ossenkopp, K.P., & Hargreaves, E.L. (1993). Spatial learning in an enclosed eight-arm radial maze in rats with sodium arsenilate-induced labrynthectomies. *Behavioral and Neural Biology, 59,* 253–257.

Packard, M.G., & White, N.M. (1991). Memory facilitation produced by dopamine agonists: Role of receptor subtype and mnemonic requirements. *Pharmacology, Biochemistry and Behavior, 33,* 511–518.

Packard, M.G., Winocur, G., & White, N.M. (1992). The caudate nucleus and acquisition of win-shift radial-maze behavior: Effect of exposure to the reinforcer during maze adaptation. *Psychobiology, 20,* 127–132.

Pallage, V., Orenstein, D., & Will, B. (1992). Nerve growth factor and septal grafts: A study of behavioral recovery following partial damage to the septum in rats. *Behavioral Brain Research, 47,* 1–12.

Pearson, D.E., Raskin, L.A., Shaywitz, B.A., Anderson, G.M., & Cohen, D.J. (1984). Radial arm maze performance in rats following neonatal dopamine depletion. *Developmental Psychobiology, 17,* 505–517.

Peele, D.B., & Baron, S.P. (1988). Effects of selection delays on radial maze performance: Acquisition and effects of scopolamine. *Pharmacology, Biochemistry and Behavior, 29,* 143–150.

Petrinovich, L., & Bolles, R.C. (1954). Deprivation states and behavioral attributes. *Journal of Comparative and Physiological Psychology, 47,* 45–453.

Phelps, M.T., & Roberts, W.A. (1989). Central-place foraging foraging by *Rattus norvegicus* on a radial maze. *Journal of Comparative Psychology, 103,* 326–338.

Pico, R.M., & Davis, J.L. (1984). The radial maze performance of mice: Assessing the dimensional requirements for serial order memory in animals. *Behavioral and Neural Biology, 40,* 5–26.

Pico, R.M., Gerbrandt, L.K., Pondel, M., & Ivy, G. (1985). During stepwise cue deletion, rat place behaviors correlate with place unit responses. *Brain Research, 330,* 369–372.

Picq, J.L. (1993). Radial maze performance in young and aged grey mouse lemurs (*Microcebus murinus*). *Primates, 34,* 223–226.

Potegal, M. (1969). Role of the caudate nucleus in spatial orientation in rats. *Journal of Comparative and Physiological Psychology, 69,* 756–764.

Poucet, B., Lucchessi, H., & Thinus-Blanc, C. (1991). What information is used by rats to update choices in the radial-arm maze? *Behavioural Processes, 25,* 15–26.

Rauch, S.L., & Raskin, L.A. (1984). Cholinergic mediation of spatial memory in the preweanling rat: Application of the radial maze paradigm. *Behavioral Neuroscience, 98,* 35–43.
Rawlins, J.N.P., & Deacon, R.M.J. (1993). Further developments of maze procedures. In A. Saghal (Ed.), *Behavioural neuroscience: A practical approach* (Vol. 1). Oxford: IRL Press.
Reinstein, D.K., DeBoissiere, T., Robinson, N., & Wurtman, R.J. (1983). Radial maze performance in three strains of mice: Role of the fimbria/fornix. *Brain Research, 263,* 172–176.
Roberts, W.A. (1979). Spatial memory in the rat on a hierarchical maze. *Learning and Motivation, 10,* 117–140.
Roberts, W.A. (1981). Retroactive inhibition in rat spatial memory. *Animal Learning and Behavior, 9,* 566–574.
Roberts, W.A. (1984). Some issues in animal spatial memory. In H. L. Roitblat, T.G. Bever, & H.S. Terrace (Eds.), *Animal cognition.* Hillsdale, NJ: Lawrence Erlbaum Associates.
Roberts, W.A., & Dale, R.H.I. (1981). Remembrance of places lasts: proactive inhibition and patterns of choice in rat spatial memory. *Learning and Motivation, 12,* 261–281.
Roberts, W.A., & Van Veldhuizen, N. (1985). Spatial memory in pigeons on the radial maze. *Journal of Experimental Psychology: Animal Behavior Processes, 11,* 241–260.
Robinson, S.E., Hambrecht, K.L., & Lyeth, B.G. (1988). Basal forebrain carbachol injection reduces cortical acetylcholine turnover and disrupts memory. *Brain Research, 445,* 160–164.
Roitblat, H.L. (1982). The meaning of representation in animal memory. *Behavioral and Brain Sciences, 5,* 353–406.
Roitblat, H.L., Tham, W., & Golub, L. (1982). Performance of *Betta splendens* in a radial arm maze. *Animal Learning and Behavior, 10,* 108–114.
Roullet, P., & Lassalle, J.M. (1992). Behavioural strategies, sensorial processes and hippocampal mossy fibre distribution in radial maze performance in mice. *Behavioral Brain Research, 48,* 77–85.
Ryabinskaya, E.A., & Ashikhmina, O.V. (1988). Volume of reinforcement and structure of rats' behaviour in the radial maze. *Journal of Higher Nervous Activity, 38,* 675–683. (In Russian)
Schenk, F., & Grobety, M.C. (1992). Interactions between directional and visual environmental cues in spatial learning in rats. *Learning and Motivation, 23,* 80–98.
Seamans, J.K., & Phillips, A.G. (1994). Selective memory impairments produced by transient lidocaine-induced lesions of the nucleus accumbens in rats. *Behavioral Neuroscience, 108,* 456–468.
Seymoure, P., Dou, H., & Juraska, J.M. (1996). Sex differences in radial maze performance: Influence of rearing environment and room cues. *Psychobiology, 24,* 33–37.
Shors, T.J., & Dryver, E. (1992). Stress impeded exploration and the acquisition of spatial information in the eight-arm radial maze. *Psychobiology, 20,* 247–253.
Simon, H., Taghzouti, K., & Le Moal, M. (1986). Deficits in spatial memory tasks following lesions of septal dopaminergic terminals in the rat. *Behavioral Brain Research, 19,* 7–16.
Spetch, M.L., & Honig, W.K. (1988). Characteristics of pigeons' spatial working memory in an open-field task. *Animal Learning and Behavior, 16,* 123–131.
Spetch, M.L., & Wilkie, D.M. (1980). A programme that simulates random choice in radial arm mazes and similar choice situations. *Behavioral Research Methods and Instrumentation, 12,* 377–378.
Staddon, J.E.R. (1983). *Adaptive behavior and learning.* New York: Cambridge University Press.
Stevens, R. (1981). Scopolamine impairs spatial maze performance in rats. *Physiology and Behavior, 27,* 385–386.
Stokes, K.A., & Best, P.J. (1988). Mediodorsal thalamic lesions impair radial maze performance in the rat. *Behavioral Neuroscience, 102,* 294–300.

Strupp, B.J., Bunsey, M., Levitsky, D., & Kesler, M. (1991). Time-dependent effects of post-trial amphetamine treatment in rats: Evidence for enhanced storage of representational memory. *Behavioral and Neural Biology, 56*, 62–76.

Sutherland, N.S. (1957). Spontaneous alternation and stimulus avoidance. *Journal of Comparative and Physiological Psychology, 50*, 358–362.

Sutherland, N.S., & Mackintosh, N.J. (1971). *Mechanisms of animal discrimination learning.* New York: Academic Press.

Sutherland, R.J., & Rudy, J.W. (1989). Configural association theory: The role of the hippocampal formation in learning, memory, and amnesia. *Psychobiology, 17*, 129–144.

Suzuki, S., Augerinos, G., & Black, A.H. (1980). Stimulus control of spatial behaviour on the eight-arm maze. *Learning and Motivation, 11*, 1–18.

Takai, R.M., & Wilkie, D.M. (1985). Foraging experience affects gerbils' (*Meriones unguiculatus*) radial arm maze performance. *Journal of Comparative Psychology, 99*, 361–364.

Tees, R.C., Midgley, G., & Nesbit, J.C. (1981). The effect of early visual experience on spatial maze learning in rats. *Developmental Psychobiology, 14*, 425–438.

Thinus-Blanc, C. (1996). *Animal spatial cognition: Behavioral and neural approaches.* Singapore: World Scientific.

Timberlake, W., & White, W. (1990). Winning isn't everything: Rats need only food deprivation and not food reward to efficiently traverse a radial arm maze. *Learning and Motivation, 21*, 153–163.

Tolman, E.C., Ritchie, F.B., & Kalish, D. (1946). Studies in spatial learning: I. Orientation and the short cut. *Journal of Experimental Psychology, 36*, 13–24.

Tomonaga, M., Kinoshita, M., & Ohta, H. (1990). Performance of common tree shrews (*Tupaia glis*) in an enclosed radial-arm maze: Preliminary research. *Japanese Journal of Animal Psychology, 40*, 26–43. (In Japanese)

Toumane, A., Durkin, T., Marighetto, A., Galey, D., & Jaffard, R. (1988). Differential hippocampal and cortical cholinergic activation during the acquisition, retention, reversal and extinction of a spatial discrimination in an 8-arm radial maze by mice. *Behavioral Brain Research, 30*, 225–234.

Uster, H.J., Bättig, K., & Nageli, H.H. (1976). Effects of maze geometry and experience on exploratory behavior in the rat. *Animal Learning and Behavior, 4*, 84–88.

Van Gool, W.A., Mirmiran, M., & Van Haaren, F. (1985). Spatial memory and visual evoked potentials in young and old rats after housing in an enriched environment. *Behavioral and Neural Biology, 44*, 454–469.

Van Haaren, F., Wouters, M., & Van de Poll, N.E. (1987). Absence of behavioral differences between male and female rats in different radial-maze procedures. *Physiology and Behavior, 39*, 409–412.

Van Luijtelaar, E.L., Van der Staay, F.J., & Kerbusch, J.M. (1989). Spatial memory in rats: A cross validation study. *Quarterly Journal of Experimental Psychology, 41B*, 287–306.

Ward, L., Mason, S.E., & Abraham, W.C. (1990). Effects of the NMDA antagonists CPP and MK-801 on radial arm-maze performance in rats. *Pharmacology, Biochemistry and Behavior, 35*, 785–790.

Wathen, C.N., & Roberts, W.A. (1994). Multiple-pattern learning by rats on an eight-arm radial maze. *Animal Learning and Behavior, 22*, 155–164.

Watson, C.D., Hewitt, M.J., Fone, K.C.F., Dickenson, S.L., & Bennett, G.W. (1994). Behavioural effects of scopolamine and the TRH analogue RX77368 on radial arm maze performance in the rat. *Journal of Psychopharmacology, 8*, 88–93.

Watts, J., Stevens, R., & Robinson, C. (1981). Effects of scopolamine on radial maze performance in rats. *Physiology and Behavior, 26*, 845–851.

Whishaw, I.Q., & Tomie, J. (1996). Of mice and mazes: Similarities between mice and rats on dry land but not water mazes. *Physiology and Behavior, 60*, 1191–1197.

Wilkie, D.M., Mumby, D.G., Needham, G., & Smeele, M. (1992). Sustained arm visiting by non-deprived, non-rewarded rats in a radial maze. *Bulletin of the Psychonomic Society, 30*, 314–316.

Wilkie, D.M., & Slobin, P. (1983). Gerbils in space: performance on the 17-arm radial maze. *Journal of the Experimental Analysis of Behavior, 40*, 301–312.

Wilkie, D.M., Spetch, M.L., & Chew, L. (1981). The ring dove's short-term memory capacity for spatial information. *Animal Behaviour, 29*, 639–641.

Williams, C.L., Barnett, A.M., & Meck, W.H. (1990). Organizational effects of early gonadal secretions on sexual differentiation in spatial memory. *Behavioral Neuroscience, 104*, 84–97.

Winocur, G. (1982). Radial-arm-maze behavior by rats with dorsal hippocampal lesions: Effects of cuing. *Journal of Comparative and Physiological Psychology, 96*, 155–169.

Wirsching, B.A., Beninger, R.J., Jhamandas, K., & Boegman, R.J. (1989). Kynurenic acid protects against the neurochemical and behavioral effects of unilateral quinolinic acid injections into the nucleus basalis of rats. *Behavioral Neuroscience, 103*, 90–97.

Wishart, T.B., Ijaz, S., & Shuaib, A. (1994). Differential effects of amphetamine and haloperidol on recovery after global forebrain ischemia. *Pharmacology, Biochemistry and Behavior, 47*, 963–968.

Woodruff, M.L., Baisden, R.H., & Cannon, R.L. (1993). Transplant-induced working memory deficits in hippocampectomized rats. *Physiology and Behavior, 54*, 579–587.

Yoerg, S.I., & Kamil, A.C. (1982). Response strategies in the radial maze: Running around in circles. *Animal Learning and Behavior, 10*, 530–534.

Zoladek, L., & Roberts, W.A. (1978). The sensory basis of spatial memory in the rat. *Animal Learning and Behavior, 6*, 77–81.

5 The Morris Water Maze (is not a Maze)

Françoise Schenk
University of Lausanne, Switzerland

A TIMELY TASK DESIGN

Imagine that you are swimming in a lake, at a certain distance from shore. Somewhere, between a diving board and the mouth of a small river, you find a submerged rock on which you take some rest. I guess that you will remember this episode and the general spatial context in which it happened. But how much practice will you need to remember the rock's exact location, so that you can swim straight towards it from different positions in the lake? Besides, how often would you swim over its exact location after a dredger had removed the rock in your absence? These very severe criteria, however, are met by laboratory rats after a dozen training trials in the Morris place navigation task, which, represents the laboratory version of this holiday game.

This task, invented by Richard Morris (Morris, 1981), was a perfectly timed answer to the methodological needs generated by the publication of a theory (O'Keefe & Nadel, 1978) that, after Tolman (1948), claimed: (1) that rats were using spatial mental maps, and (2) that the hippocampus was the substrate of such cognitive maps. According to this theory, rats remember space as a set of ordered places, defined by their relations with major landmarks in an environment. Thus, the main property of this so-called locale memory system is that no individual cue is necessary to define a place, providing that other related cues are present. This implies that a place should be recognised even in the absence of any local cue that might have been associated with it in the past. Localising an invisible submerged platform in a perfectly circular pool of opaque water thus appeared to be an ideal task for studying place memory. It was also ideal for practical reasons: it required no pretraining of any sort, as most rats seemed to catch the aim of the task in two or three trials.

At that time, most of the research about place learning was conducted in complex mazes, or in T- or cross-mazes (see Olton, 1977). These apparatuses generally have one or many choice points, offering a minimum of two choices, one of which leads to a food reinforcement. Wrong choices are punished since the subjects enter dead ends, and find no food during an incorrect trial. In mazes, the food reinforcement is obtained when the rat follows a specific route from the last choice point. Theoretically, it is thus unnecessary for subjects to remember a place: they can memorise a direction relative to a major landmark, a strategy that some individuals seem to develop spontaneously (Poucet, 1985). Systematic egocentric responses, such as body turns, can also be remembered instead of places (see Barnes, Nadel, & Honig, 1980). In these cases, the mere frequency of correct choices does not indicate which strategy has been followed. In contrast, the path to the submerged platform is in no way constrained and different strategies are possible, each with a different escape delay, until subjects develop straight approach paths from any start position. An additional advantage of the Morris task is that a probe trial can be conducted following removal of the training platform, which provides a combined measure of accuracy and motivation in each subject. The Morris maze is thus not a maze, but an open field for place learning. Direct approaches are spontaneously developed by normal subjects, but they are not determined by the structure of the apparatus.

The "Morris maze" has been widely used since its first description. More than 350 references with this single key phrase can be found for the last 5 years! It appeared ideal for the study of spatial memory *per se*, or to evaluate cognitive abilities in a broader sense. It would be pointless to try to review all these experiments, as they are related to a great diversity of research fields. A purely methodological description may be found in Morris (1984), Sutherland and Dyck (1984), Stewart and Morris (1993) or Hodges (1996). The basic features and uses of the Morris task for the study of normal and brain-lesioned subjects have been integrated in an exhaustive review by Brandeis, Brandys, and Yehuda (1989). A review of the neuropharmacological and neurochemical basis of place learning using the Morris task is provided by McNamara and Skelton (1993). However, three related issues may be clarified by the results of experiments using this methodological approach: what are the properties of spatial memory in rodents, which brain structures are specifically involved in spatial memory, and what is the ecoethological relevance of the cognitive abilities involved in the acquisition of the Morris task?

Therefore, I will discuss three main aspects of the research using this task. First, I will review some of the experimental procedures aimed at analysing which mental mechanisms might account for the observed performance, and whether they provide evidence as to the nature of spatial representations in rats. Second, I will discuss spatial memory as it appears to be dissociated by some brain lesions, by age-related changes, and by reductions in the activity of the cholinergic system. Besides providing information as to the participation of brain structures in spatial abilities, these experiments revealed some elementary modules of spatial memory. These modules develop at different ages during ontogeny. They might also help to clarify the nature of spatial skills and representations. In the third part, I will briefly mention some ethologically relevant findings in experiments with various rodent species in the Morris navigation task. Studying how other species solve this task might help us to understand whether and how spatial abilities have evolved. Finally, I will discuss some hypotheses to account for the apparently contradictory results obtained from the performance of rats with hippocampal lesions in some experimental designs with the Morris task.

THE ANALYSIS OF SPATIAL REPRESENTATIONS

In the classic design, rats are repetitively released in a circular pool filled with opaque water, in which the only escape opportunity is provided by a small submerged platform. The platform is invisible and occupies a fixed position throughout training, hereafter referred to as the "training position". To escape, rats must learn the platform distance from the pool wall and/or remember its position relative to visible or otherwise perceptible landmarks in the distant environment. Not all these landmarks are simultaneously visible from all parts of the water surface due to the presence of a relatively high opaque wall (15–40cm) encouraging the test subjects to rely on a configuration of landmarks.

It is assumed that rats cannot perceive the platform, an assumption that is readily confirmed by direct observations. Naïve rats may swim along the platform, and even touch it with their flank, without obvious reaction. Moreover, if the escape position is randomly changed before each trial, the rats must touch the platform with their head or shoulder to detect its position.

On the very first trial, naïve rats spend most of their time swimming by the wall, but do occasionally swim across the pool. Most often, they are left on the platform for 10–30 seconds before being picked up by the experimenter, to be released after a delay (most often in the next minute) at another start position. During the first trials, various aspects of the task must be learned simultaneously: that there is no escape along the wall, that there is a platform somewhere in the central area of the pool, and that climbing on to this platform represents the best available escape in the circumstances. In one or two trials, normal rats learn to swim away from the wall and appear to search all over the surface of the water. The most striking feature of the normal performance is how rapidly the test subjects appear to learn about the spatial position of the invisible platform and insist on searching around the correct location if it has been displaced or removed from the pool.

Which measures must be recorded?

The training and test trials are usually video-recorded so that various behavioural parameters can be manually or automatically measured to evaluate how well the subjects remember the spatial position of the platform. Most often, the video image is analysed through a videotracking system, on line during acquisition, or off line. This provides a sequence of coordinates of the rat's locations in the pool, from which all secondary measures can be obtained.

Escape latencies show a clear decrease throughout training, indicating an adaptation to task requirements, whatever strategy develops. A decrease in escape latency is also observed when the platform is at an unpredictable position in each trial (see Morris, 1981). Rats must be severely impaired to show no learning curve. The slowest subjects are old rats that have remained untested during their whole life and persevere in swimming by the wall even though they have been given help in finding the platform. Normal young adults show the earliest decrease in latency and the highest frequency of very short escape trials, as illustrated in Fig. 5.1.

However, the absolute value of the latency does not indicate which learning strategy has been adopted. A more rapid decrease in escape latency is expected from the subjects' developing a place memory, as compared to the subjects who merely remember the distance between the platform and the wall and swim at a certain distance from the wall, usually in the same clock- or counter-clockwise direction—as illustrated in Fig. 5.3. Escape latency is also very sensitive to a change in the position of the escape platform, as normal rats spend some time searching in the previously reinforced position before visiting other parts of the pool. Even rats treated with a selective antagonist of NMDA (N-methyl-D-aspartate) receptors, AP5 (D(-)-2-amino-5-phosphonopentanvic acid) that appear to have little memory of the escape position during probe trials, show an increase in latency when the escape position is changed (see Morris, Anderson, Lynch, & Baudry, 1986a).

If they have a spatial representation of the training position, the subjects must be able to swim straight to the platform from any starting position. This implies that starting directions and swim distances must be computed. Swim distance can then be divided by the length of the straight escape

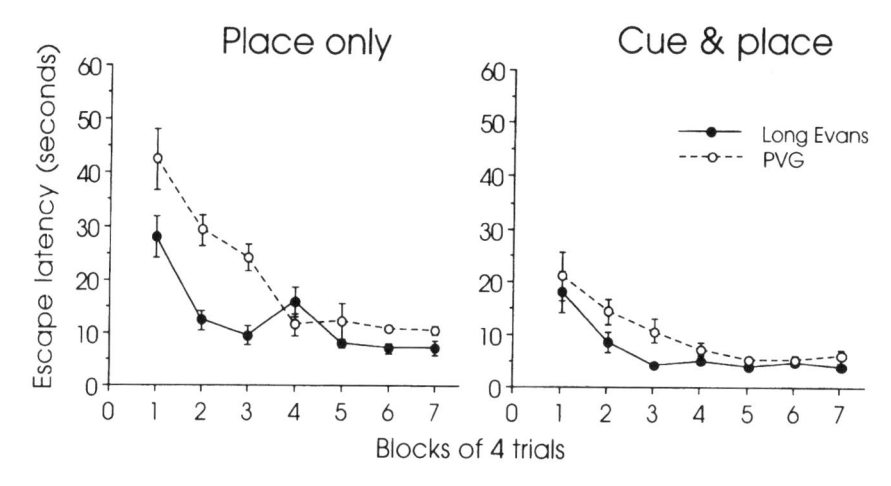

FIG. 5.1 Mean escape latency by groups of adult rats in two separate designs of the Morris task. The place only condition is the classical design with the invisible platform. In the cue-and-place condition, a black cylinder (diameter 4.5cm, length 10cm) was suspended 30cm above the platform for a cueing procedure. Rats of the Long Evans strain are more rapid than rats of the PVG (Madorin) strain in catching up with task requirements in both conditions. The differences in latencies correspond to significant differences in path length.

path. This ratio is close to 1 when rats follow nearly straight approaches. It might quantify the capacity to follow a short-cut from any starting position. The usual variability in normal adult subjects provides a mean escape distance that is about 130–150% of the straight approach path (see Fig. 5.2).

The most convincing measures of how much the rats remember about the training position is obviously provided by the behaviour expressed in probe trials, during which they are left in the pool in the absence of the escape platform. Normal rats make numerous attempts to find the platform, swimming repetitively over its usual position, sometimes diving underwater on this exact spot. This very convincing demonstration is most often quantified as the time spent in the training sector (quadrant of the pool in which the training platform is situated, or a circular area around it) as compared to three other equivalent untrained sectors. This measure is usually accompanied by an accuracy score represented by the frequency with which rats crossed the annulus indicating the virtual position of the absent platform. A high accuracy score might be due to swimming in circles at a certain distance of the wall. Thus, the accuracy score is most often compared to the crossing of three equivalent annuli in the centre of the other sectors. The animal's degree of preference for the quadrant in which the training platform was located (disproportionate amount of time in the training sector and high accuracy score) is referred to as "spatial bias", an index of spatial memory.

Other parameters can help to describe the subjects' behaviour during acquisition and during test trials as well. The proportion of time spent swimming by the side wall has been shown to reflect stress or helplessness (Stewart & Morris, 1993). Errors, counted as the proportion of time spent outside from an 18cm corridor from the start point to the platform, can also be recorded (Whishaw, 1985b). Other more sophisticated measures of the quality of the swim path have been developed in some groups. Gallagher, Burwell, and Burchinal (1993) proposed a platform proximity index based on the cumulative distance between the rat and the position of the platform, during the probe trials. Granon and Poucet (1995) recorded a cumulative search error and search error dispersion (from Benhamou, 1989) to quantify inaccuracy in searching for the platform during the training trials.

Comparison with a task on solid ground

We have developed another place learning task in order to reduce the possible stress of prolonged escape latencies in subjects with poor spatial abilities (Schenk, 1989). The task was inspired from Barnes (1979) with several modifications to make it a homing task and to reduce the possible reliance on visual guidances. Rats are trained to find a hole at a fixed position on a hexagonal or circular (Schenk, Contant, & Werffeli, 1990) table surrounded by a 40cm wall. This escape hole is connected to the test subject's home cage, which provides an ethologically relevant motivation and allows the testing of very immature rats. Additional food reinforcement can be provided in the cage, to maintain a high motivation in adult subjects. Non-connected holes are closed by a foam rubber plug, and all the holes are covered by a small disk of light plastic that is easily removed by trained rats. Rats must thus uncover a hole at the correct location to find the escape hole. Different release positions are used for successive trials. Olfactory cues on the table are made irrelevant by rotating the table and connecting another hole, at the fixed training location, before each trial. However, it is also possible to demonstrate whether, and under which conditions, rats rely on olfactory traces for spatial learning using this set-up (Lavenex & Schenk, 1997; Lavenex, 1995).

Escape trials and probe trials are quantified as in the water maze, with similar measures of latencies, distances, and time spent in various areas. Figure 5.2 shows that the development of accurate escape performance in juvenile PVG rats is very similar in both tasks. However, the homing task is more akin to a spatial discrimination task (choose the item at the correct location), as there are several unconnected holes and rats must merely

5. THE MORRIS WATER MAZE (IS NOT A MAZE)   151

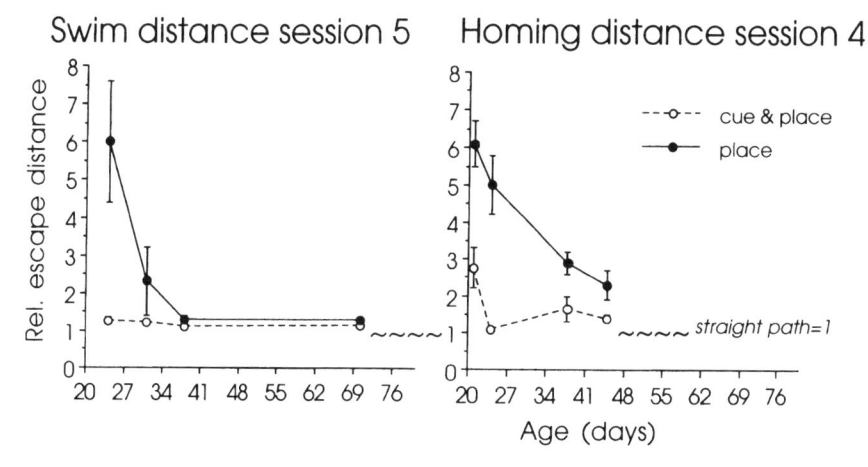

Swim distance session 5   **Homing distance session 4**

FIG. 5.2 Escape performance as measured by the relative distance (measured path length related to the distance of the straight path from start to goal) in different age groups of PVG rats. This performance was measured following 28 training trials over 5 days in the Morris task and following 20 trials over 4 days in the homing task.

discriminate which hole is in the training position. The accuracy requirement for the spatial definition of the target area is thus lower than when an invisible target must be reached along a swim trajectory in the conventional water task.

The homing task thus provides a reference experimental design to study place learning on a solid ground in conditions as similar as possible to those of the Morris maze. It might thus allow us to obtain results that complement the water maze, but avoiding difficulties due to swimming stress.

What do rats actually know: Place memory or alternative strategies?

The Morris navigation task was designed to test place learning with the expectation that the subjects had to rely exclusively on distant landmarks to localise the goal platform (Morris, 1981). The best learning curves and the shortest escape paths are shown by rats trained with a visible platform, or by those trained with an invisible platform at a fixed position in space. The memory of the training position is particularly evident in the probe trials without platform or with a displaced platform. Alternative strategies, such as following a stereotyped approach route, at a certain distance from the wall, are easily discriminated from the normal performance. In the experiment shown in Fig. 5.3, the lesioned rats were obviously capable of adjusting their swim trajectories relative to cues outside the pool since they

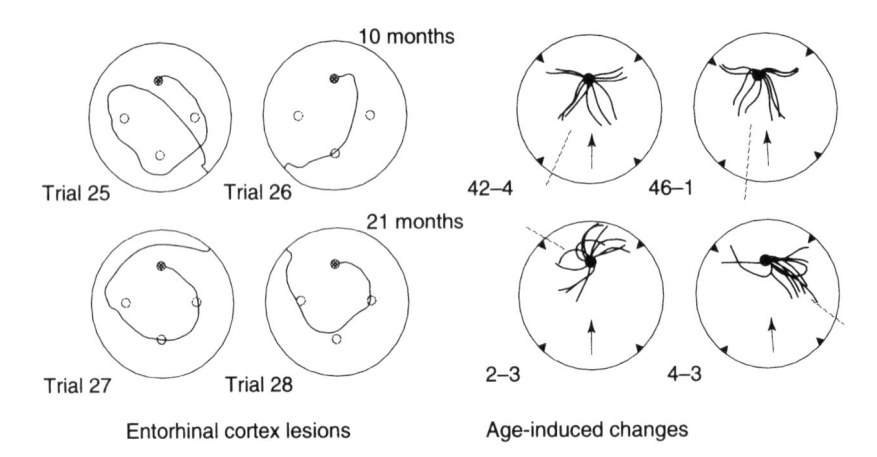

FIG. 5.3 Samples of training trials by rats with impaired spatial performance. The rat with a lesion of the entorhinal cortex develops a typical circular approach with occasional attempts to more direct approach. For the senescent rats, only the last segment of each of eight trials (session 4) is reported. The mean angle computed in each subject for the theoretical straight path is indicated by an arrow. The mean angle on the recorded paths is indicated by a dotted line. In the two younger rats (10 months), both vectors have the same orientation. In the two aged subjects (21 months), the landing orientations are clustered and deviate from the expected approach. (From Contant-Åström, 1994, with permission.)

had a higher annulus score in the training position during the probe trials (see Schenk & Morris, 1985). However, they had higher overall annulus scores than controls because they tended to swim at a fixed distance from the wall. Rats with fornix lesion were considerably impaired in reaching the training platform after a change in the location of some highly salient cues outside the pool (Eichenbaum, Stewart, & Morris, 1990). Their disturbance is an indication that they might have relied on these cues to adjust their swim path, whereas the undisturbed control subjects behaved as if their memory representation of the landscape could compensate for the absent cues. This suggests that rats with lesions might recover a good performance in learning accurate sensorimotor adjustments relative to the main landmarks around the pool.

Obviously, normal subjects are capable of discriminating among places in the pool, and they do not need to follow constrained approach routes, while such a strategy is sometimes developed by subjects with impaired spatial memory. However, normal rats do also develop the strategy that appears to be the most appropriate in specific training conditions (Whishaw & Mittleman, 1986). They remember all the different starting positions and, in the absence of the training platform, they develop search patterns that associate these starting positions and the training position. Moreover, the same authors have shown that if trained with several possible positions of the platform, rats do systematically visit all these positions during a probe trial.

Critical steps in task acquisition: Platform pre-exposure and instantaneous transfer tests

It was commonly assumed that rats used distant room cues to locate the invisible platform. However, it was not clear at which time they recorded this information. In the classic water maze procedure, rats are allowed a full view of the environment at all times throughout a trial: before they are released (in their home cage or in a waiting cage, in the hand of the experimenter), during their approach to the target platform, or when standing on it. It was soon clear that rats given experience of the escape location before training showed a more rapid acquisition than rats that had been placed on the same platform, but at another place in the pool (Sutherland & Linggard, 1982). This indicated that a part of the representation of the spatial relationships between landmarks is independent of the behaviour by which this knowledge is expressed. Later on, Keith and McVety (1988; see also Keith, 1989) showed that rats acquired knowledge about the platform's spatial position when exposed on it for 120 seconds, but only if they had already been trained to locate a platform in another room. During this pre-exposure to the platform, they also seemed to learn more general information, such as the availability of a platform in the new environment, since rats that were pre-exposed to a platform in the quadrant 180 degrees diagonally opposite from the future training position also showed better performance than nonpre-exposed rats. Thus, to be relevant for future training, the information provided by "being on the platform" had to be associated with an already familiar motor behaviour, but actual escape behaviour and information about the escape position had not to be simultaneous.

It is, however, not surprising that the placement-induced improvement is less important than the improvement that results from a single swim trial (Whishaw, 1991). Accordingly, Sutherland, Chew, Baker, and Linggard (1987) showed that illumination of the room during the middle segment of the swim plays a critical role in reducing escape latency and in protecting against the disruptive effect of a change in the position of auditory beacons. Devan, Blank and Petri (1992) confirmed that rats acquired distal cue information during approach trajectories and were not impaired if they had no visual access to the environment around the escape platform.

This suggests that most learning occurs during escape at a given position. It might be that staying on the platform provides accurate information about its position, while reaching the platform and climbing on to it reinforces the motivation to approach this location. This might be particularly evident in water tasks. Swimming seems to encourage a win-stay strategy, whereas a win-shift strategy is more likely to appear in an equivalent food reinforced task on a solid ground (Goodlett, Nonneman, Valentino, & West, 1988; Means, 1988). The reinforcement value of the approach strategy might thus develop from the opportunity to escape.

If rats rely on a memory representation of the different places in the pool, then experienced subjects must head toward the escape platform from any starting position, even one that has never been used before. In his original paper, Morris (1981) demonstrated instantaneous transfer, a further indication that the memory of the training position was not linked with a specific starting position. Indeed, rats trained with a unique starting position were easily capable of reaching the platform from a new release point and showed accurate searching behaviour if the platform had been removed. In order to control for possible access to the transfer starting position during the training trials, Sutherland et al. (1987) restricted the rats' access to half of the pool, using transparent or opaque partitions, finding that rats were severely impaired in reaching the platform when released from this new half following removal of the partition. For the authors, this indicated that rats are not capable of instantaneous transfer, even after having been allowed to swim—but not to find escape—in the second area. Indeed, the fact that the swim trajectories were then very inaccurate, even in the familiar half of the pool near the platform, suggests that this situation was highly disturbing for all the animals. What seems to have been new, and therefore disturbing to the animals was the novelty of having a complete view of the pool environment along a given swim trajectory. This suggests that being able to experience the general visual panorama is of a high importance during swim paths. Thus, whatever the form of spatial representation that guides approach to the escape platform, it must be reupdated when swim trajectories providing new visual input are made possible. This appears to be a typical reaction of rats to a change in movement opportunity in a familiar environment (see Thinus-Blanc, Save, and Poucet, Chapter 3, this volume).

A role for vestibular and other idiothetic cues

A fundamental question for spatial orientation is the degree to which kinaesthetic cues are involved. The mechanism of path integration or dead reckoning (Gallistel, 1990) by which route-based information, including vestibular signals (Etienne, Maurer, & Saucy, 1988), are processed to provide positional information, plays a critical role in the acquisition of a spatial representation, or in its use in restricted cue conditions. During an exploratory phase, rats learn the spatial relationships between visible and other perceptible landmarks in the environment, and the mechanism of path integration allows an efficient encoding of these relationships. During patrolling in a familiar environment, the spatial representation might enable the subject to anticipate landmark configurations that could be seen from a future position. This capacity has an important survival value. It allows partial independence relative to environmental landmarks, and thus a higher degree of tolerance to environmental changes, especially when illumination level is reduced, or the direction of light is changed (e.g. morning vs. evening, light and shadows). If the number of available landmarks is reduced, or if they are invisible in darkness, the mechanism of path integration might again play a critical role for correct orientation in a familiar area (Alyan & Jander, 1994). It is assumed that subjects need just to identify their position once, and can thereafter rely on the integration of the directional and distance information collected during their movements, in order to know their position relative to the start of their journey, or to reupdate a hypothetical representation of the environment (McNaughton, Chen, & Markus, 1991). The resetting of the path integration vector in the presence of a familiar cue might correspond to a readjustment of the preferred firing direction of units in the subiculum or in the anterior thalamus (Goodridge & Taube, 1995), as if the familiar visual cues had gained control of the spatial reference framework during spatial learning. The capacity of visual cues to control or reupdate head direction or place units must be acquired during the first exploratory phases in which subjects learn to associate the position of the cue with a reference direction. This coupling is not possible, or less consistent, if the rat is systematically disorientated before being introduced into the field (Knierim, Kudrimoti, & McNaughton, 1995). In this case, visual cues appear to have only a weak control over the place fields and head direction cells.

The specific effect of vestibular stimulation on place navigation using the Morris task has been studied by Semenov and Bures (1989) showing that "agreement between vestibular and visual signals is a prerequisite of efficient navigation". Pre-trial vestibular stimulation in overtrained rats affects mainly swimming efficiency, since the after effects of the imposed rotation induce compensatory body movements which result in swimming in circles. However, when the rats recover normal swim capacities, they show accurate location of the platform. In a working memory design, the same rats had to learn a new location in each daily pair of acquisition and retrieval trials with a five-minute inter-trial interval in a waiting cage, before or after which they were rotated for one minute. The performance in the retrieval trial remained unaffected by the rotation, which indicates that it did not interfere with the persistence of spatial memory. During pre-exposure learning tests, in which rats were only informed about the platform position while being deposited on it for 30 seconds, vestibular stimulation for 1 minute interfered with normal acquisition only if the after effect of the rotation (post-rotatory state) coincided with the stay on the platform. Moreover, being exposed to a rotatory platform interfered with latent learning even if the rotation was imposed after the rats had been allowed a period of exposure to a stable platform, or had stayed in the waiting cage. In this case, a second stay on the now rotatory platform had a disrupting effect. This might be due to the fact that the remote cues on which spatial memory was based lost their stability. Indeed, a period of vestibular stimulation which did not confuse the critical visual cues (rotation on the platform with no access to visual cues) had no effect.

One can thus assume that a mechanism based on path integration, and thus one that might strongly depend on the integrity of vestibular information, is highly crucial for the acquisition of new spatial information, because it leads to the association of a reference vector with the visual landmarks, whereas a disturbance of the vestibular input has only transient effects on orientation in an environment with familiar landmarks.

More generally, these results raise the question about the effects of selective motor deficits on spatial memory. Mutant mice with cerebellar degeneration have been trained in a Morris maze (actually, a small rectangular tank), a task in which they were considerably impaired relative to normal littermate controls (Lalonde & Boetz, 1986). The extent to which this impairment is the consequence of poor motor coordination is not clear, since the performance in the spatial task is correlated with motor coordination, both in normal and in mutant mice (Lalonde & Thifault, 1994). At this stage, only complex task designs, that should compensate for the motor deficit, might give an indication as to whether there is a specific involvement of the cerebellum in the cognitive spatial abilities of mice. From a theoretical point of view, this is an important question, because it might shed some light on to the role of active motor strategies in the acquisition of spatial representations.

What is the minimum visual information required for place memory?

Vision plays a critical role in the rapid acquisition of efficient escape by normal rats (e.g., see Sutherland et al., 1987). This does not mean that rats show no learning when tested in darkness (Sutherland & Dyck, 1984) although place discrimination is very poor in this condition (see Fenton, Arolfo, Nerad, & Bures, 1994; Arolfo, Nerad, Schenk, & Bures, 1994) and vision appears necessary for normal accuracy in finding and remembering the platform's position.

The minimal visual cue requirement has been studied by Fenton et al. (1994) with a limited number of controlled remote visual landmarks. This study was conducted with rats pretrained in a normal illumination condition, which precludes any conclusion as to the value of these sets of cues in the acquisition of the task by naïve rats. In the test condition, the acquisition of a new training location was similar, whether four visible cues, or only two cues were available. Indeed, the latency was higher in the first training trials when four cues were available, than it was with only two cues. This suggests that learning an escape position with a set of four cues requires more behavioural adjustments than when only two cues are present. Correspondingly, adding two new cues to a set of two familiar cues delayed escape in a significant manner, whereas removing two cues from a set of four cues had no detectable effect. Thus, new cues seem to induce either a phase of re-exploration of the pool, or the learning of the relations between platform location and the location of each new cue. Finally, the optimal performance reached during training with two cues was abolished when one of these cues was replaced by a new one, suggesting that both cues contributed to the spatial adjustment.

This set of experiments demonstrates in a convincing manner that escape performance is related to the position of the controlled cues. But it also indicates that escape performance is systematically disturbed when rats have to process new visual cues, which suggests that they need first to discriminate among the cues, and/or learn the relation between the training location and the position of each cue. This might explain why the removal of any one cue does not affect the optimal performance reached with a set of four redundant cues. Moreover, as admitted by the authors, it is not yet clear whether and how rats learn about the relationship between an extramaze position (where they wait between trials) and the positions of the cues and platform. This necessity, to anchor the learned set of landmarks to a general spatial reference, seems to characterise spatial behaviour in various spatial tasks (see discussion in Lavenex & Schenk, 1996) and might represent an important mechanism in learning about the relative position of different places (McNaughton, Knierim, & Wilson, 1995).

The importance of good vision has often been neglected in studies with senescent rats. Recently, it has been shown that the behavioural impairment of aged Sprague Dawley rats was related to the amount of retinal degeneration (O'Steen, Spencer, Bare, & McEwen, 1995; Spencer, O'Steen, & McEwen, 1995). This does not demonstrate a causal link between retinal changes and cognitive decrease, as both effects could be due to a common cause. Nevertheless, it emphasises the need to control for the visual capacities of old subjects.

Finally, the importance of visual cues from the environment suggests that auditory cues play a less important role in spatial orientation by rats. Nevertheless, there are indications that rats can rely on some auditory beacons to compensate for the restriction of visual input (Sutherland & Dyck, 1984; Sutherland et al., 1987). Moreover, rats can use nondirectional auditory signals to approach the platform (Buresova, Homuta, Krekule, & Bures, 1988).

How critical is the memory of the panorama from the training sector?

The above experiments suggest that rats rely strongly on the memory of visual images of the landmarks around the pool and reach an optimal performance in visually adjusting accurate approach trajectories with memory images. This type of associative learning ought to be relatively simple, as demonstrated by Wilkie and Palfrey (1987) with a model that adjusts the current position relative to two major cues in the environment. It thus seems critical to develop tasks which cannot be solved by direct sensory adjustment, since, according to the cognitive mapping theory, animals encode the spatial relationships between landmarks perceived at different times during an exploratory bout and can thereafter anticipate the landmark configuration that will be perceived from a future position.

If visual input is critical for task acquisition, it might be that rats rely on a snapshot memory (Cartwright & Collett, 1983; Collett, 1987) to recognise the training position. In this case, they should be unable to learn the platform location if the light is switched off before they reach the training position. The condition in which illumination was contingent upon the position of the animal in the apparatus (hereafter termed "conditional illumination") implies that rats discriminate the training position without direct visual feedback from the landmarks around it. This was tested in an experiment in which rats were trained in the Morris task in which the light was switched off each time they entered the inner circular area of the pool where the training platform was situated (Arolfo et al., 1994). Rats were thus allowed access to the visual environment when in the periphery of the pool, but had to find the escape platform in complete darkness. This was a difficult training condition. Escape performance was impaired and the spatial bias measured during probe trials was reduced. Pretraining with full access to visual room cues followed by prolonged training in conditional illumination led to an improved, although still impaired, escape performance. Nevertheless, these rats seemed to rely on the visual information obtained from the pool periphery, since their impairment was enhanced when they were tested in darkness.

In the conditional illumination condition, rats must rely on the mechanism of path integration in order to find the platform while the light is switched off. However, it is not known how accurate this mechanism might be during swimming in the absence of vision. It might be that kinaesthetic information in the absence of visual flux is more accurate when walking on solid ground than when swimming. The same type of experimental design has thus been used with a homing board task such as that described above (Schenk, Grobéty, Lavenex, & Lipp, 1995). In this experiment, rats were trained to find one escape hole out of four at a fixed position in space. The room lights were switched off each time the rat entered the central area of the table, and were switched on when it returned to the periphery. The absence of the table wall provided vision of more numerous room cues from the periphery. Rats trained under conditional illumination showed a bias comparable to that observed in the usual situation, when the room lights remained on. In contrast, rats trained in permanent darkness showed no bias toward the training sector. The discrepancy between these results and those obtained by Arolfo et al. (1994) in the Morris task might be due to the fact that path integration mechanisms are more accurate when subjects are walking on solid ground, instead of swimming. But it might also be due to the fact that, in the homing task, rats were not prevented from returning from the connected hole to explore the illuminated periphery, which they did frequently during the first learning trials. This allows for a dual action of the path integration mechanism, first when rats enter the dark zone and second when they leave it after having found the escape hole. The second vector generated by the path integration mechanism is related to the escape position, and should provide complementary opportunities to relate this location to the environmental room cues seen in the periphery.

Another reason why vision might play a more important role in the Morris task than it does in place tasks on solid ground comes from the different effects of a perturbation of the path integration mechanisms in these two conditions (see, for example, Dudchenko, Goodridge, Seiterle, & Taube, 1997; Martin, Harley, Smith, Hoyles, & Hynes, 1997). From these studies, it appears that place learning is less affected by prior disorientation in the Morris task than in tasks on 8-arm or 4-arm mazes. This might be due to the strong aversive motivation in the swimming task, combined with a higher dependence on a visual snapshot memory of the escape position. This is confirmed by a more disturbing effect of cue removal in the Morris maze after pre-test disorientation (Martin et al., 1997).

To study whether and how the relations between different visual cues are used for orientation on the homing table, it is necessary to work with controlled cues. Three identical light-emitting diodes placed in three different positions, at the same distance and height relative to the centre of the homing table, were used as controlled visual cues. The brightness of the diodes was too low to illuminate the table. As expected, discrimination of the training sector in total darkness remained accurate even when only the three diodes were on. When the position of the three diodes was changed by 90 degrees, the rats spent most of their time at the new spatial position, as defined by the set of diodes. For a conditional illumination design, the cues were switched on or off, depending on the rat's position on the table, using a videotracker coupled with a computer (Rossier, Grobéty, & Schenk, 1996). A single cue was on when the rat was in the central part of the table where they could escape. The two other cues were on when the rat was in the periphery. Accurate discrimination of the training position was observed in this condition, but not when only one or two cues were visible. This confirmed that rats were capable to rely on a representation of the relations between visual cues that were not perceived simultaneously.

Assessing the importance of local cues in place-learning designs

The resistance of place memory to the removal of a local cue is a major issue in cognitive mapping theory. It is generally observed in our homing task, where rats persist in attempting to enter a hole at a trained location, despite the absence of a very salient object or olfactory trace that was signalling the escape hole during training (see Schenk et al., 1995). In the experimental designs in which memory for object location is assessed (see Chapter 3, this volume), the focused searching in the place where a salient object was previously located is confirmation that object memory is related to a spatial reference and that rats expect specific objects in the places where they had been previously.

In the classical Morris task, the training position can only be discriminated on the basis of its relations with distant room cues. Escape on to a visible platform thus represents an easy task. It is often used to demonstrate that rats with supposed specific impairment of spatial memory are otherwise capable of complying with all the nonspatial requirement of this task (i.e., swimming ability, motivation, climbing on the platform; e.g., see Morris, Garrud, Rawlins, & O'Keefe, 1982). In some cases, cued tasks have been used to assess spatial memory *per se*, or to reveal attentional deficits, indicating that these two capacities are tightly coupled. Three types of designs can be considered: a salient cue placed on the escape platform, a distractor platform distant from the escape platform, a salient cue out of reach.

In the simplest condition, rats are trained to escape on to a visible platform at a fixed position (cued place task). Experimental designs in which the position of the invisible platform is indicated by a salient cue suspended above it can also be regarded as cued place tasks. The question is then to what extent the subjects rely on the distant or on the local cue to solve the task. This is classically assessed during probe trials with no local cue, nor escape platform. A weaker spatial bias following cued than following noncued training, might be due to the fact that the presence of the salient local cue impairs subjects' attention toward the more distant contextual cues. In classic conditioning experiments, this effect depends on the salience of the overshadowing stimulus (Dickinson, 1980; Mackintosh, 1985).

In the second type of experimental design, a salient cue is placed in the water at a certain distance from the escape platform and plays the role of a distractor during training. This may reduce escape efficiency and might also result in a poorer discrimination of the training position. The classic design consists of a simultaneous spatial and discrimination task (see Schenk & Morris, 1985; Morris, Hagan, & Rawlins, 1986b). Two platforms are placed in the pool, but only the one at the training location provides escape, while the second one is a floating platform that sinks when the rats try to climb. Since platforms with different visual appearances were interchangeably used as the escape platform, the rats could not perform the task as a simple visual discrimination, they had to rely on a spatial memory of the escape location. For a visual discrimination, the floating cue and the cue fixed on the platform can be discriminated on the basis of their visual appearances. In this case, rats cannot rely on a spatial memory, since the escape and distractor platforms are at unpredictable positions in the pool. It is thus possible to compare the performances of groups of rats trained in a spatial or in a visual discrimination task in the same apparatus. This provides an experimental design to demonstrate double dissociations induced by different brain lesions (see Packard & McGaugh, 1992).

In a third category of experimental design, a salient cue is present in the pool, but cannot be reached. It might thus help the organisation of efficient escape routes. Among these designs is one in which a black rectangle is placed on the pool wall (Pelleymounter, Smith, & Gallagher, 1987), or a hanging cue can be placed at some distance from the platform (Rudy, Stadler-Morris, & Albert, 1987; Chevalley & Schenk, 1987, 1991; Schenk & Brandner, 1995).

Typical results obtained with each category of design will be briefly reviewed. Training with a visible platform was described in the original reference paper of Morris (1981). The group trained with a visible platform showed a less pronounced spatial bias toward the training sector than did rats trained with an invisible platform. This might be due to an overshadowing effect of the highly salient visible platform. Another evaluation of the relative importance of local and environmental cues can be based on a conflict experiment in which these cues are dissociated (McDonald & White, 1994). Rats trained to reach a visible platform at a fixed position were given a transfer test in which the visible platform was at a new position in the pool. In this test, 50% of the control rats swam to the cued platform first (cue response) while the others swam toward the familiar training location (place response). This suggests that in normal rats, both the local cue and the environmental landmarks have control over performance. In this experiment, the rats showing a place response in the conflict trial had also developed an accurate place memory at an earlier phase, during acquisition.

Similar results were found when the escape platform was signalled by a cylindrical cue hanging 40cm above it (see a description in Schenk & Brandner, 1995). As expected, this cue had a significant overshadowing effect on the spatial memory of juvenile rats as assessed in the probe trial, but not on that of the adult rats. Approach to the platform was facilitated by the cue, but final adjustments to actually land on the platform had to be related to distant cues, since the cylindrical hanging cue apparently became less salient when rats were near the platform. The optimal strategy was thus to combine a guidance, for rapid approach to the target area, with an increased attention to distant room cues when in the vicinity of the training platform. Some groups of subjects such as immature rats or rats with lesions of the medial septum (Brandner & Schenk, in press) showed an accurate heading toward the cued platform when leaving the start area, but they seemed to be less attentive to the distant cues and were impaired both in escape performance (they often missed the platform and made a long detour before heading again toward the cued area) and in searching when the platform and hanging cue were absent. Chronic perinatal supplementation with choline has been observed to improve this type of combined cue-and-place learning (Schenk & Brandner, 1995).

The simultaneous spatial discrimination tasks with a lure platform, representative of the second type of cue design, have been observed to aggravate spatial deficits (see Schenk & Morris, 1985). Lesioned rats, although they were showing efficient escape and discrimination of the training position in probe trials in the place condition, performed at chance when required to discriminate between two platforms. Morris et al., (1986b) trained rats with aspiration of the hippocampus over a long period in a simultaneous discrimination design. They reported that some of the lesioned rats achieved a criterion of 85% correct choices, which suggests that the difficulty may be overcome by prolonged training. Interestingly, the presence somewhere in the pool of a floating platform during classic place training with the invisible platform interfered with escape efficacy in immature mice, but did not affect place discrimination during the probe trial with no intra-pool cue (Chapillon & Roullet, 1996).

The visual discrimination in which the escape platform can be discriminated from the floating one on the basis of its visual properties (e.g., black, grey or striped) is not an easy task. In some cases, the discrimination shown by control rats remains very poor (Selden, Cole, Everitt, & Robbins, 1990). Moreover, when rats have been pretrained to reach a randomly positioned platform signalled by a salient cue, they appear very disturbed by the addition of a second distractor cue, even one very different from the training cue (Schenk & Brandner, 1995). Environmental cues must be masked and the discriminanda chosen very carefully to be equally attractive (see Whishaw & Petrie, 1988; Paylor & Rudy, 1990). Otherwise, normal rats can develop a persistent egocentric bias, most often heading for the visible cue to their right or to their left. The best evaluation of cue attractiveness is to conduct probe trials with two floating cues.

The third category of cued task has often been used to improve the performance of impaired rats (e.g., immature or aged rats). Rudy et al. (1987) suspended a cup in the centre of the quadrant adjacent to the training sector and observed good performance in immature rats. The same condition has also been found to facilitate the performance of juvenile rats (Chevalley & Schenk, 1987, 1988) and to increase the frequency of direct escape paths. Moreover, a greater bias toward the training sector was found following removal of the cue and platform, as if the hanging cue had facilitated spatial learning relative to more distant room cues.

In summary, the risk of overshadowing is thus inversely proportional to the distance between the visual cue and the target platform. A salient cue distant from the target platform appears to improve spatial learning because it promotes fixed approach routes which appear to facilitate learning of the platform position relative to the distant room cues. A salient cue confounded with the target is likely to overshadow the distant cues.

Reversal learning and learning sets

As discussed above, rats do spontaneously develop a win-stay strategy in water tasks (Means, 1988; Goodlett et al., 1988), suggesting that they remember the last place where they have found escape, and tend to return to this spot. This explains why the classic task is mastered so rapidly, and why rats learn a new position in few trials. This does also suggest that the water maze might be ideal for delayed matching tasks in which rats would be trained to learn a new position in each trial block.

Whishaw (1985a,b) used a task design in which rats were trained to reach a new position each day, in sessions of 16 trials. Each of the four possible starts was used for two pairs of massed trials. The difference in latency between the first and the second trial of each pair was taken as an indication that rats remembered the relationship between the start position and the escape position in the first trial. In addition, the latencies in both the first and second paired trials showed a steady decrease over days, as the rats remembered the possible escape positions. Indeed, during a probe trial without a platform, the swim paths had a very typical shape as if the rats were sequentially swimming in each possible location (see also Whishaw & Mittleman, 1986). Two aspects of the learning set task are learned throughout training: rats memorise the possible escape locations in the same time as the win-stay strategy.

This experimental design can be used to reveal impaired spatial abilities in rats which have been allowed to overcome an acquisition impairment with an overtraining procedure. It can help in further dissociating short- and long-term memory impairments (see Morris, Schenk, Tweedie, & Jarrard, 1990). Rats with long-term memory impairments might be less efficient when tested for retention over days, but not in a matching task with short intertrial intervals.

Long-term retention: Memory consolidation or primacy effect?

Long-term retention following the acquisition of the water maze task is usually tested in probe trials. Indeed, it does not seem convenient to test long-term retention of the task using a retraining design, because all the procedural aspects of the task (existence of the platform, distance from the wall, etc.) might be confounded with a more specific spatial memory of the environment around the pool, and of the platform position itself. Many retraining conditions should be compared, in the same and in a different pool, using same or different escape positions. The spatial bias recorded during a probe trial thus seems to be the best available indication of long-term retention. However, this bias is sometimes very weak, after only a few days interval (see the control in the experiment by Brioni, Arolfo, Jerusalinsky, Medina, & Izquierdo, 1991, or in the experiment by Warren, Castro, Rudy, & Maier, 1991), and even in overtrained rats (from Morris et al., 1990). It might thus be suitable to use a particular training procedure during original acquisition in order to enhance the spatial bias during probe trials. This can be done by using an on-demand platform design (Buresova, Krekule, Zahalka, & Bures, 1985) in which the rat is progressively trained to swim on the correct spot before the rest platform is raised near the surface. A similar result can be obtained by conducting variable duration probe trials (when the escape platform is made available after a variable delay) in each block of training trials (Markowska, Long, Johnson, & Olton, 1993).

Morris and Doyle (1985) analysed various aspects of forgetting and retrieval following place training across 30 trials. In this case, retention was measured as the difference between the time spent in the training and in the opposite quadrants. This ratio decreased in each of four successive probe trials at 12-day intervals, but was still greater than 1 up to 36 days following training. Other groups of rats were given 8 trials with a new escape position in the diagonal quadrant before the series of probe trials started. This reversal training was either massed in 1 day, or dispersed over 8 days. The difference between the time in the quadrant of the first trained position and the opposite quadrant (new position) provided information about the balance between the biases toward the first and the most recently learned positions. The rats given dispersed training on the second position showed a clear-cut bias toward this latter quadrant, that decreased very much like the bias shown by rats trained on a single position. They showed no bias toward the original training position at any retention interval. The group having received massed training showed a more complex feature. In the first probe trial, they had a clear bias toward the most recently learned position. This bias vanished progressively through successive probe trials, while the bias for the originally learned position regained significance to reach the same level as that of the rats trained on a single position.

This short overview suggests that the memory trace of past training in a specific environment undergoes a complex long-term maturation. This might be studied in greater detail in order to clarify some memory consolidation processes involved. Indeed, there is evidence that this long-term memory can be enhanced, either by a particular training procedure, or by a pharmacological treatment, such as flumazenil, an antagonist of benzodiazepines (Brioni et al., 1991). Prior cueing with the presentation of some aspects of the learning context results in a better expression of learning. For an avoidance brightness discrimination task, different cues are active at different retention intervals (Gisquet-Verrier, Dekeyne, & Alexinsky, 1989). It might thus be profitable to find out which type of cue can induce an improvement in the retention of place learning, or, more precisely, in the behavioural expression of this memory.

The ambiguity of stress effects

It is common to assume that a high level of stress reduces memory. In the Morris task, in particular, stress might play a complex role, more so than in other spatial tasks. A stress response is certainly triggered by the first introduction in the pool, when rats rely on the inappropriate strategy of swimming by the wall, but it is likely to be rapidly reduced as rats learn to find the escape platform. Consequently, any cognitive impairment is likely to amplify the stress effects as escape is delayed. There is thus a reciprocal and somewhat circular link between stress and spatial memory.

Exposure to inescapable shocks is known to induce deficits in the ability to learn a new escape response (see Maier, Albin, & Testa, 1973). In addition, the treated rats appear to develop different strategies from rats given no shocks (Lee & Maier, 1988). A perseverative response style is also induced by exposure to uncontrollable foot shocks in some strain of mice (Francis, Zaharia, Shanks, & Anisman, 1995). However, this treatment had no effect on the acquisition or the retention of the classical place learning task by rats (Warren et al., 1991). It is true that the Holtzman rats tested in this experiment are not frequently used in the Morris task and showed an unusual combination of good asymptotic performance in escape (latencies below 10 seconds in the last blocks of training) with a rather weak spatial bias during the probe trial. However, the lack of treatment effect is not likely to be attributable to a ceiling or floor effect. In comparison, changes in the activity of the noradrenergic system, as induced by lesions of the dorsal noradrenergic bundle, appear to improve escape performance when the water is cold (Selden et al., 1990; see also Szuran, Zimmermann, & Weltzl, 1994), suggesting that the effect of stress might not appear in the usual training conditions in the Morris task, which uses water temperatures between 23 and 26 degrees Celsius. It is also regrettable that Warren et al. (1991) did not study whether the shocked rats were equally able to learn about another position in the pool. They might also have tried a design likely to induce overshadowing.

In contrast with the above results, adult rats subjected to chronic restraint stress showed significant although reversible impairment in a radial maze task (Luine, Villegas, Martinez, & McEwen, 1994). The absence of adrenal hormones might also have a direct effect on spatial performance in the radial maze (Vaher, Luine, Gould, & McEwen, 1994) and in the Morris task (Oitzl & de Kloet, 1992). In the latter study, antagonists of the glucocorticoid receptors or of the mineralocorticoids seemed to have different effects, depending on the time of injection during task acquisition. In fact, a prolonged absence of glucocorticoid caused cell loss in the hippocampal dentate gyrus and impaired the acquisition of navigation tasks (Conrad & Roy, 1995). In this case, the hormonal effects on task performance appeared dual: an indirect effect via cell destruction in the dentate gyrus, plus a possibly more direct effect on performance as a result of a short-term influence of glucocorticoids on hippocampal plasticity (Diamond, Bennett, Fleshner, & Rose, 1992). However, prolonged exposure to a social stress, or to corticosterone treatment (over 3 months) produced an impairment in mid-aged, but not in young rats (Bodnoff, Humphrey, Lehman, Diamond, Rose, & Meaney, 1995).

These data suggest that corticosteroid level, before or during the conduction of an experiment in the Morris maze (Conrad & Roy, 1995), might interfere with task performance. It is, however, difficult to predict exactly in which conditions and to what extent the performance should be reduced. These effects seem to follow a U-shaped relationship with corticosterone level, a low level affecting hippocampal plasticity via mineralocorticoid receptors, and a high level via glucocorticoid receptors (Diamond et al., 1992). However, even though it is tempting to associate the age-linked deficits in spatial memory with alteration in hippocampal corticosteroid receptor gene expression, behavioural impairments can be found in the absence of changes in the expression of these genes (Yau, Morris, & Seckl, 1994).

In addition to corticosteroids, opiates are also known to modulate memory in a general manner, and might be involved in the deficits associated with age, as the highest amount of dynorphin A(1-8)-like immunoreactivity in CA3 pyramidal cells of the hippocampus is found in impaired aged rats (Gallagher & Nicolle, 1993). Moreover, stressful procedures are known to induce the release of endogenous peptides (Barta & Yashpal, 1981). The Morris task may be considered as stressful for subjects with impaired orientation capacities. Old subjects have exaggerated sympathetic-adrenal medullary responses to acute swim stress at 20 and 25 degrees Celsius (Mabry, Gold, & McCarty, 1995) and prior habituation to water immersion reduces some of their deficit in the Morris task (Mabry, McCarty, Gold, & Foster, 1996). An indication that the activity of opiates may reduce normal performance in the Morris task is provided by the result that the opiate antagonist, naloxone, improves task acquisition and retention (Decker, Introini-Collison, & McGaugh, 1989; Galea, Saksida, Kavaliers, & Ossenkopp, 1994). However, contrary results have been found (for a review, see McNamara & Skelton, 1993).

Taken together, these results provide a somewhat complex picture as to how the hormonal changes induced by stress might affect spatial abilities. On one hand, subjecting rats to inescapable shocks seems not to affect task acquisition (Warren et al., 1991), whereas on the other, changes in the blood level of hormones known to be affected by stress appear to have an effect on spatial abilities. Most likely, this is due to the diversity of central effects of stress. Nevertheless, it calls for a better understanding of which aspects of the Morris task are most sensitive to stress. This should reveal some of the rules (if any) following which emotions and stress affect cognitive capacities in a more general manner.

DISSOCIATIONS BETWEEN MEMORY COMPONENTS FOR PLACE LEARNING

The hypothesis that the hippocampus is involved in the elaboration of spatial maps implied that rats with hippocampal dysfunctions should be severely impaired in the acquisition of the Morris navigation task. This prediction was frequently verified, although considerable residual capacities were observed on many occasions. Residual capacities were also shown by immature rats and by aged rats, which might help in dissecting out subcomponents of spatial memory. It is not the purpose of this section to discuss whether and how the hippocampus and other brain structures are involved in spatial orientation (see Hodges, Chapter 6, this volume). Instead, I will concentrate on changes in spatial skills and on the analysis of residual capacities, in order to identify some of the elementary mechanisms involved in spatial memory.

Maturational steps

The late maturation of the hippocampus and of the neocortex in most rodents has led to the commonly accepted view that spatial abilities might be the last to develop. As it is likely that juvenile rats leaving the nest area must rely on their spatial memory to survive, these capacities were expected to reach a certain efficiency shortly after weaning. Thus, besides indicating when brain maturation was sufficient to sustain normal spatial capacities, studies of the ontogeny of spatial memory in the Morris task could provide an interesting model as to how these capacities develop.

We have conducted developmental studies with hooded rats of the PVG strain. As is clear from Fig. 5.2, there is a dramatic improvement in place learning in both the Morris and the homing tasks in the beginning of the fourth postnatal week. In this strain, a stable adult-like performance based on straight escape paths is not reached until the sixth week (Schenk, 1985, 1987a). This slow development of the behavioural performance revealed several functional steps. Similar changes were found in the water and in the dry land escape tasks, suggesting general properties of the orientation mechanisms in developing animals that are not specific to swimming tasks.

At 4 weeks old, (PND 23–26) juvenile rats are particularly responsive to the presence of local cues. If the cue is placed on the target spot, the young rats develop strong guidance. In the homing board task, in which they have time to orient toward the cue before starting to walk, their starting directions were more precisely and systematically oriented toward the cue than those of more mature subjects (Chevalley & Schenk, 1987). In this case, however, removing the cue, or placing a second identical cue in the field at another position, reduced the spatial bias, suggesting an important overshadowing of distant landmarks (Schenk, 1987a, 1989). In this design, it was the accuracy score, rather than the time in the training quadrant, that indicated the most pronounced effects in rats up to 7 weeks old (Schenk, 1985). Figure 5.4 illustrates the overshadowing effect produced by the addition of a small object on the escape platform during training in different groups of juvenile rats in a large pool (diameter 160cm). In contrast, the presence of a similar cue at some distance from the target improved both escape efficiency and memory for the spatial position of the target even after removal of the cue (Chevalley & Schenk, 1987). These results suggest that during this period, juvenile rats are particularly attentive to local cues, which might play an important role in organising spatial movements. The risk of overshadowing of the more distant cues by the various local cues in the immediate vicinity of the nest entrance is of little relevance in a natural situation, since the simultaneous disappearance of all these cues is not likely.

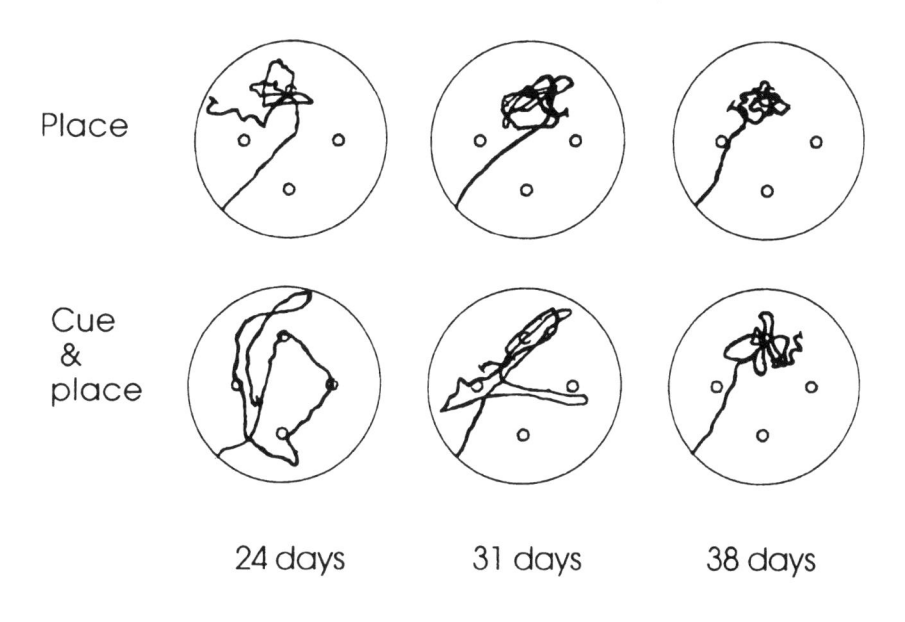

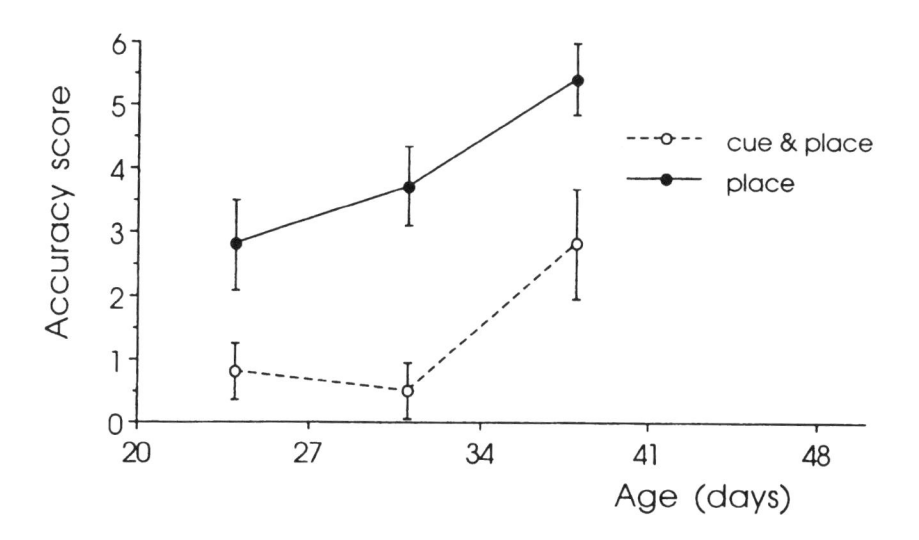

FIG. 5.4 Representative paths of rats in different age groups during the first 20 seconds of a probe trial (trial 29) following training with (cue & place) or without (place) a cue on the training platform. The accuracy score is computed from the annulus score in the training position minus the mean number of scores over the three irrelevant positions.

This description does not fit with the results of a more abrupt development of spatial abilities reported by Rudy et al. (1987). This might be due to three related causes. First, Rudy et al. studied Long Evans rats, whereas our study was based on hooded rats of the PVG Madorin strain. Long Evans rats are more rapid in their acquisition of the Morris task (see Fig. 5.1). In addition, Rudy et al. introduced a salient cue in the pool, a condition known to improve escape acquisition. Finally, they did not seem to qualify the behaviour of post weanling rats relative to adults, but relative to the more immature ones, whereas we tried to undertake a more complete analysis of the performance of juvenile rats as compared to that of fully mature ones.

The study of ontogenetic changes in other species of rodent confirms that escape in the Morris task shows a dramatic improvement shortly after weaning. Chapillon and Roullet (1996) showed that 22-day-old mice showed the same performance as adults during acquisition and probe trials without any overshadowing in the case of cued training. However, unlike adults, they showed an impaired escape performance when a distracting cue was present in the pool, suggesting that they relied to a larger extent on salient local cues.

Another study by Galea, Ossenkopp, and Kavaliers (1994) confirms the existence of a marked improvement in spatial skills around the time of weaning in meadow voles. This study indicates further improvement during the following week. Later steps in the maturation of spatial abilities thus appear to be more or less pronounced depending on the strain or species, on the task design, and on whether or not the comparison is made with an adult reference group. More difficult testing procedures, such as the conditional illumination design on a homing board, might reveal whether there is greater dependence on visual information during the first week after weaning.

The development of adult-like capacities does not appear as a sudden emergence of a fully mature function, and seems to rely on a coupling of different processes such as a dependence on olfactory gradients, path integration, guidances based on visual cues, memory for places expressed by a selective place reactivity, and anticipatory processes based on a spatial representation. This coupling of different functional modules develops in several steps during ontogeny, which might at the same time provide adaptive strategies for homing or exploration needs at different ages. Some steps are likely to be characterised by an imbalance between the use of different types of information, which might indicate that already mature modules participate in the calibration of more recently developed ones.

Performance in the Morris task following damage of the hippocampal formation

During the last ten years, many papers reported that lesions of the hippocampus caused a dissociation between memory components (reference vs. working memory: Olton, Becker, & Handelman, 1979; declarative vs. nondeclarative: Squire & Zola-Morgan, 1988; configural vs. associative: Rudy & Sutherland, 1989). The demonstration was usually based on the comparisons of impaired and intact capacities in different task designs. Selective lesions of the hippocampus were reported to induce a dissociation between memory components in the Morris task (Schenk & Morris, 1985; Morris et al., 1990). A clear recovery of a normal accuracy score was obtained with pretraining or prolonged postoperative training in rats with lesions of the entorhinal cortex, provided that the lesions spared the subiculum (Schenk & Morris, 1985). However, the recovered rats made no prolonged visits to the training quadrant and they did not display the normal searching pattern, which consisted of abrupt changes in direction when swimming about the position of the absent platform. Their trajectories appeared highly repetitive, albeit accurate, as if they remembered a limited number of approach paths. No recovery was observed during the same period if the lesion was more invasive.

In a second study, rats with selective lesions of either the hippocampus or the subiculum (Morris et al., 1990) showed a very high degree of recovery following a prolonged training phase of alternate blocks of visible or invisible platform training. In this case, rats showed a nearly normal performance in terms of time and accuracy scores, and their swim pattern was not apparently different from that of control rats. A procedure in which cued and noncued trials are alternated might induce a more complete type of recovery because it provides the opportunity to reach the escape platform along a straight line during the cued trials. During straight approaches of the cued platform, rats can learn sensorimotor adjustments to visual environmental cues. This might overcome the stereotypical approach tendency shown in Fig. 5.3, and facilitate an efficient approach of the platform in the absence of the cue. However, the recovered rats were markedly impaired when they had to learn a new escape position every day, following a matching-to-place procedure. Again, the group with mixed lesions of the subiculum and hippocampus showed no recovery, although their latencies decreased significantly with training.

A similar degree of recovery has been observed following lesions of the fornix. In one study (Eichenbaum et al., 1990), the lesioned rats showed a normal spatial bias following a prolonged training phase with a fixed start position. However, the bias was considerably reduced when major room landmarks were removed or when the start position was changed. More recently, Whishaw, Cassel, and Jarrard (1995) trained rats with fornix lesions to acquire a place response in cued and noncued trials. After 40 trials, the spatial bias exhibited by the lesioned rats was extremely high, as they spent 80% of their time in the training quadrant and showed accurate starting orientations. However, their swim paths remained less accurate than those of the control rats and a careful analysis showed that they made characteristic adjustment loops as they approached the platform (see Figure 7 in Whishaw et al., 1995). This indicates that rats with fornix lesions had to adjust their position in the vicinity of the training platform, whereas control rats generated trajectories that anticipated the finding of the platform at a given position in the pool.

Studying rats with lesions of the fornix or of the caudate nucleus revealed a double dissociation between the effects of these lesions (Packard & McGaugh, 1992), the former affecting spatial discrimination, and the latter cue discrimination. The latency data for the spatial task indicated that the rats with caudate lesions were slightly slower in reaching the correct platform, maybe because, unlike control rats, they could not rely on a combined "guidance and place" strategy. As discussed in Part 1, rats with lesions of the striatum persisted in searching in the previous escape position despite the displacement of the visible platform, indicating that they relied exclusively on place memory to reach a visible platform at a fixed position in space (McDonald & White, 1994). Place memory can thus develop almost normally in spite of an impaired capacity for cue guidance.

Like selective lesions of the hippocampus, chronic treatment with a selective antagonist of NMDA receptors, AP5, was shown to induce marked impairments in place learning in the Morris task (Morris et al., 1986a). A high degree of recovery in spatial performance could be promoted by various pretraining procedures. However, this protective effect was very variable. It was obvious when the rats had been pretrained to find an invisible platform at a fixed position in space in another spatial environment (Bannerman, Good, Butcher, Ramsay, & Morris, 1995). In the same series of experiments, there was almost no protective effect if the pretraining was conducted with an unpredictable escape position. However, another pretraining procedure with a cued platform at unpredictable locations led to a significant recovery when the treated rats were trained in a classic place paradigm (Saucier & Cain, 1995). In contrast, pretraining in the same pool at a fixed position did not protect against the disruptive effect of AP5 perfusion when the treated rats had to learn a new escape position in the same pool (Morris, Davis, & Butcher, 1991). This indicates that in certain circumstances, rats with NMDA-receptor blockade can learn one escape position in a given environment, but not a second position in the same environment. In this sense, one can consider that "spatial learning can be divided into pharmacological dissociable components"

(Bannerman et al., 1995, p.185). However, the most clearly marked dissociation is between acquiring a first escape position in a given environment and learning new positions in this environment. The second, more obscure dissociation, might be related to a protective effect from the knowledge that the task is an easy one, due to past experience with either a spatial task (Bannerman et al., 1995) or an easy cued task (Saucier & Cain, 1995). This protection seems ineffective for reversal learning in a familiar environment (Morris et al., 1991).

This short review suggests that learning processes in cases of dysfunction of the hippocampus do certainly allow the use of more than just the elementary taxon strategies proposed by O'Keefe and Nadel (1978). The logical suggestion, from these results, is that the training procedure must compensate for the inappropriate behavioural strategies that lesioned animals develop in order to acquire information about the escape platform position. Training in an easy task in which even impaired rats can follow discrete direct paths might facilitate learning about the visual environment. It might also reduce some of the stress induced by a prolonged escape latency. Both effects are likely to enhance performance.

Multiple aspects of cholinergic functions

The acquisition of the classical Morris task is sensitive to anticholinergic drugs (Sutherland, Whishaw, & Regehr, 1982; Hagan, Tweedie, & Morris 1986; Whishaw, 1985b). The performance in a cued task is also affected by the treatment, suggesting that the impairment is not specifically spatial. Conversely, a chronic treatment enhancing cholinergic activity might increase place learning in normal rats via an increase in the number of nicotinic cholinergic receptors (Abdulla, Calaminici, Stephenson, & Sinden, 1993).

Overshadowing by a salient proximal cue might account for the results shown by Whishaw and Tomie (1987), when rats treated with atropine sulphate (50mg/kg, intraperitoneally) did not react to the introduction of novel cues and showed a poor spatial bias toward the training quadrant after removal of the visible escape platform. In addition, they were not capable of discriminating between two identical visible platforms, one of which was at the correct location. A systematic analysis confirmed that, in a variety of dry land and water tasks, atropine-treated rats made less use of distal cues for place learning than did control rats (Whishaw, 1989). In fact, atropine appears to affect strategy selection, not discrimination learning, an effect which is obviously incompatible with normal acquisition of the Morris task (Whishaw & Petrie, 1988).

The Morris task is also affected by lesions of the basal forebrain cholinergic nuclei (Kelsey & Landry, 1988; Hagan, Salamone, Simpson,

Iversen, & Morris, 1988). More specifically however, recent work with neurotoxic lesions of the medial septum have shown that the acquisition of the Morris navigation task might be relatively insensitive to a loss of septohippocampal cholinergic fibres, while performance in the radial maze was severely affected (Decker, Radek, Majchrzak, & Anderson, 1992). We have found that quisqualic lesions of the medial septum affect the performance in the apparently easy cued task in which a salient cue is hanging above the escape platform, at a fixed position in space (Brandner & Schenk, in press). In this case, the deficit seems to be due to the presence of the cue and to its subsequent overshadowing effect. The lesioned rats were obviously impaired each time their first approach trajectory missed the platform. In such cases, most of the normal rats showed immediate correction and searched in the vicinity of the platform. Rats with septal lesions did little correction in this area, suggesting that they were impaired in combining a guidance strategy with the use of distant room cues. The overshadowing effect might be due to the fact that, like rats with selective lesions of the hippocampal region, rats with septal lesions are more dependent on being able to view the panorama around the platform to compensate for a spatial deficit, and that such a strategy is much more sensitive to overshadowing by a salient visual cue.

The fact that cholinergic blockade acted mainly when administered prior to training (see Hagan et al., 1986; Aigner, Walker, & Mishkin, 1991) indicates that this treatment might interfere with the initial storage of information, or, more generally, with the process by which ongoing information is integrated prior to the selection of an appropriate strategy. In line with this, a cholinergic dysfunction also affects attention to environmental stimuli (Muir, Dunnett, Robbins, & Everitt, 1992; Callahan, Kinsora, Harbaugh, Reeder, & Davis, 1993). This system might thus be involved in attentional processes and in the selection of an appropriate strategy as well. This does not necessarily exclude a participation in memory processes.

In the Morris task, the most specific effect seems to be the overshadowing by a goal-associated cue, particularly in immature and in senescent subjects (Schenk et al., 1990). Age effects might be related to a change in cholinergic activity, since juvenile rats (Hess & Blozovski, 1987) like senescent rats (Lindner & Schallert, 1988; Markowska, Olton, & Givens, 1995), show an increased sensitivity to cholinergic antagonists, even though reduced activity in the cholinergic system is not likely to be the only cause of age-related deficits in the Morris maze (Abdulla, Abu-Bakra, Calaminici, Stephenson, & Sinden, 1995).

In general, rats with lesions of the cholinergic nuclei show an exaggerated tendency to develop a strategy based on a guidance to salient cues. Most experimental environments are poorly controlled, they are merely described. They may or may not contain highly contrasted elements in the vicinity of the pool that remain permanently visible when approaching the platform. Thus, depending on the visual environment, and perhaps on the position of the platform within the pool (in a "minimal cue quadrant" or not, as discussed by Hodges, 1996) the deficit of the lesioned rats can be compensated by an efficient use of salient visual cues. The prediction is thus that the deficit should reappear when training to a new quadrant, or when major room cues are masked.

ECO-ETHOLOGICAL RELEVANCE OF THE MORRIS TASK

Is it a valid task for the evaluation of adaptive spatial abilities?

It has always been tacitly assumed that most spatial tasks used in the laboratory provide an indication of a subject's spatial skills, as required by its life habits in the field. But even though *Rattus norvegicus* was known for being a good swimmer, there was no demonstration that the Morris swim task might indeed evaluate a subject's capacity to orient itself in a natural context. Nor was this demonstration established for most laboratory tasks.

In an attempt to assess the validity of laboratory data in the "real world", Gaulin and Fitzgerald (1986, 1989) put forward the hypothesis that species with a large home range might profit from an increase in spatial skills, and that, for the same reasons, males of polygynous species should also benefit from enhanced spatial memory that could facilitate mobility. This prediction leads to three sub-hypotheses. First, the difference in spatial skills between males and females of a polygynous species, and between males of species with different home ranges, might be evident in classic laboratory tasks. Second, the difference in spatial abilities might be related to a difference in the architecture of the hippocampus. Third, seasonal changes in reproduction should be related to changes in male/female differences in their spatial skills and in the development of the hippocampus. More recently, Jacobs (see Jacobs, 1995, for a review) showed that mammalian species that were scatter hoarders might also benefit from enhanced spatial memory and hippocampal volume.

Experimental work with wild rodent species has brought confirmation to some of these hypotheses. As predicted, voles of polygamous species exhibited significant sex-related differences in spatial abilities, while monogamous species lacked such differences (see Gaulin, 1992 for a review). In addition, differences in hippocampus size were found, as predicted, in bird and in mammal species (see Jacobs, Gaulin, Sherry, & Hoffman, 1990).

As in other laboratory tasks, the performance in the Morris swim task revealed effects that appeared relevant from an ethological perspective. Sexually dimorphic spatial abilities in task acquisition were observed in two subspecies of deer mice, *Peromyscus maniculatus artemisae* and *P. m. angustus* (Galea, Kavaliers, Ossenkopp, Innes, & Hargreaves, 1994). Moreover, spatial performance of males and females is differentially affected by changes in reproductive status (Galea, Kavaliers, & Ossenkopp, 1996). This was related to changes in life habits in both sexes, since the difference between males and females was significant under long photoperiods, when the females were least mobile. A similar approach (Sawrey, Keith, & Backes, 1994) failed to reveal sex differences in two polygynous vole species (*Microtus ochrogaster* and *M. pennsylvanicus*). Although many factors, including stress, might account for the discrepancy between these two studies, a careful reading of the description of the apparatus used by Sawrey et al. (1994) suggests that females might have benefited from the presence of two salient cues "mounted over the rim of the pool (which) were clearly visible from within the pool". This might have reduced the hypothetical deficit of the females, which, like juveniles, appear to rely more systematically on visible cues around a maze than do males, which, in turn, seem to rely more on the geometrical properties of the room (Williams, Barnett, & Meck, 1990).

Another question is whether the cues that are available in the field might also be used in a laboratory task. In recent work, Kavaliers and Galea (1994) demonstrated that meadow voles (*Microtus pennsylvanicus*) use the position of the sun and that of other celestial cues available during the day, in a Morris maze placed in a large open field. The acquisition of the task was comparable to that obtained in laboratory conditions with uncontrolled visible cues. The demonstration that subjects relied on the position of the sun was indirect. There was a reduction in performance when the sky was overcast and when the subjects were tested in the afternoon, instead of the morning. In this latter condition, a rapid improvement in escape suggests that the use of the sun compass might require experience at various times of the day, as if for calibration purposes.

It is tempting to relate between-species differences in the performance in the Morris task to differences in spatial memory. However, it should first be demonstrated that they are not due to different swimming capacities. Brazilian short-tailed Oppossums (*Monodelphis domestica*) are much less efficient than rats in the Morris task (Kimble & Whishaw, 1994), which might be due to a poorer swimming ability. It is our experience that, although they do acquire good escape performance, wild woodmice (*Apodemus sylvaticus*) are not very good subjects for testing in the Morris task (Schenk, 1987b). Interestingly, woodmice show a good spatial memory in probe trials following place learning in both the Morris and the homing task, but do not develop straight approaches in either task. Another point is thus related to species-specific habits in planning trajectories: it may be that the more or less straight trajectory, expected as an indication of a capacity to take a shortcut, is of limited adaptive value because it might be too predictable by a predator. As for immature subjects, this issue calls for the use of other ethologically relevant dry land tasks.

SHALL WE REJECT OR RATHER REFINE OUR HYPOTHESES?

Do we need to reject the spatial map hypothesis?

It might be tempting to discuss whether the experiments reviewed above provide evidence about the existence of a quasi-mythical cognitive spatial map in rats. However, it might be more profitable, as already suggested by Thinus-Blanc (1984), to concentrate first on an analysis of the cognitive processes underlying spatial skills.

Obviously, rats "know" where the platform ought to be, and how to get there. The work reviewed above provides a general agreement that this behaviour is based on views of the environment, together with the efficient integration of route-based information. Moreover, access to visual cues appears most critical a few seconds before subjects land on the platform (Sutherland et al., 1987; Arolfo et al., 1994). Apparently, the accuracy required by the swimming task exceeds the capacities of path integration, except in particular training conditions (Save & Moghaddam, 1996). On solid ground, where the accuracy requirement is low, due to the presence of holes, the discrimination appears less dependent on a specific view of the environment (Schenk et al., 1995). Thus, the hypothesis put forward by Wilkie and Palfrey (1987) that rats need to rely on the permanent visibility of some major cues to adjust their position might be valid in the case of the Morris task because of its high accuracy requirements. A mechanism based on a snapshot memory (Cartwright & Collett, 1983; Benhamou, Bovet, & Poucet, 1995) might thus play a critical role in the development of optimal performance in the Morris task. This does not mean, however, that vision of the goal or of the panorama expected in its vicinity is essential for place-learning in other tasks, such as the homing board (Schenk et al., 1995).

Another set of evidence which attests to the particular importance of vision is derived from experiments in which the capacity for instantaneous transfer is tested. In these tests, performance appears to be disrupted each time new visual information is provided in such a manner that either a familiar swim path meets new visual cues (Whishaw & Tomie, 1987; Fenton et al., 1994) or a new trajectory can be followed relative to known visual cues (Sutherland et al., 1987). In these circumstances, rats do not instantaneously adopt a new optimal approach. This might be in part because the reupdating of a familiar spatial representation is of higher priority and interferes with task performance. If an adjustment of path trajectory to visual cues plays a critical role in the performance of the Morris task, then it is not surprising that an unexpected change in visual input during swimming triggers an exploratory reaction that increases escape latency.

These experiments do not appear to really challenge the cognitive mapping hypothesis. Rather, they emphasise the fact that experimental designs are based on a crude human logic that does not fit with the constraints of the spatial representation(s) guiding a rat's behaviour. This might be due to the difficulty to remember that, as stated by Neisser (1976, p.131), maps "are not pictures in the head, but plans for obtaining informations from potential environments". A more objective analysis of what rats do is obviously necessary to allow a better understanding of how a spatial representation contributes to optimal adaptation.

Do we need to reconsider which brain structures are involved in spatial memory?

The traditional view is to consider the capacity to develop straight approaches and focused searching behaviour in the Morris task as indications of a spatial representation. Rats with hippocampal or fornix lesions are severely impaired in the acquisition of a normal performance in the Morris task. However, a significant but incomplete recovery could be observed if they were trained in particular conditions such as alternate blocks of trials with a visible and an invisible platform, or progressive adaptation to the invisible platform (Eichenbaum et al., 1990; Morris et al., 1990; Whishaw et al., 1995). In all cases, there was a significant residual deficit.

Clearly, rats with selective hippocampal lesions are capable of significant spatial learning. However, it does not mean that the hippocampus is not critically involved in the development of spatial representations. These results indicate that hippocampal lesions have a double effect on spatial abilities, in slowing down place-learning rate and in leading to a qualitatively poorer spatial representation (Schenk et al., 1995). The slow development of efficient escape is due to an organisational impairment. This deficit is due to the lack of an important property of the spatial representation, the testing of hypotheses. A cognitive mapping process should allow for the planning of novel shortcuts, and for the anticipation of what should be met along this path (Poucet, 1993). If the expectation is confirmed, this might, in turn, consolidate the spatial representation. This type of behavioural hypothesis might be the product of an emergent representation, as in perceptual cycles guided by orientation schemata (Neisser, 1976). It should accelerate the development of optimal escape paths. Our interpretation is that rats with lesions of the hippocampus are unable to generate and test such hypotheses. Instead, they tend to repeat successful approach paths, hence their stereotyped behaviour. What they learn is thus embedded in complex and interfering trajectories. When the platform is visible, they develop straight approach paths. If progressively trained to escape onto the invisible platform, they are able to express what might be taken as an evidence of place memory. But this memory has considerable limitations. It appears as a short-term memory and it requires prolonged training to express nearly normal escape and searching of the platform (Morris et al., 1990). It is also highly susceptible to the removal of salient environmental cues (Eichenbaum et al., 1990) because it relies on a possibly complex set of sensorimotor associations, not on a memory of the spatial relation between different places. In this case, a visual snapshot memory from the exact platform location might compensate for the lack of spatial memory. This would explain in the same time a slower acquisition than in normal subjects, and an incapacity to learn about platform relocation that is common in all the experiments indicating an apparent recovery from the effects of hippocampal lesions on the acquisition of the Morris task (see Whishaw et al., 1995).

This opens up the double question of what type of spatial memory-representation is thus developed and by which brain structures it is supported. Our data suggest that this might not be a property of extrahippocampal structures, but of spared hippocampal components, since we could not find such recovery in rats with lesions to more of one component of the hippocampal formation (Schenk & Morris, 1985; Morris et al., 1990).

We propose that the recovered rats have learned highly specific visuomotor adjustments which help them in reaching a given place and searching around it during the probe trials. Indeed, the first time one observes rats with selective hippocampal lesions in the Morris task, one is struck by the accuracy with which they swim repeatedly along the same track, crossing the central region of the pool at a given place. Clearly, they are capable of adjusting their path relative to distant room cues, most likely visual ones. If the organisational deficit is compensated by an adequate training procedure that guides the rats along simple approach paths, lesioned rats can memorise, albeit at a slow rate, the visual landscape around the target. The amount of spatial information about the escape position thus seems normal. However, the residual inaccuracy and typical incapacity to learn rapidly a new position in the same environment indicate that the lesioned rats have not memorised the relations between places and cannot use their spatial representation to generate predictions for a direct approach path or shortcut. Rather, they seem to adjust their movements relative to sensory feedback in a reactive rather than in a predictive mode (Paillard, 1994). To account for this difference, we consider that the lesioned rats express place reactivity, a capacity close to that shown by immature rats. Another characteristic of place reactivity is the high vulnerability to overshadowing, as discussed above. It indicates that visual adjustments in reference to memorised images are based on the salience of each visual cue, not on the spatial relations among these cues. Thus, the deficit induced by hippocampal lesions concerns both spatial strategies (i.e. the development of spatial "hypotheses" as suggested above) and the quality of spatial representations. As a consequence, the lesioned rats are expected to be unable to solve a conditional illumination task in which the spatial position must be learned on the basis of an integration of different visual cues perceived in different parts of the environment.

Obviously, extrahippocampal regions such as the parietal cortex (DiMattia & Kesner, 1988; Save & Moghaddam, 1996) or the prefrontal cortex (Sutherland, Kolb, & Whishaw, 1982; Granon & Poucet, 1995) participate in spatial learning, but a more analytical approach should go beyond the mere demonstration that lesions produce spatial deficit. A dissociative approach has already confirmed that spatial memory is primarily affected by lesions of the hippocampal region, whereas each of three different behavioural components of spatial abilities (remaining in a place that had been positively reinforced, approaching a cue predicting the availability of a reinforcement, or patrolling on the basis of a spatial representation) is dependent on the active participation of different brain structures, respectively, the amygdala, the striatum and the hippocampus (McDonald & White, 1993). Other procedures aimed at qualifying the deficit should be developed.

Does it raise some experimental issues?

This review reveals some specific aspects of the Morris task, in particular its dependence on visual adjustments, as compared to dry land tasks. It suggests also that residual learning capacities, leading, for example, to a quasi-normal performance by rats with lesions of the hippocampus or of the cholinergic nuclei, and by rats treated with NMDA blockers, are difficult to interpret because there is no consensus as to whether the observed performance relies on the same learning mechanisms as in control rats. The hypothesis that visuomotor adjustments might support the recovery shown by rats with hippocampal or septal lesions and that this type of strategy is more susceptible to overshadowing by salient environmental cues calls for a better control of environmental landmarks. Conducting a probe trial following training with controlled salient environmental cues allows us to assess the amount of overshadowing and the capacity to associate a guidance strategy with a memory of the relative position of different places. Using a conditional illumination design with controlled visual cues allows us to demonstrate whether the subjects rely on specific visuomotor adjustments supporting selective place reactivity, or whether they have a representation of the spatial relation among different places in the environment.

This review shows also that some important issues in the study of memory, as, for example, the long-term change in spatial memory, are not given enough attention and might be developed, especially as a model to study retrograde amnesia. It also strengthens the need for an extensive comparative approach, both based on the use of different species and adapted task designs.

ACKNOWLEDGEMENTS

This research was supported by a grant from the Fonds National Suisse de la Recherche Scientifique No. 3100–039754.93/1. Thanks are due to Marianne Gafner for skilled technical assistance.

REFERENCES

Abdulla, F.A., Abu-Bakra, M.A.J., Calaminici, M.R., Stephenson, J.D., & Sinden, J.D. (1995). Importance of forebrain cholinergic and GABAergic systems to the age-related deficits in water maze performance of rats. *Neurobiology of Aging, 16*, 41–52.

Abdulla, F.A., Calaminici, M.R., Stephenson, J.D., & Sinden, J.D. (1993). Chronic treatment with cholinoceptor drugs influences spatial learning in rats. *Psychopharmacology, 111*, 508–511.

Aigner, T.G., Walker, D.L., & Mishkin, M. (1991). Comparison of the effects of scopolamine administered before and after acquisition in a test of visual recognition memory in monkeys. *Behavioral and Neural Biology, 55*, 61–67.

Alyan, S., & Jander, R. (1994). Short-range homing in the house mouse, *Mus musculus*: stages in the learning of directions. *Animal Behaviour, 48*, 285–298.

Arolfo, M.P., Nerad, L., Schenk, F., & Bures, J. (1994). Absence of snapshot memory of the target view interferes with place navigation learning by rats in the water maze. *Behavioral Neuroscience, 108*, 308–316.

Bannerman, D.M., Good, M.A., Butcher, S.P., Ramsay, M., & Morris, R.G.M. (1995). Distinct components of spatial learning revealed by prior training and NMDA receptor blockade. *Nature, 378*, 182–186.

Barnes, C.A. (1979). Memory deficits associated with senescence: a neurophysiological and behavioral study in the rat. *Journal of Comparative and Physiological Psychology, 93*, 74–104.

Barnes, C.A., Nadel, L., & Honig, W.K. (1980). Spatial memory deficits in senescent rats. *Canadian Journal of Psychology, 34*, 29–39.

Barta, A., & Yashpal, K. (1981). Regional distribution of β-endorphin in the rat brain: the effect of stress. *Progress in Neuropsychopharmacology, 5*, 595–598.

Benhamou, S. (1989). An olfactory orientation model for mammals' movement in their home range. *Journal of Theoretical Biology, 139*, 379–388.

Benhamou, S., Bovet, P., & Poucet, B. (1995). A model for place navigation in mammals. *Journal of Theoretical Biology, 173*, 163–178.

Bodnoff, S.R., Humphreys, A.G., Lehman, J.C., Diamond, D.M., Rose, G.M., & Meaney, M.J. (1995). Enduring effects of chronic corticosterone treatment on spatial learning, synaptic plasticity, and hippocampal neuropathology in young and mid-aged rats. *Journal of Neuroscience, 15*, 61–69.

Brandeis, R., Brandys, Y., & Yehuda, S. (1989). The use of the Morris water maze in the study of memory and learning. *International Journal of Neuroscience, 48*, 29–69.

Brandner, C., & Schenk, F. (in press). Septal lesions impair the acquisition of a cued place navigation task: Attentional or memory deficit? *Neurobiology of Learning & Memory.*

Brioni, J.D., Arolfo, M.P., Jerusalinsky, D., Medina, J.H., & Izquierdo, I. (1991). The effect of flumazenil on acquisition, retention, and retrieval of spatial information. *Behavioral and Neural Biology, 56*, 329–335.

Buresova, O., Homuta, L., Krekule, I., & Bures, J. (1988). Does nondirectional signalization of target distance contribute to navigation in the Morris water maze? *Behavioral and Neural Biology, 49*, 240–2480.

Buresova, O., Krekule, I., Zahalka, A., & Bures, J. (1985). On-demand platform improves accuracy of the Morris water maze procedure. *Journal of Neuroscience Methods, 15*, 63–72.

Callahan, M.J., Kinsora, J.J., Harbaugh, R.E., Reeder, T.M., & Davis, R.E. (1993). Continuous infusion of scopolamine impairs sustained attention of rhesus monkeys. *Neurobiology of Aging, 14*, 147–151.

Cartwright, B.A., & Collett, T.S. (1983). Landmark learning in bees: experiments and models. *Journal of Comparative Physiology, A151*, 521–543.

Chapillon, P., & Roullet, P. (1996). Use of proximal and distal cues in place navigation by mice change during ontogeny. *Developmental Psychobiology, 26*, 529–545.

Chevalley, A.-F., & Schenk, F. (1991). The ontogeny of spatial orientation in the rat. In *Ontogenesis of the brain* (Vol. 5, pp.191–194). S. Trojan & M. Langmeier (Eds.), Karolinum, Praha.

Chevalley, A.-F., & Schenk, F. (1987). Immature processes of spatial learning in hooded rats. *Society for Neurosciences Abstracts, 17*, 184–185.

Chevalley, A.-F., & Schenk, F. (1988). Ontogenèse du comportement d'orientation spatiale chez le rat. Utilisation d'indices proximaux et distaux. *Sciences et Techniques de l'Animal de Laboratoire, 13*, 49–52.

Collett, T.S. (1987). The use of visual landmarks by gerbils: reaching a goal when landmarks are displaced. *Journal of Comparative Physiology, A160*, 109–113.

Contant-Åström, B. (1994). *Assessment of age-related alterations of spatial memory in rats.* Paul Åströms Förlag, Jonsered.

Conrad, C.D., & Roy, E.J. (1995). Dentate gyrus destruction and spatial learning impairment after coritcosteroid removal in young and middle-aged rats. *Hippocampus, 5*, 1–15.

Decker, M.W., Introini-Collison, I.B., & McGaugh, J.L. (1989). Effects of naloxone on Morris water maze learning in the rat: enhanced acquisition with pretraining but not posttraining administration. *Psychobiology, 17*, 270–275.

Decker, M.W., Radek, R.J., Majchrzak, M.J., & Anderson, D.J. (1992). Differential effects of medial septal lesions on spatial-memory tasks. *Psychobiology, 20*, 9–17.

Devan, B.D., Blank, G.S., & Petri, H.L. (1992). Place navigation in the Morris water task: effects of reduced platform interval lighting and pseudorandom platform positioning. *Psychobiology, 20*, 120–126.

Diamond, D.M., Bennett, M.C., Fleshner, M., & Rose, G.M. (1992). Inverted U-relationship between the level of peripheral corticosterone and the magnitude of hippocampal primed burst potentiation. *Hippocampus, 2*, 421–430.

Dickinson, A.K. (1980). *Contemporary animal learning theories. Problems in the Behavioural Sciences.* Cambridge: Cambridge University Press.

DiMattia, B.D., & Kesner, R.P. (1988). Spatial cognitive maps: differential role of parietal cortex and hippocampal formation. *Behavioural Neuroscience, 102,* 471–480.

Dudchenko, P., Goodridge, J., Seiterle, D., & Taube, J.S. (1997). Effects of repeated disorientation on the acquisition of spatial tasks in rats: Dissociation between the appetitive radial arm maze and aversive water maze. *Journal of Experimental Psychology: Animal Behaviour Processes, 23,* 194–210.

Eichenbaum, H., Stewart, C., & Morris, R.G.M. (1990). Hippocampal representation in place learning. *Journal of Neuroscience, 10,* 3531–3542.

Etienne, A.S., Maurer, R, & Saucy, F. (1988). Limitation in the assessment of path integration. *Behaviour, 106,* 81–111.

Fenton, A.A., Arolfo, M.P., Nerad, L., & Bures, J. (1994). Place navigation in the Morris water maze under minimum and redundant extra-maze cue conditions. *Behavioral and Neural Biology, 62,* 178–189.

Francis, D.D., Zaharia, M.D., Shanks, N., & Anisman, H. (1995). Stress-induced disturbances in Morris water-maze performance: interstrain variability. *Physiology and Behavior, 58,* 57–65

Galea, L.A., Kavaliers. M., & Ossenkopp, K.P. (1996). Sexually dimorphic spatial learning in meadow voles *Microtus pennsylvanicus* and deer mice *Peromyscus maniculatus. Journal of Experimental Biology, 199,* 195–200

Galea, L.A.M., Kavaliers, M., Ossenkopp, K.P., Innes, D., & Hargreaves, E.L. (1995). Sexually dimorphic spatial learning varies seasonally in two populations of deer mice. *Brain Research, 635,* 18–26.

Galea, L.A.M., Ossenkopp, K.P., & Kavaliers, M. (1994). Developmental changes in spatial learning in the Morris water-maze in young meadow voles, *Microtus pennsylvanicus. Behavioural Brain Research, 60,* 43–50.

Galea, L.A.M., Saksida, L., Kavaliers, M., & Ossenkopp, K.P. (1994). Naloxone facilitates spatial learning in a water-maze task in female, but not in male, adult non-breeding meadow voles. *Pharmacology, Biochemistry and Behavior, 47,* 265–271.

Gallagher, M., Burwell, R., & Burchinal, M. (1993). Severity of spatial learning impairment in aging: development of a learning index for performance in the Morris watermaze. *Behavioral Neuroscience, 107,* 618–626.

Gallagher, M., & Nicolle, M.M. (1993). Animal models of normal aging: relationship between cognitive decline and markers in hippocampal circuitry. *Behavioural Brain Research, 57,* 155–162.

Gallistel, C.R. (1990). *The organization of learning.* Cambridge, MA: Bradford Books/MIT Press.

Gaulin, S.J.C. (1992). Evolution of sex differences in spatial ability. *Yearbook of Physical Anthropology, 35,* 125–151.

Gaulin, S.J.C., & FitzGerald, R.W. (1986). Sex differences in spatial ability: An evolutionary hypothesis and test. *American Naturalist, 127,* 74–88.

Gaulin, S.J.C., & FitzGerald, R.W. (1989). Sexual selection for spatial-learning ability. *Animal Behavior, 37,* 322–331.

Gisquet-Verrier, P., Dekeyne, A., & Alexinsky, T. (1989). Differential effects of several retrieval cues over time: Evidence for time-dependent reorganization of memory. *Animal Learning and Behavior, 17,* 394–408.

Goodlett, C.R., Nonneman, A.J., Valentino, M.L., & West, J.R. (1988). Constraints on water maze spatial learning in rats: implications for behavioral studies of brain damage and recovery of function. *Behavioural Brain Research, 28,* 275–286.

Goodridge, J.P., & Taube, J.S. (1995). Preferential use of the landmark navigational system by head direction cells in rats. *Behavioral Neuroscience, 109,* 49–61.

Granon, S., & Poucet, B. (1995). Medial prefrontal lesions in the rat and spatial navigation: evidence for impaired learning. *Behavioral Neuroscience, 109,* 474–484.

Hagan, J.J., Salamone, J.D., Simpson, J., Iversen, S.D., & Morris, R.G.M. (1988). Place navigation in rats is impaired by lesion of medial septum and diagonal band but not nucleus basalis magnocellularis. *Behavioural Brain Research, 27,* 9–20.

Hagan, J.J., Tweedie, F., & Morris, R.G.M. (1986). Lack of task specificity and absence of posttraining effects of atropine on learning. *Behavioral Neuroscience, 100,* 483–493.

Hess, C., & Blozovski, D. (1987). Hippocampal muscarinic cholinergic mediation of spontaneous alternation and fear in the developing rat. *Behavioural Brain Research, 24,* 203–214.

Hodges, H. (1996). Maze procedures: the radial-maze and water-maze compared. *Cognitive Brain Research, 3,* 167–181.

Jacobs, L. (1995). The ecology of spatial cognition: adaptive patterns of space use and hippocampal size in wild rodents. In E. Alleva, A. Fasolo, H-P. Lipp, L. Nadel, & L. Ricceri (Eds.), *Behavioral brain research in naturalistic and semi-naturalistic settings: Possibilities and perspectives.* Kluwer: Dordrecht.

Jacobs, L.F., Gaulin, S.J.C., Sherry, D.F., & Hoffman, G.E. (1990). Evolution of spatial cognition: Sex-specific patterns of spatial behavior predict hippocampal size. *Proceedings of the National Academy of Science USA, 87,* 6349–6352.

Kavaliers, M., & Galea, L. (1994). Spatial watermaze learning using celestial cues by the meadow vole, *Microtus pennsylvannicus. Behavioural Brain Research, 61,* 97–100

Keith, J.R. (1989). Does latent learning produce instantaneous transfer of place navigation? *Psychobiology, 17,* 210–211.

Keith, J.R., & McVety, K.M. (1988). Latent place learning in a novel environment and the influences of prior training in rats. *Psychobiology, 16,* 146–151.

Kelsey, J.E., & Landry, B.A. (1988). Medial septal lesions disrupt spatial mapping ability in rats. *Behavioral Neuroscience, 102,* 289–293.

Kimble, D., & Whishaw, I.Q. (1994). Spatial behavior in the brazilian short-tailed opossum (*Monodelphis domestice*): comparison with the norway rat (*Rattus norvegicus*) in the morris water maze and radial arm maze. *Journal of Comparative Psychology, 108,* 148–155.

Knierim, J.J., Kudrimoti, H.S., & McNaughton, B.L. (1995). Hippocampal place fields, the internal compass, and the learning of landmark stability, *Journal of Neuroscience, 15,* 1648–1659.

Lalonde, R., & Thifault, St. (1994). Absence of an association between motor coordination and spatial orientation in lurcher mutant mice. *Behavior Genetics, 24,* 497–501.

Lalonde, R., & Boetz, M.I. (1986). Navigational deficits in weaver mutant mice. *Brain Research, 398,* 175–177.

Lavenex, P. (1995). *Importance des informations olfactives dans les comportements spatiaux chez le rat de laboratoire (Rattus norvegicus).* Thèse de doctorat des Sciences, Université de Lausanne.

Lavenex, P., & Schenk, F. (1996). Integration of olfactory information in a spatial representation enabling accurate arm choice in the radial arm maze. *Learning and Memory, 2,* 299–319.

Lavenex, P., & Schenk, F. (1997). Olfactory cues potentiate learning of distant visuospatial information. *Neurobiology of Learning and Memory, 68,* 140–153.

Lee, R.K.K., & Maier, S.F. (1988). Inescapable shock and attention to internal versus external cues in a water discrimination escape task. *Journal of Experimental Psychology: Animal Behavioral Processes, 14,* 302–310.

Lindner, M.D., & Schallert, T. (1988). Aging and atropine effects on spatial navigation in the Morris water task. *Behavioral Neuroscience, 102,* 621–634.

Luine, V., Villegas, M., Martinez, C., & McEwen, B. (1994). Repeated stress causes reversible impairments of spatial memory performance. *Brain Research, 639,* 167–170.

Mabry, T.R., Gold P.E., & McCarty, R. (1995). Age-related changes in plasma catecholamine responses to acute swim stress. *Neurobiology of Learning and Memory, 63,* 260–268.

Mabry, T.R., McCarty R., Gold P.E., & Foster, T.C. (1996). Age and stress history effects on spatial performance in a swim task in Fischer-344 rats. *Neurobiology of Learning and Memory, 66,* 1–10.

Mackintosh, N.J. (1985). Varieties of conditioning. In N.M. Weinberger, J.L. McGaugh, & G. Lynch (Eds.), *Memory systems of the brain* (pp. 335–350). New York: Guilford Press.

Maier, S.F., Albin, R.W., & Testa, T.J. (1973). Failure to learn to escape in rats previously exposed to inescapable shock depends on the nature of the escape response. *Journal of Comparative and Physiological Psychology, 85,* 581–592.

Markowska, A.L., Long, J.M., Johnson, C.T., & Olton, D.S. (1993). Variable-interval probe test as a tool for repeated measurements of spatial memory in the water maze. *Behavioral Neuroscience, 107,* 627–632.

Markowska, A., Olton, D.S., & Givens, B. (1995). Cholinergic manipulations in the medial septal area: age-related effects on working memory and hippocampal electrophysiology. *Journal of Neuroscience, 15,* 2063–2073.

Martin, G.M., Harley, C.W., Smith, A.R., Hoyles, E.S., & Hynes, C.A. (1997). Spatial disorientation blocks reliable goal location on a plus maze but does not prevent goal location on the Morris maze. *Journal of Experimental Psychology: Animal Behaviour Processes, 23,* 183–193.

McDonald, R.J., & White, N.M. (1993). A triple dissociation of memory systems: hippocampus, amygdala, and dorsal striatum. *Behavioral Neuroscience, 107,* 3–22.

McDonald, R.J., & White, N.M. (1994). Parallel information processing in the water maze: Evidence for independent memory systems involving dorsal striatum and hippocampus. *Behavioral and Neural Biology, 61,* 260–270.

McNamara, R.K., & Skelton, R.K. (1993). The neuropharmacological and neurochemical basis of place learning in the Morris water maze. *Brain Research Reviews, 18,* 33–49.

McNaughton, B.L., Chen, L.L., & Markus, E.J. (1991). "Dead reckoning", landmark learning, and the sense of direction: a neurophysiological and computational hypothesis. *Journal of Cognitive Neuroscience, 3,* 190–202.

McNaughton, B.L., Knierim, J.J., & Wilson, M.A. (1995). Vector encoding and the vestibular foundations of spatial cognition: Neurophysiological and computational mechanisms. In M.S. Gazzaniga (Ed.), *The Cognitive Neurosciences* (pp.585–595). Cambridge, MA: MIT Press.

Means, L.W. (1988). Rats acquire win-stay more readily than win-shift in a water escape situation. *Animal Learning and Behavior, 16,* 303–311.

Morris, R.G.M. (1981). Spatial localization does not require the presence of local cues. *Learning and Motivation, 12,* 239–260.

Morris, R.G.M. (1984). Developments of a water-maze procedure for studying spatial learning in the rat. *Journal of Neuroscience Methods, 11,* 47–60.

Morris, R.G.M., Anderson, E., Lynch, G., & Baudry, M. (1986a). Selective impairment of learning and blockade of long term potentiation by an N-methyl-D-aspartate receptor antagonist, AP5. *Nature, 319,* 774–776.

Morris, R.G.M., Davis, S., & Butcher, S.P. (1991). Hippocampal synaptic plasticity and N-methyl-D-aspartate receptors: a role in information storage. In M. Baudry & J.L. Davis (Eds.), *Long-term potentiation: a debate of current issues* (pp.267–300). Cambridge, MA: Bradford/MIT Press.

Morris, R.G.M., & Doyle, J. (1985). Successive incompatible tasks: Evidence for separate subsystems for storage of spatial knowledge. In G. Buszaki and C. Vanderwolf (Eds.), *Electrical Activity of the Archicortex.* (pp.281-293) Budapest: Kiado.

Morris, R.G.M., Garrud, P., Rawlins, J.N.P., & O'Keefe, J. (1982). Place navigation impaired in rats with hippocampal lesions. *Nature, 297,* 681-683.

Morris, R.G., Hagan J.J., & Rawlins, JN. (1986b). Allocentric spatial learning by hippocampectomised rats: a further test of the "spatial mapping" and "working memory" theories of hippocampal function. *Quarterly Journal of Experimental Psychology: B. Comparative Physiological Psychology, 38,* 365-395.

Morris, R.G.M., Schenk, F., Tweedie, F., & Jarrard, L. (1990). Ibotenate lesions of hippocampus and/or subiculum: dissociating components of allocentric spatial learning. *European Journal of Neuroscience, 2,* 1016-1029.

Muir, J.L., Dunnett, S.B., Robbins, T.W., & Everitt, B.J. (1992). Attentional functions of the forebrain cholinergic systems: effects of intraventricular hemicholinium, physostigmine, basal forebrain lesions and intra cortical grafts on a multiple-choice serial reaction time task. *Experimental Brain Research, 89,* 611-622.

Neisser, U. (1976). *Cognition and reality: principles and implications of cognitive psychology.* New York: Freeman.

O'Keefe, J., & Nadel, L. (1978). *The hippocampus as a cognitive map.* Oxford: Clarendon Press.

O'Steen, W.K., Spencer, R.L., Bare, D.J., & McEwen, B.S. (1995). Analysis of severe photoreceptor loss and Morris water-maze performance in aged rats. *Behavioural Brain Research, 68,* 151-158.

Oitzl, M.S., & De Kloet, E.R. (1992). Selective corticosteroid antagonists modulate specific aspects of spatial orientation learning. *Behavioral Neuroscience, 106,* 62-71.

Olton, D.S. (1977). Spatial memory. *Scientific American, 236,* 82-98.

Olton, D.S., Becker, J.T., & Handelman, G.E. (1979). Hippocampus, space and memory. *Behavioral and Brain Sciences, 2,* 313-365.

Packard, M., & McGaugh, J.L. (1992). A double dissociation of fornix and caudate nucleus lesions on acquisition of two water maze tasks: further evidence for multiple memory systems. *Behavioral Neuroscience, 106,* 439-446.

Paillard, J. (1994). L'intégration sensori-motrice et idéo-motrice. In M. Richelle, J. Requin, & M. Robert (Eds.), *Traité de psychologie expérimentale* (pp.925-961). Paris: PUF.

Paylor, R., & Rudy, J.W. (1990). Cholinergic receptor blockade can impair the rat's performance on both the place learning and cued versions of the morris water task: the role of age and pool wall brightness. *Behavioural Brain Research, 36,* 79-90.

Pelleymounter, M.A., Smith, M.Y., & Gallagher, M. (1987). Spatial learning impairments in aged rats trained with a salient configuration of stimuli. *Psychobiology, 15,* 248-254.

Poucet, B. (1985). Spatial behaviour of cats in cue-controlled environments. *Quarterly Journal of Experimental Psychology, 37B,* 155-179.

Poucet, B. (1993). Spatial cognitive maps in animals: new hypotheses on their structure and neural mechanisms. *Psychological Review, 100,* 163-182.

Rossier J., Grobéty, M.-C., & Schenk, F. (1996). Place learning under discontinuous access to a limited number of visual cues. *European Journal of Neuroscience, Suppl. 9,* 119.

Rudy, J.W., Stadler Morris, S., & Albert, P. (1987). Ontogeny of spatial navigation behaviors in the rat: Dissociation of "proximal"- and "distal"-cue based behaviors. *Behavioral Neuroscience, 101,* 62-73.

Rudy, J.W., & Sutherland, R.J. (1989). The hippocampal formation is necessary for rats to learn and remember configural discriminations. *Behavioural Brain Research, 34,* 97-109.

Save, E., & Moghaddam, M. (1996). Effects of lesions of the associative parietal cortex on the acquisition and use of spatial memory in egocentric and allocentric navigation tasks in the rat. *Behavioral Neuroscience, 110,* 74-85.

Sawrey, D.K., Keith, J.R., & Backes, R.C. (1994). Place learning by three vole species (*Microtus ochrogaster, M. montanus,* and *M. pennsylvanicus*) in the Morris swim task. *Journal of Comparative Psychology, 108,* 179–188.

Saucier, D., & Cain, D.P. (1995). Spatial learning without NMDA receptor-dependent long-term potentiation. *Nature, 378,* 186–189.

Schenk, F. (1985). Development of place navigation in rats from weaning to puberty. *Behavioral and Neural Biology, 43,* 69–85.

Schenk, F. (1987a). Dissociation between components of spatial memory in the rat during ontogeny. In P. Ellen & C. Thinus-Blanc (Eds.), *Cognitive processes and spatial orientation in animal and man.* (Vol. I.) *Animal psychology and ethology* (pp.160–167). Dordrecht: Martinus Nijhoff.

Schenk, F. (1987b). A comparison of spatial learning in woodmice (*Apodemus Sylvaticus*) and hooded rats. *Journal of Comparative Psychology, 101,* 150–159.

Schenk, F. (1989). A homing procedure for studying spatial memory in immature and adult rodents. *Journal of Neuroscience Methods, 26,* 249–258.

Schenk, F., & Brandner, C. (1995). Enhanced visuospatial memory following pre- and postnatal choline treatment. *Psychobiology. 23,* 302–312.

Schenk, F., Contant, B., & Werffeli, P. (1990). Intrahippocampal cholinergic grafts in aged rats compensate impairments in a radial maze and in a place learning task. *Experimental Brain Research, 82,* 641–650.

Schenk, F., Grobéty, M.-C., Lavenex, P., & Lipp, H.-P. (1995). Dissociation between basic components of spatial memory in rats. In E. Alleva, A. Fasolo, H-P. Lipp, L. Nadel, & L. Ricceri (Eds.), *Behavioral brain research in naturalistic and semi-naturalistic settings: Possibilities and perspectives.* Dordrecht: Kluwer.

Schenk, F., & Morris, R.G.M. (1985). Dissociation between components of spatial memory in rats after recovery from the effects of retrohippocampal lesions. *Experimental Brain Research, 58,* 11–28.

Selden, N.R.W., Cole, B.J., Everitt, B.J., & Robbins, T.W. (1990). Damage to coeruleocortical noradrenergic projections impairs locally cued but enhances spatially cued water maze acquisition. *Behavioural Brain Research, 39,* 29–51.

Semenov, L.V., & Bures, J. (1989). Vestibular stimulation disrupts acquisition of place navigation in the Morris water tank task. *Behavioral and Neural Biology, 51,* 346–363.

Spencer, R.L., O'Steen, W.K., & McEwen, B.S. (1995). Water maze performance of aged Sprague-Dawley rats in relation to retinal morphologic measures. *Behavioural Brain Research, 68,* 139–150.

Squire, L.S., & Zola-Morgan, S. (1988). Memory: brain systems and behavior. *Trends in Neurosciences, 11,* 170–175.

Stewart, C.A., & Morris, R.G.M. (1993). The watermaze. In A. Sahgal (Ed.), *Behavioural neuroscience: a practical approach: Vol. I. The practical approach series.* Oxford: IRL Press.

Sutherland, R.J., Chew, G.L., Baker, J.C., & Linggard, R.C. (1987). Some limitations on the use of distal cues in place navigation by rats. *Psychobiology, 15,* 48–57.

Sutherland, R.J., & Dyck, R.H. (1984). Place navigation by rats in a swimming pool. *Canadian Journal of Psychology, 38,* 322–347.

Sutherland, R.J., Kolb, B., & Whishaw, I.Q. (1982). Spatial mapping: definitive disruption by hippocampal or medial frontal cortical damage in the rat. *Neuroscience Letters, 31,* 271–276.

Sutherland, R.J., & Linggard, R.C. (1982). Being there: a novel demonstration of latent spatial learning in the rat. *Behavioral and Neural Biology, 36,* 103.

Sutherland, R.J., Whishaw, I.Q., & Regehr, J.C. (1982). Cholinergic receptor blockade impairs spatial localization by use of distal cues in the rat. *Journal of Comparative and Physiological Psychology, 96,* 563–573.

Szuran, T., Zimmermann, E., & Weltzl, H. (1994). Water maze performance and hippocampal weight of prenatally stressed rats. *Behavioural Brain Research, 65*, 153–155.

Thinus-Blanc, C. (1984). A propos des cartes cognitives chez l'animal: l'hypothèse de Tolman. *Cahiers de Psychologie Cognitive, 4*, 537–558.

Tolman E.C. (1948). Cognitive maps in rats and men. *Psychological Review, 55*, 189–208.

Vaher, P.R., Luine, V.N., Gould, E., & McEwen, B.S. (1994). Effects of adrenalectomy on spatial memory performance and dentate gyrus morphology. *Brain Research, 656*, 71–78.

Warren, D.A., Castro, C.A., Rudy, J.W., & Maier, S.F. (1991). No spatial learning impairment following exposure to inescapable shock. *Psychobiology, 19*, 127–134.

Whishaw, I.Q. (1985a) Formation of a place-learning set in the rat: a new procedure for neurobehavioral studies. *Physiology and Behavior, 35*, 139–143.

Whishaw, I.Q. (1985b). Cholinergic receptor blockade in the rat impairs locale but not taxon strategies for place navigation in a swimming pool. *Behavioral Neuroscience, 99*, 979–1005.

Whishaw, I.Q. (1989). Dissociating performance and learning deficits on spatial navigation tasks in rats subjected to cholinergic muscarinic blockade. *Brain Research Bulletin, 23*, 347–358.

Whishaw, I.Q. (1991). Latent learning in a swimming pool place task by rats: evidence for the use of associative and not cognitive mapping processes. *Quarterly Journal of Experimental Psychology, 43*, 83–103.

Whishaw, I.Q., Cassel, J.-C., & Jarrard, L. (1995). Rats with fimbria-fornix lesions display a place response in a swimming pool: a dissociation between getting there and knowing where. *Journal of Neuroscience, 15*, 5779–5788.

Whishaw, I.Q., & Mittelman, G. (1986). Visits to starts, routes, and places by rats (*Rattus norvegicus*) in swimming pool navigation tasks. *Journal of Comparative Psychology, 100*, 422–431.

Whishaw, I.Q., & Tomie, J.-A. (1987). Cholinergic receptor blockade produces impairments in a sensorimotor subsystem for place navigation in the rat: evidence from sensory, motor, and acquisition tests in a swimming pool. *Behavioral Neuroscience, 101*, 603–616.

Whishaw, I.Q., & Petrie, B.F. (1988). Cholinergic blockade in the rat impairs strategy selection but not learning and retention of nonspatial visual discrimination problems in a swimming pool. *Behavioral Neuroscience, 120*, 662–677.

Wilkie, D.M., & Palfrey, R. (1987). A computer simulation model of rats' place navigation in the Morris water maze. *Behavior Research Methods, Instruments and Computers, 19*, 400–403.

Williams, C.L., Barnett, A.M., & Meck, W.H. (1990). Organizational effects of early gonadal secretions on sexual differentiation in spatial memory. *Behavioral Neuroscience, 104*, 84–97.

Yau, J.L.W., Morris, R.G.M., & Seckl, J.R. (1994). Hippocampal corticosteroid receptor mRNA expression and spatial learning in the aged wistar rat. *Brain Research, 657*, 59–64.

6 Testing for Spatial Brain Dysfunction in Animals

Helen Hodges
Institute of Psychiatry

INTRODUCTION: APPROACHES TO THE ASSESSMENT OF SPATIAL BRAIN DYSFUNCTION

Tests involving spatial learning and/or memory are used in animal models more extensively than any other type of cognitive assessment. There are two obvious reasons for this. First, most experimental work is carried out in the rat, a species endowed with formidable spatial abilities, and second, the use of spatial tasks, the majority of which are rapidly learned, may obviate the need for an extensive period of behavioural training. Two broad approaches have been adopted; studies in which the primary emphasis lies in elucidating the brain mechanisms involved in spatial navigation, and factors contributing to successful place learning, and studies which use spatial tasks as a method of assessing cognitive function in impaired animals. Studies designed to investigate factors and mechanisms involved in spatial navigation have used environmental and apparatus manipulations to control the visual and sensory information available to animals, such as the distribution of extra- or intra-maze cues and their relationship to reinforcement. In addition, lesion, pharmacological, and electrophysiological methods have been employed to elucidate the brain regions and neuronal systems involved in spatial learning and memory, often in combination. Some approaches have been broad-based, as for example, the investigations by McNamara and Skelton (1993) into the neuronal systems involved in place learning in the water maze which involved assessment of treatments reported to exert facilitatory (cholinergic, glutamatergic and peptidergic), detrimental (GABAergic, benzodiazepines and alcohol) or null (catecholaminergic and serotonergic) effects on acquisition. Other approaches have been more restricted, for example comparisons of the effects of lesions of the hippocampus or its subfields in spatial and nonspatial tasks (see Eichenbaum et al., 1994, for a review), or use of intra-cranial drug infusions at selected sites to examine effects of receptor-specific compounds on behavioural or electrophysiogical response. These studies rest on two assumptions, first, that spatial abilities measured in different spatial tasks are broadly comparable, and second, that it is impairment of spatial rather than nonspatial mechanisms engaged in these tasks that is responsible for any observed deficits after lesion or drug treatments.

Studies designed to assess cognitive impairment in animal models of human clinical pathology have placed particularly heavy reliance on the use of spatial tasks. These models include cholinergic deficiency induced by lesions, chronic alcohol consumption or ageing as partial analogues of the neuronal loss and cognitive impairment seen in Alzheimer's disease, alcoholic dementia, Korsakoff's disease, and old age (Bartus et al., 1982; Arendt et al., 1989; Hodges et al., 1991a,b,c.); striatal or nigral lesions as models for deficits seen in Huntington's chorea or Parkinson's disease (Bjorklund et al., 1994a,b) and focal and global ischaemia typically brought about middle cerebral artery occlusion (MCAO) or four vessel occlusion (4 VO: permanent occlusion of the vertebral arteries combined with transient occlusion of the carotids), as models for deficits induced by stroke and heart attack, respectively (Hodges et al., 1994). Use of spatial tasks to assess cognitive dysfunction in these models is frequently made on the assumption that these tasks tap underlying processes that are engaged in learning and memory in both spatial and nonspatial procedures. Thus radial or water maze deficits are presented as examples of 'memory impairment' and the double Y-maze or use of subsets of rewarded arms in the radial maze have explicitly been advocated as a method of dissociating long-term 'reference' memory deficits from short-term 'working' memory deficits (Mallet & Beninger, 1993). In terms of specificity of effects, 'clinical model' approaches face an even more difficult task than 'spatial mechanism' approaches in showing how far deficits in spatial tasks are related to problems in spatial information processing, and what brain mechanisms are involved, since the brain damage can be quite extensive, for example, following use of toxins or alcohol that are not specific to cholinergic neurones, or MCAO that results in large infarcts in cortex and striatum. This chapter will examine procedures designed to isolate processes involved in spatial learning, and those aimed to detect improvements or impairments in task performance following brain damage and/or drug treatments, bearing four main questions in mind: (1) What process and brain mechanisms contribute to spatial navigation? (2) How far do different spatial procedures assess the same processes? (3) How far do spatial tasks tap specific spatial processes, as opposed to those involved in nonspatial learning and memory? (4) What light do patterns of recovery from brain damage in animals throw on the organisation and flexibility of spatial information processing?

VARIETY OF PROCEDURES FOR THE ASSESSMENT OF SPATIAL LEARNING AND MEMORY

There is an astonishing variety of spatial tests (see Fig. 6.1 and earlier chapters), ranging from featureless arenas such as the open field (OF) or water maze (WM), where search is unconfined, to the restricted routes available in the radial or Y-maze. Channelled mazes can offer a smaller or greater amount of choice from the unidirectional runway to the eight or more arms of the radial maze, and complex sequences available in the Stone maze (a multiple T-maze with restricted availability of cues at the extremities). Ostensibly, spatial tasks are designed to assess an animal's capacity to learn which locations provide food, safety, interesting objects or some other objective, using mainly visuospatial environmental cues. However, by design or accident, mazes are likely to tap a variety of cognitive processes, including associative learning, short- and long-term spatial and nonspatial memory, temporal order, conditional discrimination and anxiety. The elevated plus maze, hole board and open field, for example, have explicitly been developed as tests of anxiety, that are sensitive to the effects of anxiolytic and anxiogenic drugs, but they do not preclude spatial learning (File, 1993). Some spatial tasks are designed to assess components of memory, rather than navigational skills. For example, Jarrard's tasks in the radial arm maze (Jarrard, 1986) involve two versions, a place task, in which food rewards are always found at the ends of the same four arms, their location denoted by invariant extra-maze visuospatial cues, and a cue task in which textured inserts in each arm are moved to different arms on each trial. Only four textures are rewarded, so that the rat must learn the association between textured cue and food, regardless of the position of the cue. Furthermore, by rewarding only 4/8 arms, it is possible to dissociate long-term 'reference memory' errors (entries into never-rewarded arms) from short-term 'working memory' errors (re-entries into arms that have already been visited within a trial) in both associative and spatial modes. However, if the same rats are trained on both place and cue tasks, a conditional discrimination is also built into the procedure, since the rats must recognise that if textures are present, their position is irrelevant, whereas if they are absent, the position of the arms is essential for locating the food. Reference versus working memory dissociations can be accomplished by simpler procedures, with only two choices at each stage, so that the probability of making reference and working memory errors is equated. In the split stem T-maze (Volpe et al., 1992) the correct reference choice of

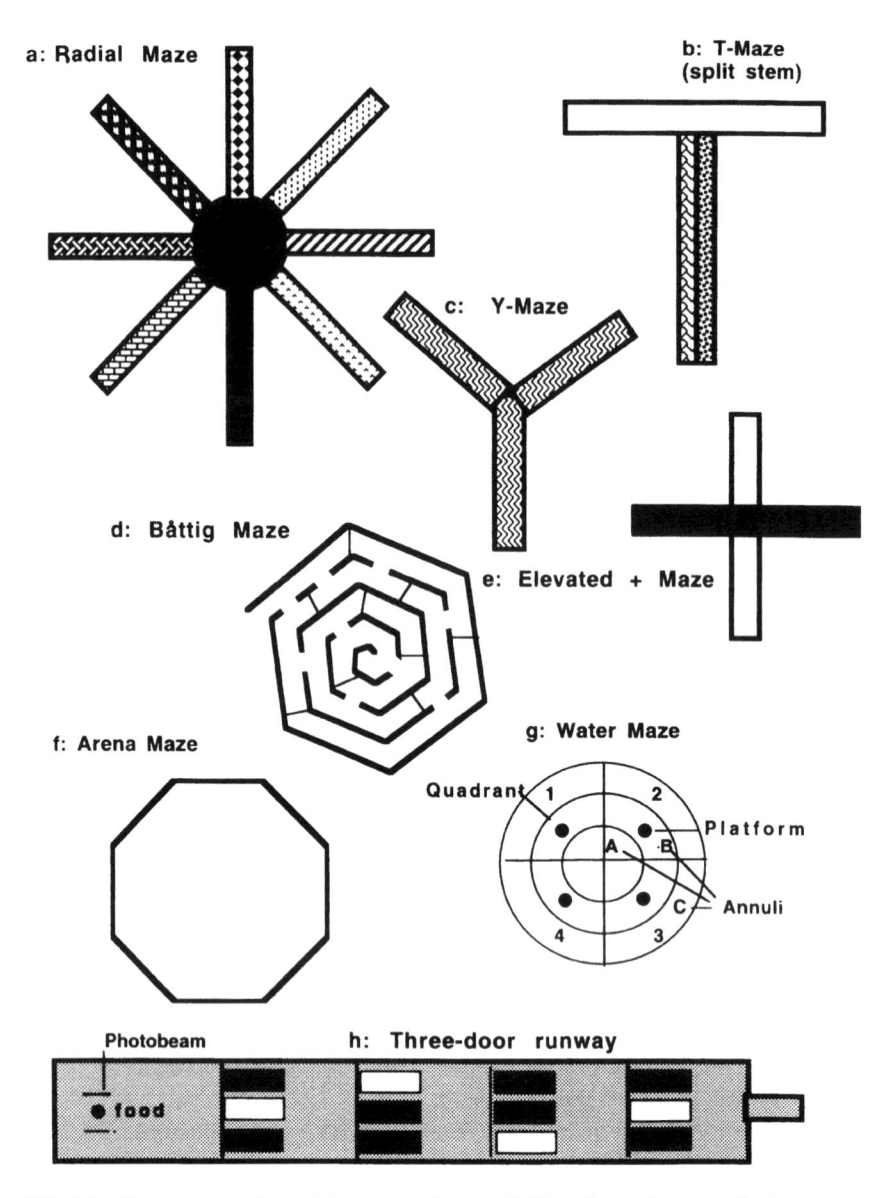

FIG. 6.1 Mazes commonly used to assess rodent spatial learning and memory (not drawn to scale). In the radial-arm maze (a) rats learn to retrieve food from the ends of all, or a subset, of the arms. Working memory errors occur when rats re-enter arms within a trial, and reference memory errors occur when they visit arms that are never rewarded. By including moveable textured inserts within the arms, the maze can be used to measure associative as well as spatial learning. The T-maze (b) and Y-maze (c) measure spontaneous alternation and working memory by the rat's ability to retrieve food from the arm not visited on the previous trial. Textured inserts in the stem of the T-maze, or a double Y-maze configuration, can be used

the stem route is denoted by invariant position or texture cues, but working memory is assessed by choice of arm, where entry into the arm not visited on the previous trial is correct. In the double Y-maze (Mallet & Beninger, 1993) the type of choice, as well as probability of error, is identical in the reference and working memory components. For the reference task the rat must proceed from a side arm to the same arm on each trial, namely, the arm leading to the centre of the maze and giving access to the identical second half. For the working memory task the rat must select one of two arms, with the correct choice based on the start position in the reference task. In addition to memory storage, spatial tasks can be used to assess the characteristics of storage, for example by looking at proactive interference via manipulation of the number of previous trials in the T-maze (Beracochea et al., 1987), or by controlling the temporal order of arm visits in the radial maze, in order to measure serial position effects (Harper et al., 1993).

The variety of procedures for assessing dysfunction in 'spatial' tasks means that there are substantial differences in the cognitive processes that they require, in the incentives used to motivate performance, and in the motor requirements for successful task completion (e.g. in lever pressing as opposed to swimming). Animals may therefore perform well in one type of spatial task, and poorly in another apparently similar one. For example, we have found (Nunn et al., 1991; Nunn & Hodges, 1994) that rats with ischaemic damage to the CA1 field of the hippocampus following 4 VO show substantial impairment of spatial learning in the water maze, but no impairment in learning Jarrard's place or cue tasks in the radial maze (see Fig. 6.2). Other workers (Davis & Volpe, 1990; Kiyota et al., 1991) have reported a profound impairment in radial maze, but not in water maze tasks following ischaemic brain damage, so the pattern of deficit varies according to laboratory and precise testing conditions (see Nunn & Hodges, 1994, for a review). However, the conclusion clearly emerges

---

to measure reference as well as working memory. The enclosed Bättig maze (d) measures spatial learning in the absence of extra-maze visuospatial cues. The elevated plus maze (e) is typically used to assess anxiety by the percentage of time spent on the anxiogenic open as opposed to the enclosed arms. In contrast to mazes with specific routes, open field or arena mazes (f) are used to measure spontaneous exploratory activity. However, in the case of the water maze (g) active search for a safe platform is enforced. Rats learn to locate a submerged platform typically in the centre of one of the four quadrants, and within annulus B, using different start points (N, S, E, or W) on each trial. The three-door runway (h) measures working memory by requiring rats to learn which of three doors, at four barriers along the runway, will open to permit progress towards food in the goal box. Typically, a different pattern of opening doors on each day is cued by a distinctive colour (e.g. white vs. black) on Trial 1, but for matching-to-position on subsequent trials all doors are the same colour.

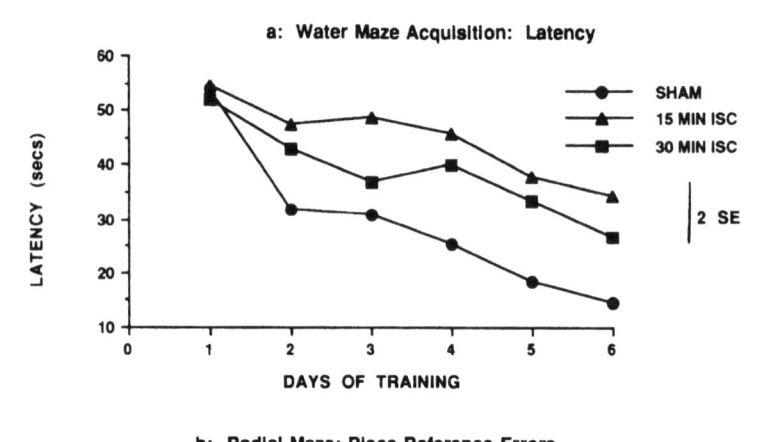

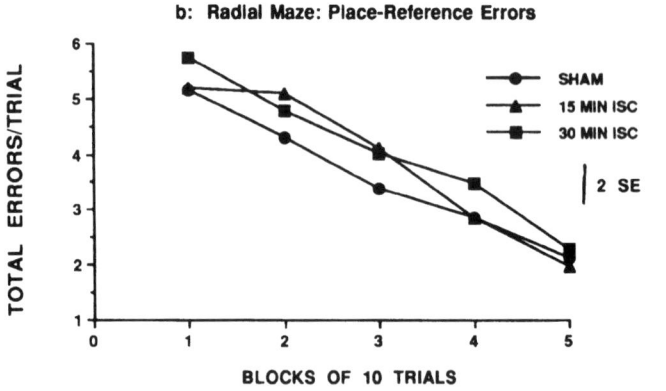

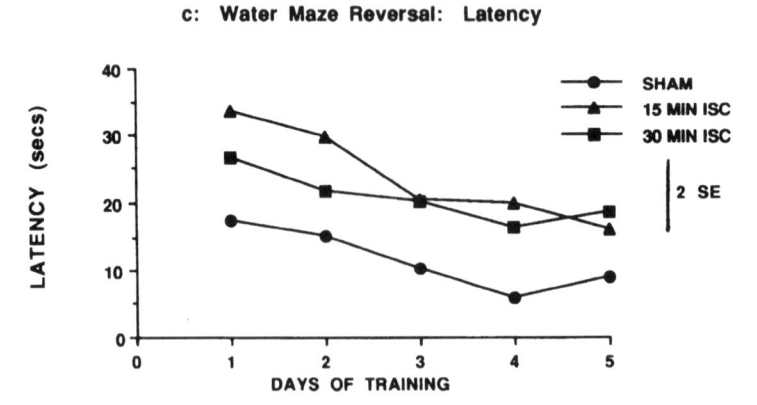

FIG. 6.2 Water maze and radial arm maze learning in rats subjected to 15 or 30 minutes of global cerebral ischaemia by four vessel occlusion (4 VO). Ischaemic rats were severely impaired in time taken to locate a submerged platform in the water maze, both initially (a) and

that performance in water and radial maze tasks is not necessarily correlated, since the same ischaemic animals can perform as well as controls in one of these tasks and significantly worse in the other. Findings of Van Luijtelaar et al. (1989) provide a striking example of the failure of both reference (visits and revisits to never-rewarded locations) and working (revisits to rewarded locations) memory errors to generalise between two spatial tasks, a hole board with 4/16 holes baited, and a radial maze with 4/8 arms baited. Although both types of apparatus showed reasonable reliability in terms of correlations between odd and even trials, neither reference nor working memory errors correlated significantly in the same rats trained on both tasks, regardless of order of training. One possible explanation is that the holeboard had transparent walls, permitting use of visual cues for allocentric spatial learning, whereas the radial maze had high opaque walls, which would foster egocentric or associative place learning, so that, as the authors pointed out 'animals seem to learn different things in the two apparatuses'. It is therefore of primary importance to try to understand and control the processes that are being assessed by different spatial tasks, in order to understand the nature of the deficits shown by impaired animals. These may arise through several factors including deficits in processing allocentric visuospatial information, impaired associative learning, heightened anxiety or stress, sensorimotor or motor deficits, which are differentially affected by different types of task, and by different types of brain damage.

VISUOSPATIAL INFORMATION PROCESSING

Concepts about how animals find their way about have been dominated for half a century by the idea that they form a 'cognitive map' relating visuospatial cues in the environment in a flexible way, so that they can follow novel routes to specific locations from any start point, and make detours if the direct route is blocked. This idea originated from Tolman's (1948) demonstrations with sunburst and detour mazes that rats can reach goals from different unfamiliar starting points and show directional rather route-specific learning, thus exhibiting navigational skills that could not be

---

four months later, after radial maze training, when the platform was in the opposite quadrant (c). However, ischaemic rats showed no impairment in radial maze reference (b) or working memory errors, relative to sham operated controls. In the 15-minute ischaemic group, cell loss occurred chiefly in the CA1 area of the hippocampus, whereas the 30-minute group showed additional hippocampal damage, and some cell loss in cortex and striatum. However, the similar profile of water maze deficits in the two groups suggested that the common loss of CA1 cells was a major determinant of this spatial impairment. (Reprinted from Hodges, 1996, with kind permission of Elsevier Science BV, Amsterdam Publishing Division, 1055 KV Amsterdam, The Netherlands.)

accounted for by the formation of stimulus-response or stimulus-stimulus associations or the chaining of motor responses. However, it was not until O'Keefe's identification of hippocampal 'place units' that this idea was firmly linked to visuospatial environmental cues, and given a physiological substrate (O'Keefe & Nadel, 1978; O'Keefe, 1983). Place units were found to fire when a rat was in a particular location, as defined by the configuration of objects in the environment, and ceased to fire when these objects were moved. O'Keefe's work provided a major input into theories of hippocampal function, by suggesting that the hippocampus constructs cognitive maps, using a flexible 'locale' strategy, as opposed to 'taxon' or cue-place associative learning. Initially, O'Keefe and co-workers argued that the hippocampus both constructed and stored cognitive maps, but more recent findings suggest that they are stored elsewhere in the brain (O'Keefe & Speakman, 1987). The work of O'Keefe's group provoked an explosion of research into how manipulations of visuospatial cues, particularly their stability, distribution and salience, affect place learning and interact with different types of brain damage or drug treatment.

CUE DISTRIBUTION AND SALIENCE

Cognitive map accounts of place learning have centred on the use of visuospatial cues distributed around the environment (allocentric spatial learning), even though visual acuity in the rat is limited, particularly in the albino strains commonly used in the laboratory (Dean, 1978; Birch & Jacobs, 1979). Contributions from olfaction, vibrissal stimulation, proprioception and body movement have been relatively neglected, although they are likely to provide highly salient information to animals adapted to tunnel dwelling that are capable of finding their way efficiently through dark enclosures such as the Bättig maze (Bättig, 1983). However, there is much evidence from studies that manipulate visuospatial cues to show that these, when accessible, play an important part in spatial navigation and place learning. O'Keefe and Conway (1980) demonstrated that the distribution of cues in the environment influences place learning, and interacts with hippocampal damage. Normal rats found it somewhat harder to learn to locate a rewarded arm in a 4-arm radial maze when cues were clustered around this arm, than when they were distributed around the room. However, after fornix lesions disrupting input to the hippocampus, rats were severely impaired in the distributed, but not the clustered condition. This suggested that animals with lesion-induced hippocampal dysfunction were capable of using an alternative strategy to locate the rewarded arm, possibly by using associative cue-reward (taxon) learning that was not disrupted by the fornix lesion. Consistent with this suggestion, Eichenbaum et al. (1991) found that rats with fimbria-fornix lesions were not impaired in learning the position of a platform in the water maze, when started at the same point on every trial, permitting associative 'guidance' learning of its location using a fixed set of cues, but were disrupted when started at a different point on each trial, which required the flexible learning of relationships between the cues around the pool. Cue distribution may therefore, intentionally or unintentionally, bias performance in spatial tasks that require the use of allocentric visuospatial cues. In the water maze, for example, we have found that rats spontaneously spent less time in the 'minimal cue quadrant', a pool sector that did not have any cues (posters, cupboards, etc.) nearby, whereas they searched at above chance level in the quadrant adjacent to where the Experimenter was standing (Hodges et al., 1995). Learning to find the submerged platform in the minimal cue as opposed to the Experimenter's quadrant was substantially retarded. Moreover, in rats with lesions to the cholinergic forebrain projection system (FCPS) involving the basal forebrain projections to cortex and the medial septal area projections to hippocampus (nucleusbasalis magnocellularis-medial septal area, or NBM-MSA lesions) this tendency was exaggerated, so that lesion deficits were far more evident in the minimal cue than the Experimenter's quadrant. These results would be consistent with proposal that animals with damage to hippocampal inputs are particularly disadvantaged when visuospatial information is sparse, but show relatively normal place learning when dominant, clustered or invariant cues encourage cue-location associative learning. The results also show, in an automated apparatus that is considered to be relatively immune to Experimenter effects, how easy it is to subtly bias the rate of spatial learning in the pool sectors by altering the distribution and/or salience of extra-maze visuospatial cues.

Cue removal

A more drastic manipulation is to remove visuospatial cues altogether. Sutherland and Dyck (1984) found that after surrounding a pool with curtains to conceal extra-maze cues, normal rats changed their strategy from search centred on the training quadrant, to circling at a fixed distance from the pool walls. This was a highly effective strategy, in that it took the animals along a trajectory that passed through the standard platform positions. Interestingly, rats with hippocampal lesions have also been reported to adopt fixed-distance circling in the water maze (Markowska et al., 1993; Nunn & Hodges, 1994; Hodges et al., 1996), suggesting that even if the ability to use extra-maze visuospatial cues is impaired, these animals are are able to use the pool wall as a guidance cue for place learning. Abrupt cue removal in maze-experienced animals can also indicate the extent to which animals' behaviour is governed by visuospatial cues in well-rehearsed task performance. For example, we (Hodges et al., 1991b) have compared the performance of control and NBM-MSA lesioned rats in Jarrard's radial maze place and cue tasks under normal lighting, and with dim lights suspended above the centre of the maze, illuminating only the entries to arms, but leaving the ends of the arms and the cues in the room obscured. Thus, if animals were using visuospatial cues in the place task, they should be disrupted in this task, but not in the cue task, where textured inserts were still accessible to touch, sight and smell. Under the fluorescent strip lighting used throughout training, control animals showed a low level of errors (<1 reference error and almost no working memory errors per pair of daily trials), whereas both types of error in both place and cue tasks were significantly and stably elevated in rats with NBM-MSA lesions When tested in dim light control rats showed a substantial increase in both working and reference memory error, relative to normal performance, in the place but not the cue task, suggesting that they used allocentric visuospatial information both for correct choices and to avoid re-entries, rather than relying, for example, on a rehearsed sequence of choices, or within-trial odour marking (see Fig. 6.3). The high error rate of lesioned rats, on the other hand, was scarcely affected by testing in dim light, suggesting that they made scant use of visuospatial cues in the place task, since it made no difference whether they could see them or not.

Cue stability

O'Keefe's findings (O'Keefe & Nadel, 1978) that place units cease to fire when objects defining an environment are moved, suggest that in order for visuospatial cues to be built into a cognitive map it is essential that they stay in the same place, even though their perspectives and interrelationships vary as an animal moves around. Biegler and Morris (1993) investigated the effects of landmark movement by training rats to search for food at fixed points relative to an object (e.g. 40cm 'south' of the landmark). For some animals, the landmark remained in a fixed location, for others it was moved around. Rats trained with the fixed landmark searched at the appropriate points relative to it, and took more direct paths to this point. Rats trained with the movable cue searched around the landmark itself, hence treating it as an associative cue, rather than as a reference point in space. Thus, landmark stability appears to be an essential prerequisite for true allocentric spatial learning of the location of a reward, but not for learning that a cue predicts a reward.

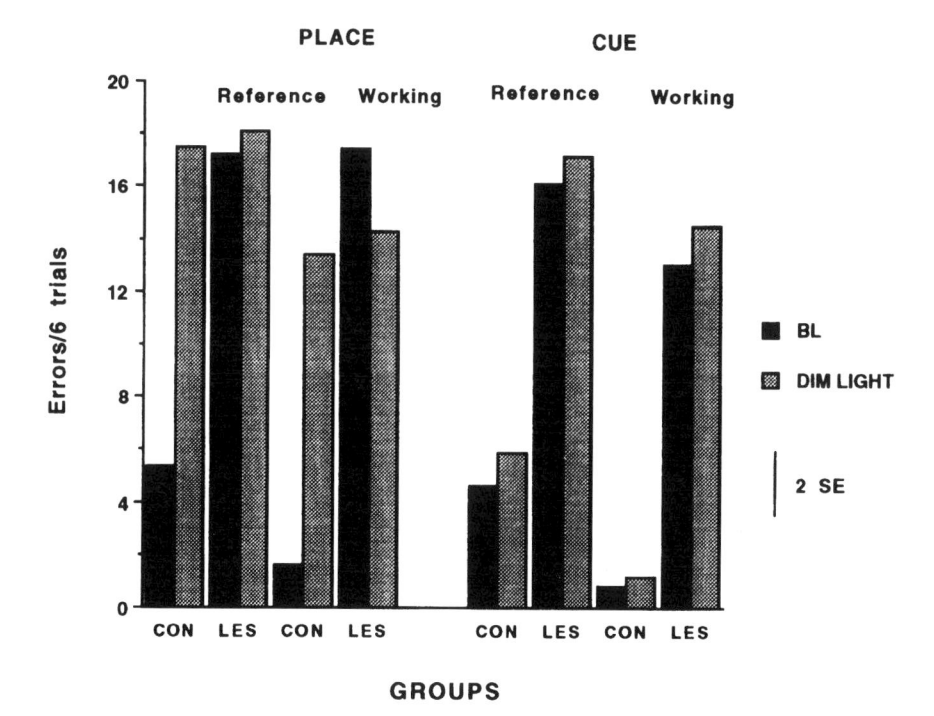

FIG. 6.3 Effects of dim light on radial maze Place and Cue task errors in controls (CON) and rats with nucleus basalis and medial septal area (NBM-MSA) cholinergic lesions (LES). Error rates in the control group increased substantially in the place task in dim light, but not in the cue task, where intra-maze cues (textured arm inserts) remained accessible. High baseline (BL) error rates were not increased in lesioned rats (LES) in dim light, suggesting that they were not using extra-maze visuospatial cues to direct their search. (Reprinted from Hodges et al., 1991b, with kind permission from Elsevier Science Ltd., The Boulevard, Langford Lane, Kidlington, OX5 1GB, UK.)

ASSOCIATIVE VS. COGNITIVE MAPPING MECHANISMS

Allocentric visuospatial cues vary in proximity to particular goal points. There is considerable debate about the extent to which an animal selects a route or learns a place using distal cues that provide pointers to an appropriate direction, or uses cues near to the goal or starting point to direct his search. Reliance on distal cues has been taken as evidence for the construction of cognitive maps, whereas use of local cues at salient sites such as start points or goal locations, has been suggested to involve associative learning of cue-goal relationships. This issue comes into prominence when mazes with fixed routes or sequential stages are used, such as the radial arm maze, Stone maze or three-door runway. Tolman's (1948) findings suggested that rats adopt a mapping strategy using distal cues, since they are able both to chose the most direct path to a goal in sunburst mazes, and to make successful detours when the direct route is blocked. However, subsequent studies have suggested than animals make use of local as well as distal cues, depending on the demands of the task, and, as implied by O'Keefe and Conway's (1980) and Eichenbaum et al.'s (1991) findings discussed above, different brain mechanisms may underlie the use of these different strategies.

Jarrard's (1986) place and cue radial maze tasks (see above) were specifically designed to separate spatial and associative processes. Findings that hippocampal lesions impaired place but not cue task performance suggested that the tasks did, indeed, tap different processes (Jarrard, 1993), and that the hippocampus was involved in spatial but not associative learning. However, working memory errors in the supposedly associative cue task were also increased following hippocampal lesions, indicating that even though cues were placed in different arms on each trial, the layout of cued arms within a trial may be acquired by rapid visuospatial mapping processes. Thus, associative and visuospatial mapping processes appear normally to act in unison, although they can be dissociated by manipulation of distal and proximal cues, as in Jarrard's tasks, or by use of the cloverleaf maze with curved arms that start in one direction and finish in another.

Brown's group (Brown, 1992; Brown et al., 1993) investigated associative and mapping processes in the search behaviour of normal rats in a 12-arm radial maze, using tasks involving working memory, with each arm baited, and a place-learning task comparable to that used by Jarrard (1986) with 8/12 arms baited to assess working (re-entries) and reference (nonrewarded arm entries) memory errors. In addition to arm choices, 'microchoices' were studied, in which rats investigated arm entrances and doors, without fully entering an arm. Although earlier studies suggested that intra-maze cues played little part in radial maze performance (Olton & Collison, 1979), Brown suggested that microchoices indexed attention to local intra-maze cues, engaged associative learning processes, and possibly governed within-trial working memory performance. A 'cognitive map' (O'Keefe & Nadel, 1978) theory of maze search would predict that, as accuracy increased, the rats would move directly from one baited arm to the next, using mainly extra-maze visuospatial cues without spending time in door investigations. A 'working memory' hypothesis (Olton et al., 1980), in which arms constitute a list of items in short-term memory, would predict that microchoices would be directed towards remaining baited but unvisited arms. Rotating the maze was predicted to increase reference memory errors, but not working memory errors, since these were hypothesised to depend on microchoice sequential processing of local cues. The data, however, showed that rotating the maze increased both working and reference memory errors, suggesting that distal visuospatial information contributed to within-trial working memory. Conversely, microchoices were not differentially associated with either baited, or unvisited arms, but increased when extra-maze cues were reduced or altered, suggesting that they were used to gain additional information when the use of distal cues was disrupted. Brown's findings therefore suggested that both intra- and extra-maze cues are involved in radial maze learning, In particular, intra-maze associative cues may be linked into a cognitive map formed via extra-maze cues and used to calibrate this map, or even to substitute for it, if extra-maze cues are reduced. Conversely, extra-maze visuospatial cues contribute to within-trial working memory choices, which do not depend solely on associative intra-maze cues, in agreement with Jarrard's (1993) findings. These results support the suggestion that visuospatial mapping and associative mechanisms are normally co-operative and additive. This conclusion is reinforced by our 8-arm radial maze study (Hodges & Green, 1986), in which cues (sandpaper strips) denoting rewarded arms were either always in the same 4 arms, providing congruent spatial and associative information, or moved randomly on each trial, so that the associative cue was relevant, but visuospatial extra-maze cues irrelevant to reward location. Rats found the constant cue condition easier to learn than the random condition, supporting the proposal that congruent spatial and associative information is facilitative. Moreover, partially trained rats treated with the benzodiazepine (BZ) chlordiazepoxide were far more impaired in the random than the constant condition, suggesting that interference between information processing demands may expose deficits in rats where cognitive mechanisms are impaired by drugs.

Manipulation of distal and proximal cues in the water maze has provided comparable evidence for the use of parallel associative and visuospatial mechanisms, and linked the hippocampus to the latter. In a direct test of the use of proximal versus distal cues, accomplished by the use of a submerged as opposed to a visible platform in the water maze, McDonald and White (1994) found that rats with dorsal striatal lesions swam preferentially to the submerged platform, even after learning to find the safe cued platform, suggesting intact visuospatial and impaired associative information processing, whereas rats with fornix lesions could not learn the location of the submerged platform, but readily learned to find the cued platform indicating the converse pattern of impairment. These results are complemented by the findings of Selden et al. (1990) that rats with dorsal noradrenergic bundle (DNAB) lesions were as efficient as controls in using visuospatial cues in standard water maze place learning. However, they were impaired in learning which of two intra-maze (proximal) cues was attached to a safe platform. Selden et al. interpret this deficit as reflecting a broadened attention span which favoured the use of distal as opposed to proximal cues. However, the results would also be consistent with the suggestion that the DNAB lesion disrupted associative spatial learning, while leaving visuospatial mapping processes intact. Taken together these findings suggest that parallel systems are involved in spatial navigation, and that animals make use of both allocentric visuospatial environmental cues, and specific cue-reinforcement associations in place learning. Furthermore, the findings suggest that the hippocampus is preferentially engaged in visuospatial information processing, whereas the dorsal striatum and DNAB are involved in cue-place associative learning.

EGOCENTRIC SPATIAL INFORMATION PROCESSING

An important source of spatial information comes from use of body position and movement to direct orientation and search. This 'egocentric' spatial information processing has been relatively neglected because of emphasis on allocentric processes, but it is likely that a large number of 'spatial' tasks, notably those involving L-R choices in the T-maze, Y-maze and Skinner box, or adjacent arm entry in more complex mazes, employ egocentric processes. Even the water maze can be used as an egocentric task by training rats in the dark. Moghaddam and Bures (1996) showed that navigation in darkness was initially less rapid and flexible than in the light with visible cues, but with overtraining rats easily learned to find the platform in novel locations, suggesting the formation of an 'egocentric map'.

Distortion of reaching for objects by prismatic lenses indicate that allocentric visuospatial and egocentric information are normally highly integrated. However, use of tasks such as adjacent arm entry and cheeseboard exploration (a black circular arena with 177 evenly distributed recessed food wells: Kesner et al., 1989) to tax ego- and allocentric processing, combined with selective lesions, have shown that the two processes can be dissociated. Cook and Kesner (1988) using two different sets of allocentric and egocentric tasks, demonstrated that bilateral caudate nucleus lesions severely disrupted retention of the egocentric tasks, but only transiently, if at all, impaired recall of the allocentric tasks. A double dissociation of function was provided by Kesner et al.'s (1989) findings that rats with medial prefrontal cortical lesions were impaired in egocentric tasks, whereas those with parietal cortical lesions were impaired in allocentric tasks. These findings were confirmed by King and Corwin (1992) with somewhat more selective uni- and bilateral lesions confined to the posterior parietal cortex (PPC) and medial agranular cortex (AGm), with some indication of functional lateralisation since only right (or bilateral) PPC lesions disrupted the performance of allocentric tasks. Corwin et al. (1994) extended the findings by showing that bilateral destruction of the ventrolateral cortex, an area connected to both the PPC and AGm and so potentially a region for integration of allocentric and egocentric information, in fact selectively disrupted allocentric task performance. These recent findings in rodent studies are consistent with results of selective prefrontal and parietal cortex lesions in nonhuman primates, and in human clinical studies, of effects of damage to these regions (Corwin et al., 1994). Thus a distinction between allocentric and egocentric information processing and its homologous organisation holds across a range of mammalian species. This distinction holds important implications for the assessment of spatial dysfunction in animal models, since it is clear that spatial tasks differ in their allocentric and egocentric information processing demands, and different types of brain damage or drug treatment may also differentially affect these processes. For example, Crowne et al. (1992) found that rats displaying a high level of amphetamine-induced rotation showed no deficits in learning to locate a submerged platform in the water maze, but were severely impaired in learning a water T-maze delayed alternation task, indicating that rotation impaired egocentric, but not allocentric localisation. Some animal models of brain damage directly target regions important for egocentric information processing notably MCAO stroke models that may damage both the caudate and prefrontal cortical areas, resulting in postural and movement asymmetry and contralateral limb weakness, together with impaired spatial learning. Gross neurological symptoms usually recede in two to three weeks, but contralateral skilled paw reaching, and deficits in spatial tasks are persistent (Grabowski et al., 1993; Hodges et al., 1994). It is therefore necessary to assess these animals in different types of spatial task, to see whether deficits extend both to allocentric and egocentric spatial task performance. Even with more subtle deficits, such as contralateral sensorimotor impairment and sensory neglect seen after large unilateral basal forebrain lesions (Abdulla et al., 1994), it is important to find out whether deficits in spatial tasks found after these lesions are influenced by egocentric spatial impairment following loss of cholinergic projections to cortex.

NONVISUAL SENSORY INFORMATION INVOLVED IN SPATIAL NAVIGATION

Relationships between visuospatial cues and their association with reinforcement have received most attention in investigations of spatial navigation and dysfunction in animal studies, possibly because of their salience to investigators rather than to their subjects. In animals, information from other sensory modalities is of critical importance for spatial localisation, with species-specific mechanisms, such as echolocation exemplifying the development of extreme specialisations. Two important sensory modalities in the rat are the olfactory and tactile systems, both augmented by mystacial vibrissae movement. Interestingly, several lines of evidence suggest that the hippocampus is involved in the processing of information from both of these sensory modalities, and hence possibly their integration into place learning.

Olfactory information

Eichenbaum et al. (1994) have suggested that the hippocampus (CA1 fields and dentate gyrus) is critically involved in the processing of relational information, from any sensory modality, whereas the parahippocampal region is important for storing information in short-intermediate (i.e. working) memory. Hence the representational demands of a task are important for detecting impairment in animals with damage to hippocampus, whereas requirements for storage in working memory will be sensitive to parahippocampal damage. Eichenbaum and co-workers (see Eichenbaum et al., 1994, for a review) demonstrated that rats with fornix lesions destroying input to the hippocampus were not impaired in discriminating between successively presented pairs of odours, hence they retained odour information in short-term storage. In contrast rats with cortical, notably rhinal cortex, damage showed working memory deficits in non-matching odour tasks (Otto & Eichenbaum, 1992). Fornix lesioned rats were substantially impaired when odours were presented simultaneously or in close approximation, which requires discrimination of the components of a mixture. Moreover, if odours were mismatched, controls were able to identify familiar odours, but rats with fornix lesions reacted as if the stimuli were novel. In dry spatial apparatus, it is virtually impossible to eliminate odours, particularly in mazes with many sections, such as the RAM, or if cues are placed within the maze. Moreover, these local odours will be successively presented, as the rat moves around, and may form one type of local cue that instigate microchoices (Brown et al., 1993). Although Olton and Collison (1979) have suggested that odour trails do not affect working memory in the radial maze with all arms rewarded, it may be that they influence choices between rewarded and nonrewarded arms in different maze tasks. Thus odour, and other local cues may contribute in an uncontrolled way to place learning, and interact with task and brain damage. As previously noted, we have found that ischaemic rats with CA1 cell loss are impaired in water maze navigation, but not in learning the Jarrard place and cue tasks in the radial maze (Nunn et al., 1991). One of the reasons for this may be that ischaemic rats were able to use odour cues as associative markers. We therefore tested ischaemic rats in an 8-arm water radial maze (Buresova et al., 1985), initially using Olton's working memory task, with a solenoid operated platform at the end of each arm, so that arms could be collapsed after that rat had found them. After a fairly extensive pre-training phase during which rats learned to visit all the arms freely, ischaemic rats made significantly more errors of re-entry in the first 8 choices, than controls (see Fig. 6.4), indicating a marked working memory deficit, that was not apparent in working memory performance in the dry radial maze (Hodges, 1996; Nelson et al., 1997a). Removal of odour cues and/or the opportunity to make microchoices may have contributed to this difference.

Tactile information and vibrissae movement

Spatial tasks differ in their requirements for finding a precise location. In a radial maze a rat has only to run down a confined path to find food at the end. In a water or holeboard arena, searching in the appropriate sector is not good enough—the rat has to find the precise target. It is possible that

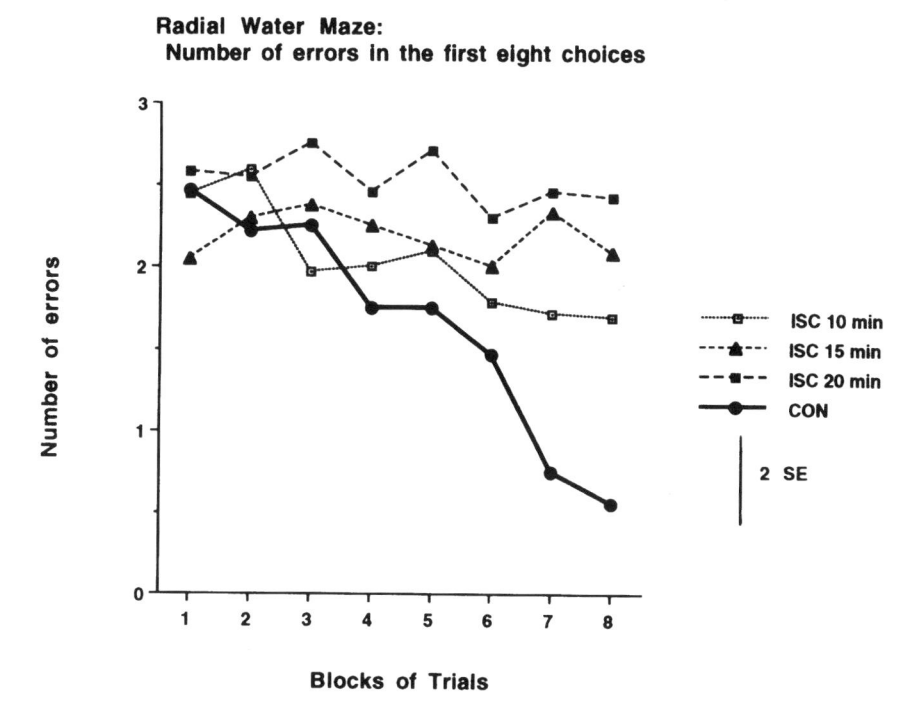

Radial Water Maze:
Number of errors in the first eight choices

FIG. 6.4 Working memory errors (arm re-entries) in the radial water maze. Control rats were superior to animals with ischaemic damage following 10, 15 or 20 minutes of global ischaemia induced by 4 VO. After pre-training, controls made fewer errors of re-entry in the first 8 choices than the ischaemic groups. (Reprinted from Hodges, 1996, with kind permission of Elsevier Science BV, Amsterdam Publishing Division, 1055 KV Amsterdam, The Netherlands.)

different cues or factors come into play when a rat has to pinpoint a location, as opposed to heading out correctly. We have observed that ischaemic rats with CA1 cell loss induced by 4 VO, when familiarised with a pool, showed accurate search in the quadrant, but spent longer than controls searching very close to the platform, even bumping directly into it, without climbing on (Netto et al., 1993; Hodges et al., 1996). This effect was picked up using the 'counter' measure which calculates the time spent in an area twice the platform size in diameter, surrounding the platform (see below).

Explanations in terms of gross motor deficits were ruled out by findings that swim speed was not impaired in ischaemic rats, and they learned to climb on to a visible platform as rapidly as controls. However, a possible explanation for this deficit is that ischaemic rats are less sensitive than controls to the presence of the platform via impaired tactile, sensorimotor or proprioceptive mechanisms induced by the CA1 damage. These deficits could arise by disruption of input from the mystacial vibrissae, which play an important part in tactile discrimination. Recent investigations of the neural pathways mediating vibrissal information in a point-to-point manner have centred on the role of the barrel field somatosensory cortex (Guic-Robles et al., 1989, 1992; Carvell & Simons, 1990). However, earlier findings (Kommisaruk, 1970; Gray, 1971) suggested that there is also a close correlation between hippocampal theta activity and vibrissal movement. Thus, ischaemic damage to the CA1 sector might impair the use of somatosensory information originating from the vibrissae. We therefore looked to see whether shaving off a rats whiskers had an effect similar to that of 4 VO, in increasing time spent in the counter area during water maze learning, and found this to be the case (J.A. Gray et al., unpublished data: see Fig. 6.5). Rats with vibrissae removed showed a significant increase in time spent in the platform counter area in two replications in maze-naïve and maze-experienced rats, though this did not necessarily lead to impairments relative to controls in other water maze measures, particularly in the maze-experienced groups. Groups with bilateral removal or removal of whiskers on the left had significantly higher counter scores than controls, but with removal of whiskers on the right counter scores were increased only on one of the two replications, leaving unresolved the question of possible laterality effects, which might be anticipated from findings that asymmetries are induced in striatal afferents by unilateral vibrissal removal (Schwarting et al., 1994). The findings suggest that even in the water maze, which has been argued to be a relatively pure measure of spatial learning using allocentric visuospatial cues, information from other modalities can play a distinct role.

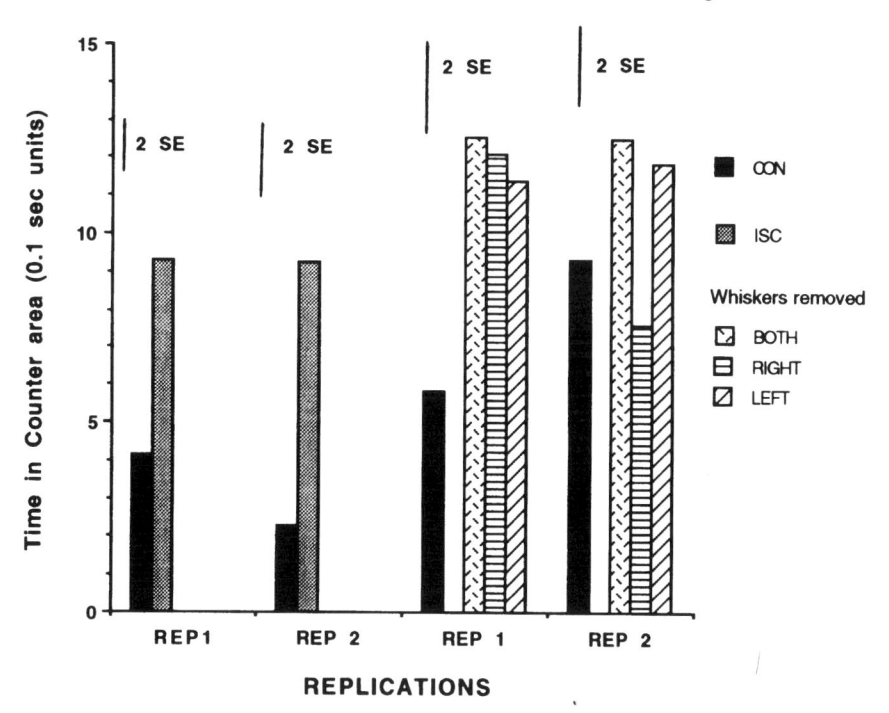

Proximity to the platform during water maze training

FIG. 6.5 Mean amount of time spent close to the water maze platform in control and ischaemic rats (15 minutes of 4 VO), and rats with whiskers shaved off on both (BOTH) sides, or on the right (RIGHT) or left (LEFT) sides only. The counter area surrounding the platform was twice the platform size in diameter. In results from two replications (REP), ischaemic rats, and animals with whiskers shaved off, spent significantly longer than controls searching round the platform in the counter area, without finding it, apart from rats with the right side shaved in the second replication. (From Gray & Grigoryan, in prep.)

SEARCH STRATEGIES

Superimposed on sensory information, egocentric and allocentric mapping and associative learning processes that animals use to construct representations of their environment, it is often possible to see regular patterns in the way that they search for safe or rewarded locations. Search strategies contribute to the efficiency in place learning, and hence to the construction of cognitive maps. Various algorithms have been developed to account for foraging strategies ranging from the simple rule 'always go to the nearest not yet visited feeder' (Bures et al., 1992) to the proposal that subjects retain a running record of previous visits, but respond randomly to locations outside the current memory set. Search strategies appear to be affected by a range of factors beyond the availability of visuospatial or associative cues. Of primary importance are differences in the constraints imposed by the apparatus. Strategies are easier to detect in open arenas than in pathway mazes, such as the RAM, which are partially designed to minimise them by confining rats to predetermined routes. Nevertheless, individual rats vary in the pattern of entries in the RAM, some choosing sequential arm entries, others entering every other arm, or opposite arms (Hodges et al., 1991b). Task requirements may further inhibit the development of search patterns in the RAM. For example, confinement of rats to the central area between choices (Olton et al., 1977), or selective reinforcement of a subset of arms (Olton & Pappas, 1979), reduces the tendency to chose adjacent arms. In contrast, sequential arm entry is a highly efficient strategy that is fostered by working memory tasks, where all arms are baited. Yoerg and Kamil (1982) suggest that it represents an optimum foraging strategy, and showed that the percentage of adjacent arm entries increased dramatically in a RAM with a large (88cm diameter) as opposed to a small (34cm diameter) central platform, with all arms rewarded. Motivational cost-benefit effects on search strategy are also implied by our findings that use of sequential arm entries was far more pronounced in rats fed for one, as opposed to two hours per day (T-A Perry, pers. comm.). Impairments in search strategy in radial mazes without doors are sometimes apparent in the form of 'error clusters' (Hodges & Green, 1986), when rats repeatedly enter the same few arms in a stereotyped and ballistic way, neglecting the rest of the maze. We have observed error clusters after cholinergic lesions, scopolamine and benzodiazepine treatments, so that this stereotyped behaviour appears to reflect a rather general response disturbance. In open arenas with automatic tracking systems such as the water maze (WM), search strategies are easier to detect and to quantify than in constrained mazes, and differing treatment and lesion effects can be examined. Normal animals initially circle round the pool wall, and then progressively spend more time in the inner annuli and the training quadrant, finally taking a direct path to the platform. During a series of repeated acquisition trials with the platform in different quadrants, thigmotaxis drops to a low level, and animals search initially around the former positions of platforms, and then appropriately in the centre of other quadrants. This pattern has been used by Morris et al. (1986) to develop a matching-to-novel-position test of working memory in the water maze. Measures provided in the WM (swim path length, latency to find the platform, the percentage of time spent in the quadrants and annuli, the accuracy of heading angle and speed of swimming) can be used to chart the progress of these changes. However, care must be taken in selecting appropriate measures. Latency and path length do not necessarily distinguish between animals that show undirected, as opposed to appropriate search. Latency, in particular, is subject to confounding factors such as speed, and may not reflect search accuracy, which detracts from the large number of studies that have used this as the sole measure of spatial learning. Gallagher et al. (1993) have suggested that average distance from the platform provides a measure that discriminates between animals that differ in accuracy of search, although showing similar latencies and distances. This problem is acute in animals that display a localisation deficit, in that they spend a longer time than controls searching very close to the platform, without climbing on to it, hence showing latency deficits, but high percentages of time in the training quadrant. Use of the 'counter' measure, computing the time spent in an area immediately around the platform position (see above) provides one method of detecting localisation deficits during training. (Netto et al., 1993; Hodges et al., 1996). Counters can therefore be used to show differences in time spent close to the platform during training, as well as differences in the accuracy of recall of the precise platform position during the 'probe trial' when the platform is removed. The WM is highly sensitive to changes on search patterns brought about by lesions or drugs, even though these are not necessarily easy to interpret. For example, cholinergic lesions and scopolamine treatment substantially increase circling close to the pool wall (Whishaw & Tomie, 1987; McNamara & Skelton, 1993), whereas hippocampal lesions have been reported to increase circling at a fixed distance from the pool wall (Markowska et al., 1993; Nunn & Hodges, 1994; Hodges et al., 1996: see above). As noted, global ischaemia resulting in damage to the CA1 field of the hippocampus and removal of vibrissae both significantly increase the time spent searching close to the platform without finding it, whereas under treatment with anxiolytics, particularly benzodiazepines, rats swim indiscriminately all over the pool (Hodges, 1996; see Schenk, Chapter 5, this volume).

NONSENSORY FACTORS AFFECTING SPATIAL NAVIGATION

Many features of tasks that are unrelated to visuospatial or sensory information processing may affect place learning in different maze procedures, and interact with brain damage. Problems engendered by sensorimotor and motor deficits have already been noted, since they interact closely with egocentric and and allocentric spatial information available to animals. Less direct, but powerful influences on spatial learning include the incentives used to motivate learning, the level of stress engendered by the task, and the schedule used to train or test the the animals.

Task motivation and stress

Different spatial tasks employ different incentives. For example the WM is aversively motivated by the desire to escape onto a safe platform, whereas dry mazes and runways are usually motivated by food reward. Differences in reinforcement may affect the speed of learning, the strategy adopted and the necessity for accurate navigation. Failure to find a food reward incurs no great penalty, except when the rats are very hungry. In the radial arm maze (RAM), as noted previously, animals fed for one, as opposed to two hours per day are more likely to visit adjacent arms (T-A Perry; see above), a strategy that is optimal in mazes with all arms rewarded, but which impedes learning when only a subset of arms are baited, as in the Jarrard tasks. In contrast, failure in aversively motivated tasks is highly stressful, indeed potentially life threatening in the WM. Fast learning in aversive tasks such as the WM in control animals may therefore more easily expose deficits in impaired animals that are masked by the slower learning of all groups in dry mazes, where animals with impaired visuospatial information processing capacity may be able to employ alternative strategies without disadvantage. Differences in incentives and stress may in part account for our failure to find deficits in rats with CA1 cell loss induced by 4 VO in the radial maze, when the same animals showed marked impairment in the water maze (Nunn et al., 1991).

Higher stress levels in aversively, as opposed to appetitively motivated tasks are likely to have differential consequences for experimental and control groups, particularly when the experimental condition itself interacts with stress or anxiety mechanisms, creating a potentially serious confounding factor. Dalrymple-Alford et al. (1985, cited in Will & Kelche, 1992) found that 10 days of housing in enriched conditions (see below) significantly improved the ability of rats with dorsal hippocampal lesions to find a food cup hidden below the surface of a dry arena, but failed to improve their location of a submerged platform in a water maze. The spatial requirements of these two tasks appear to be very similar, so that it is possible that the stressful nature of the WM task suppressed the cognition-enhancing effects of enrichment. Aversively motivated tasks may themselves differ in the levels of stress engendered. In pre-training in the water radial maze (see above) we observed that control rats visited significantly fewer arms than ischaemic groups. Errors over the first 30 trials showed no differences between ischaemic and nonischaemic groups, because of the large number of errors of omission in controls. These same controls were subsequently found to spend less time on the open arms of an elevated plus-maze than the ischaemic groups, suggesting an anxiolytic effect of ischaemic hippocampal damage, that may have contributed to their initially more widespread water radial maze exploration (A. Nelson et al., in prep.). However, once the rats were visiting all the arms freely, a significant working memory deficit was seen in ischaemic animals (see Fig. 6.4). Effects of differing stress levels within a standard WM acquisition task have been clearly demonstrated by Selden et al. (1990), who showed that control animals were impaired in cold (12°C) water in comparison with animals with dorsal noradrenergic bundle (DNAB) lesions, although no differences were seen in warm (26°C) water. Thus anxiogenic or stressful features of maze tasks, although, independent of cognitive mechanisms that underlie successful allocentric visuospatial navigation, may substantially affect initial performance and rate of spatial learning and interact with brain damage.

TRAINING AND TESTING REGIMES

Training and testing schedules can be manipulated to affect rates of acquisition and/or levels of asymptotic performance in spatial tasks, by varying the amount of information presented within training sessions, or by interpolating long or short intervals between sessions, and hence affecting retrieval from long-term memory by proactive and retroactive interference. Training schedule manipulations have been used to maximise differences between normal control animals and those with drug or lesion-induced impairments. Mandel et al. (1989) have shown that deficits in rats with cholinergic basal forebrain lesions can be exposed by use of a sparse WM training regime (a 10min, as opposed to a 5min inter-trial-interval, and/or two, as opposed to four trials/day), even though lesioned animals showed relatively normal rates of acquisition under standard training procedures. Similarly, we have found that rats with NBM-MSA lesions are more sensitive than controls to reductions in training trials (Grigoryan et al., 1994) or breaks in training (Turner et al., 1992). Indeed, Turner et al. (1992) found that RAM error rates in lesioned, but not control, animals were doubled after a three-week 'holiday break' (see Fig. 6.6), so that if stable baselines are required, care must be taken to keep to routine standardised training regimes in spatial maze tasks, as in other procedures.

Manipulation of testing regimes is commonly used to expose the sensitivity of brain-damaged animals to effects of interference in spatial and nonspatial tasks. Beracochea et al. (1987), using a T-maze spontaneous alternation task, found that mice treated with alcohol for 48 weeks or receiving mamillary body lesions, showed a marked decline in choice accuracy as a function of the number of preceding trials. When the maze context was altered on later trials, the rate of spontaneous alternation improved. Aggleton et al. (1990) report that deficits following amygdalar-fornix lesions increased with repeated use of similar goal box stimuli in a nonspatial task, and in T-maze alternation. Proactive (via an additional information trial) and retroactive (via interpolated delay) interference effects in a spatial T-maze alternation task have also been demonstrated following lesions restricted to the descending column of the fornix. (Tonkiss et al., 1990).

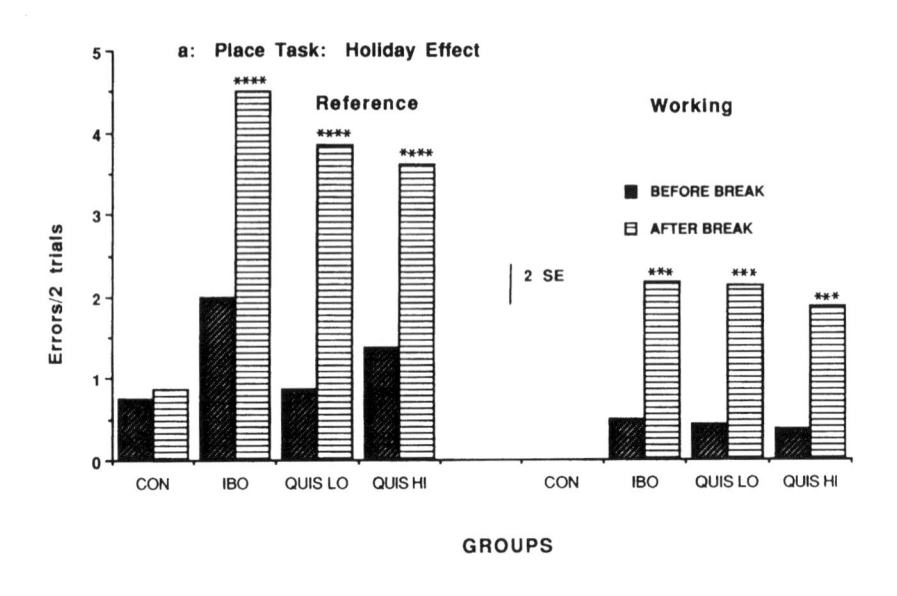

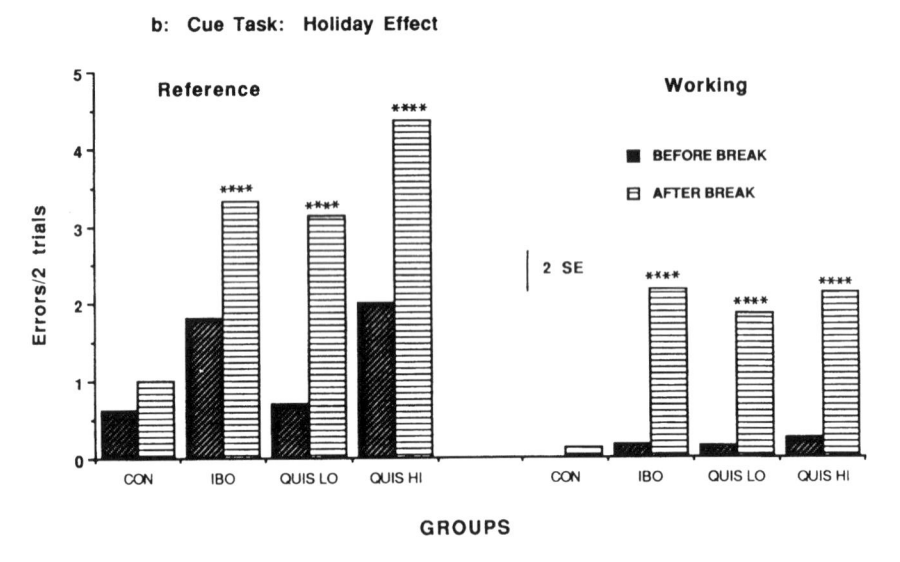

FIG. 6.6 The 'holiday effect'. Rats with cholinergic lesions (NBM-MSA) showed a substantial increase in radial maze place and cue task errors after a break in training, whereas controls were scarcely disrupted. Effects of the lesioning agents ibotenic acid (IBO) and quisqualic acid at two concentrations (QUIS LO and HI) were broadly equivalent. Increase in after-break scores: ***$P < 0.01$, ****$P < 0.001$. (Reprinted from Turner et al., 1992, with kind permission from Rapid Communications of Oxford Ltd., The Old Malthouse, Paradise Street, Oxford OX1 1LD, UK.)

ASSESSMENT OF MEMORY DEFICITS BY SPATIAL TASKS

Investigation of the processes involved in spatial navigation, and the factors affecting spatial performance outlined in the previous sections, have indicated that spatial tasks assess a range of abilities, including the ability to integrate sensory information from different modalities into a cognitive map, and to associate cues with safe or rewarded locations. It is also apparent that different spatial tasks assess a differing range of abilities, so that animals that show impairments in one procedure may not do so in another. In particular, the RAM and WM exemplify important differences between spatial tasks that permit free as opposed to constrained search, such as complexity, the extent to which a variety of strategies may be used, and task motivation. However, maze procedures are often used not simply to compare rates of place learning in brain-damaged and control groups, but as ways to access more general processes of acquisition, working memory and reference memory that are assumed to be engaged also in the processing of nonspatial information. It is therefore important to consider: (1) how far processes of learning, working memory and reference memory measured in different maze tasks will be equivalent; and (2) whether findings obtained in spatial procedures generalise to the acquisition and storage of nonspatial information.

Procedural vs. rapid acquisition tasks

Rates of acquisition differ profoundly in different spatial tasks. For example significant place learning can occur in only 10 trials in the WM, whereas around 50 trials are required in the Jarrard (1986) RAM tasks. Rapid learning can occur in some RAM procedures, as in O'Keefe and Conway's (1980) 'DESPATCH' task involving the learning of a single rewarded arm in a plus-maze. Repeated acquisition procedures with a different set of rewarded arms for each daily session also show that acquisition can occur within one session. However, as Peele and Baron (1988) found, the rats may need to be trained for up to 35 sessions (about 400 trials in all) before showing stable within-session baseline rates of acquisition. Several aspects of the tasks might contribute to differences in ease of acquisition. Use of several rewarded locations in the RAM involves both a within-trial win-shift strategy, and a between-trial win-stay strategy. In contrast, in WM tasks, there is usually only one safe platform that stays in the same place over several trials/days. Conditional discriminations that are built into several RAM tasks (e.g., the Jarrard Place and Cue tasks), and the use of intra-maze associative cues indexed by microchoices (Brown et al., 1993) will also add to task difficulty. Thus task complexity and differences in procedural demands may overlie the rapid and accurate processing of visuospatial information, and lead to markedly divergent rates of place learning in spatial tasks, which broadly speaking, can be characterised as 'procedural' and 'rapid acquisition' tasks on the basis of the length of training required. This division relates to Squire's distinction between 'declarative' and 'procedural' memory systems (Squire, 1986), in as far as procedural tasks may require mastery of routines, before the spatial learning and memory components of the task can be accessed.

As a result of the difference in rate of acquisition, 'procedural' tasks have a very different range of application from 'rapid acquisition' tasks. Procedural tasks are not well suited to assessing drug effects on learning, since acquisition is slow, involves multiple cognitive components, and drugs would have to be given over many weeks, making it difficult to distinguish between acute and chronic effects, and their interaction with the progress of learning. The notable exceptions are repeated acquisition tasks where the procedural aspects are mastered before the within-session assessment of place learning (Peele & Baron, 1988). In contrast, with relatively enduring, or slowly developing changes in brain function, such as lesions, effects of chronic alcohol consumption, old age, or growth of brain transplants (Hodges et al., 1991a,b,c), learning in procedural tasks can be used to assess the time course of changes in a way that is not feasible in rapid acquisition tasks. Once animals have mastered a procedural task, particularly one that includes both reference and working memory components such as the double Y-maze or RAM, stable asymptotic performance of the task can be used to assess effects of brain damage and drugs over an extended period, in the same animals. Impaired animals can be used to assess bidirectional drug effects, with different doses tested in the same animal against its own baseline performance, to measure dose-related effects sensitively in repeated measures designs. For example, following damage to cholinergic projections with NBM-MSA lesions or chronic alcohol treatment, we have shown reductions in RAM error rates after treatment with cholinergic agonists, cholinesterase inhibitors and BZ receptor inverse agonists (Hodges et al., 1989, 1991c,d; 1992; Turner et al., 1992). Procedural tasks, therefore, provide the opportunity to carry out a series of pharmacological probes, in order to investigate the specificity of lesions and/or deficits to particular neuronal systems, and to assess possible therapeutic agents (see Fig. 6.7). This an important advantage when neuronal loss may be widespread with nonspecific neurotoxic, ischaemic or alcohol-induced brain damage. This type of extended repeated-measures design would not typically be feasible with rapid acquisition tasks such as the WM (see below). The utility of procedural tasks does, however depend on stable performance of controls, in which choices continue to be made on the basis of current information processing, rather than solely on well-rehearsed routines. Changes in cues (e.g., by use of dim light: Hodges et al.,

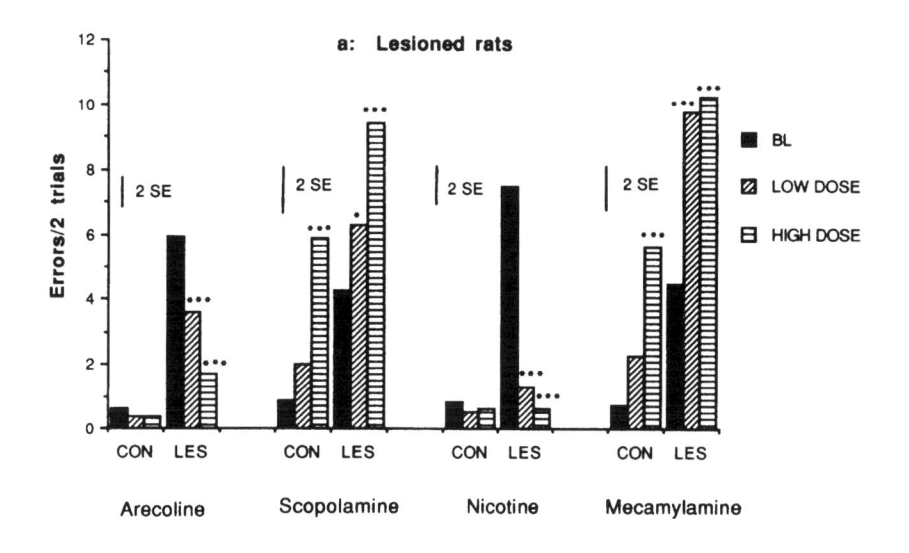

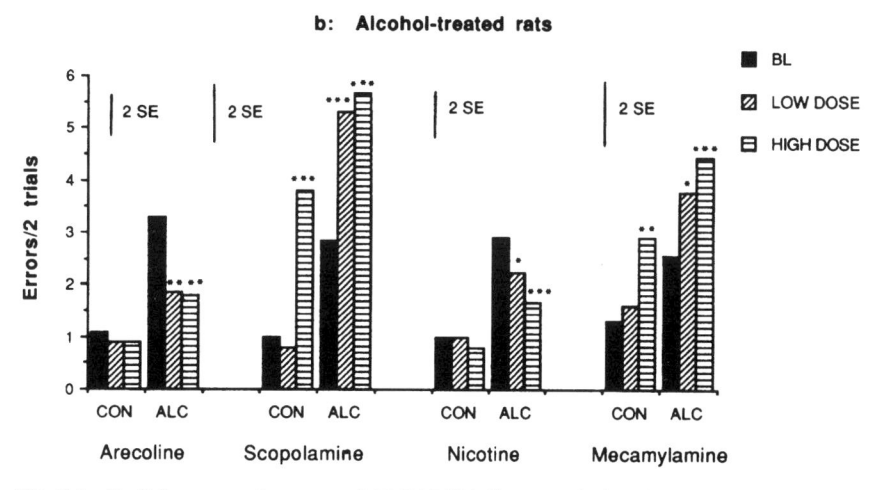

FIG. 6.7 Radial maze performance of NBM-MSA ibotenate lesioned (LES) and alcohol-treated (ALC) animals on radial maze place and cue tasks in base-line (BL) conditions and after treatment with the cholinergic receptor agonists nicotine and arecoline and antagonists scopolamine and mecamylamine (low and high doses were, respectively: 0.05mg/kg and 0.1mg/kg for nicotine and scopolamine, 0.5mg/kg and 1.0mg/kg for arecoline and 1.0mg/kg and 2.0mg/kg for mecamylamine). The data indicate that lesioned and alcohol-treated animals made more errors than controls, and were more sensitive to effects of cholinergic treatments, showing improvements with the agonists and impairments with the antagonists at low doses that did not affect the performance of controls. The stability of BL scores with these radial maze tasks enabled the comparison of several compounds and assessment dose response relationships in the same animals over an extended (about 3-month) period. Difference between baseline and treatment scores: *$P < 0.05$; **$P < 0.025$; ***$P < 0.01$. (Reprinted from Hodges, et al., 1991d, with kind permission of Birkhäuser Verlag AG., Ch-4010 Basel, Switzerland.)

1991b; see Fig. 6.3) can be used to check up on the degree of stimulus control. With standardised training regimes, the RAM or double Y-maze appear to be well-suited to measuring enduring deficits in both retrieval of information from long-term memory (reference memory errors) and particularly in working memory (see below).

Rapid acquisition tasks appear to provide a more sensitive method for assessment of drug and lesion effects on place learning than procedural tasks, but they also have their drawbacks. First, it is necessary to show that groups have a similar level of performance in the initial trials. For example, in the WM when the platform position is not known, group differences should be minimal, but then increase as learning progresses at different rates (see Fig. 6.2a). If groups exhibit differences right from the start of training, and performance improves in parallel, the findings point to factors, such as motor activity, motivation or prior experience, rather than to differences in rates of spatial learning. Figure 6.2c illustrates the occurrence of group differences right from the first day of training in maze-experienced rats, because control animals searched more efficiently than ischaemic groups. Second, deficits may be difficult to demonstrate statistically under standard multi-trial procedures in rapidly learned tasks such as the WM, because of the variability of scores. Third, once learning is asymptotic, spatial memory is highly robust so that it is difficult to retest animals in the same apparatus. Experienced rats, including those initially impaired in WM acquisition, will show good recall of pool locations when tested weeks to months after acquisition, so that differences from controls disappear (Netto et al., 1993; see Fig. 6.8). Several procedures have been developed which partially address this problem. First, as Mandel et al. (1989) have demonstrated (see above), the number of training trials can be reduced and/or the length of the ITI increased to double the period of acquisition required for asymptotic learning and so increase the opportunity to detect impairments or improvements. Alternatively, repeated short periods of acquisition may be used (Hodges et al., 1995), to decrease the opportunity for the rats to form a long-term representation of the pool during training. Variable interval probe trials (Markowska et al., 1993) may also be used to provide a sensitive index of the degree of learning at various points during training. Animals may also be moved to a new pool, or curtains placed around the pool to alter the cues available. Netto et al. (1993), for example, showed that ischaemic animals were not impaired in acquisition after 6 phases of training in a WM. However, when placed in a novel pool, deficits in ischaemic rats re-emerged, suggesting that the impairment involved spatial learning, but not recall of spatial information (see Fig. 6.8). Lesion and drug effects on acquisition may therefore be measured in the WM, provided that the testing methods are appropriate. However, independent groups must normally be used, which diminishes the power of the design, particularly as performance is quite variable from rat to rat.

In sum, the differing rates of learning in procedural and rapid acquisition tasks, in addition to affecting the type of experimental design, and the type of question that can most appropriately be investigated, may also have some bearing on the generality of the memory processes tapped by these two broad categories of spatial task. Rapid acquisition tasks such as the WM appear to make heavy, if not exclusive demands on visuospatial information processing. In contrast, in the procedural tasks, conditional discriminations and potential availability of olfactory and intra-maze cues in dry mazes all combine to influence place learning. Thus, procedural maze tasks may engage a broader spectrum of sensory processes that contribute to the formation of cognitive maps, and also engage nonspatial processes in learning and memory, to a greater extent than standard place learning in the WM. When animals show impairments in complex mazes such as the RAM, they are likely to show deficits in a variety of other tasks. In contrast, deficits in the WM do not necessarily predict impairment in nonvisuospatial tasks. For example, rats with cholinergic lesions are impaired in a range of tasks, including the RAM, WM, conditional discriminations in the Skinner box, and passive avoidance, although the profile of deficits has been shown to vary with type of lesioning agent (Dunnett et al., 1991). This broad profile contrasts with the selective WM deficits shown by 4 VO ischaemic rats (see Sinden et al., 1995, and Nunn & Hodges, 1994, for discussions). Despite the greater generality of procedural mazes and their potential for assessing 'steady state' deficits in repeated measures designs, the large number of factors likely to contribute to performance make it difficult to detect selective cognitive effects of brain damage with any precision.

Working memory tasks

Working memory involves the retention of trial specific or trial unique information for short periods (generally up to 30 seconds). Several spatial procedures have been developed to measure working memory, including within-trial re-entries in the RAM (Jarrard, 1986; Olton & Pappas, 1979; Olton et al., 1980); alternation in the Y- or T-maze (Mallet & Beninger, 1993; Aggleton et al., 1986), pushing at open/closed doors in the three-door runway (Furuya et al., 1988), learning set or matching to position tasks in WM (Morris et al., 1986; Netto et al., 1993; Hodges et al., 1995, 1996; Whishaw, 1985), delayed matching/nonmatching to position or objects in maze locations (Aggleton et al., 1991; Rafaelle & Olton, 1988; Mumby et al., 1992; Wood et al., 1993b; Kelsey & Vargas, 1993). The duration of the retention interval is variable within and between

a: **Familiar Pool**

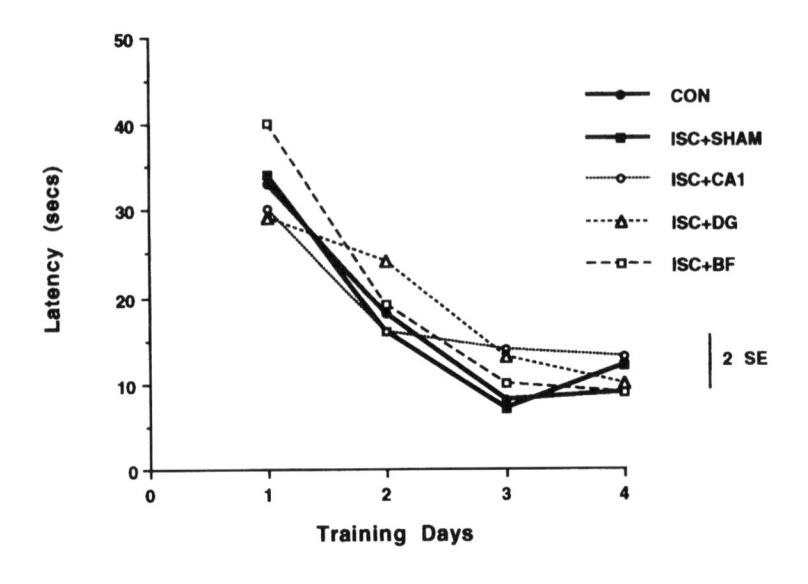

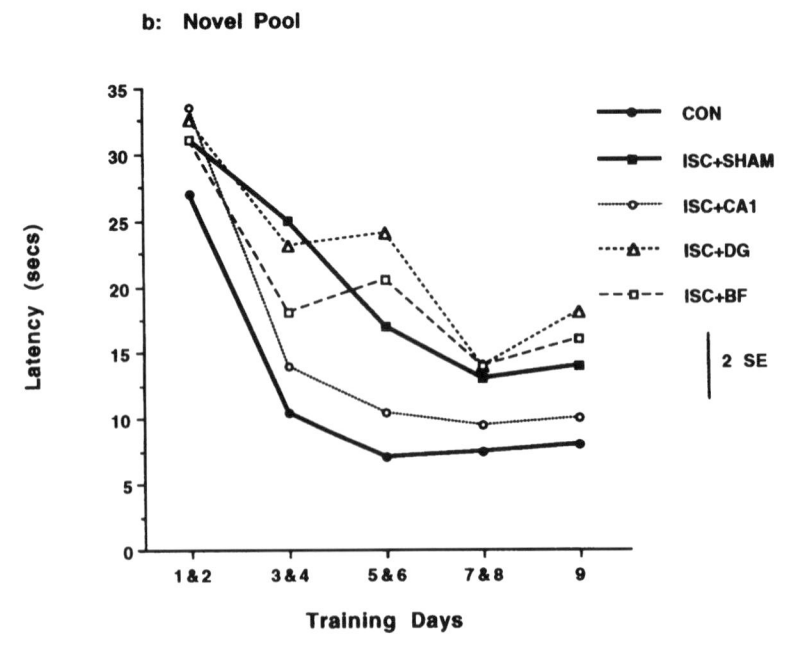

b: **Novel Pool**

FIG. 6.8 Effect of experience on water maze performance of control and ischaemic rats (15 minutes of 4 VO) and ischaemic groups with fetal grafts from the CA1 (ISC+CA1) and dentate gyrus (ISC+DG) fields of the hippocampus, and basal forebrain (ISC+BF). By the sixth

procedures and cannot be as tightly controlled in mazes as in lever press tasks in the Skinner box.

As with reference memory tasks, working memory tasks may include substantial procedural or conditional components that require extensive training, or may be rapidly learned. Delayed matching/nonmatching to position (DMTP/DNMTP) in the Skinner box and matching to position in the three door runway require pre-training, but unlike acquisition tasks in which procedural and spatial learning processes are mastered concurrently, once the procedure is learned, they provide relatively uncontaminated and stable assessments of spatial working memory. Tasks such as the three-door runway, in which the rat has to choose one (out of three) opening doors set in four or more barriers across the runway to reach food at the end, using intra-maze cues that are changed for each set of trials (see Fig. 6.1h) provide a sensitive test of spatial working memory, once the prodedure has been learned, which may take up to six weeks. Controls find this a difficult task, so the test is suitable for detecting enhancing drug or environmental effects in normal animals, that often go undetected in tasks such as DMTP in the Skinner box, where error rates are very low. The sensitivity of 'procedural' working memory tasks is increased by low inter- and intra-group variability, so that even small but systematic differences between groups reach statistical significance. Thus procedural working memory tasks tend not to suffer from the large fluctuations of scores that bedevil analysis of water maze working memory tasks, where the platform can be found rapidly on one trial, and not located at all on the next. Moreover, in some rapidly learned working memory tasks, day-to-day performance improves progressively, so that baselines are inherently unstable. For example, a popular water maze working memory task uses different platform positions on each day. On Trial 1 the platform is found 'by chance' and working memory is primarily assessed by matching to position on Trial 2. However, the same visuospatial cues are available over days, and the rats learn a lose-shift strategy and do not spend time in previously safe locations, so that over successive days the platform is found more and more rapidly. We examined spatial working memory of ischaemic rats in both a rapidly learned matching to position water maze task, and in the 'procedural' three-door runway, where rats were given 6 trials/day, with

---

period of training (a: familiar pool) 18 weeks after transplantation, all groups learned a new platform position equally rapidly. However, on transfer to a novel pool 20 weeks after grafting (b), ischaemic groups showed significant impairment in place learning relative to controls, whereas ischaemic rats with CA1 grafts learned as rapidly as controls, and were superior to the ischaemic control group. The other transplants were ineffective. These data suggest that global ischaemia impairs spatial learning, but not the recall of spatial information. (Reprinted from Netto et al., 1993, with kind permission from Elsevier Science Ltd., The Boulevard, Langford Lane, Kidlington, OX5 1GB, UK.)

a different pattern of opening doors on each day. On the first trial the opening door at each choice point was white, but on trials 2–6 all the doors were black, and the rat had to match to position in the absence of an intramaze visuospatial brightness cue. In the water maze task ischaemic rats showed marked impairment that was totally reversed by fetal grafts from the CA1, but not the CA3 hippocampal field (see Fig. 6.9a). In the three-door runway task all rats performed comparably on Trial 1, indicating that ischaemic rats were not impaired in making a simple visual discrimination. However, ischaemic rats showed a substantial increase in errors on Trial 2, indicating poorer retention of the door positions in working memory than controls (see Fig. 6.9b). Rats with CA1 grafts showed substantial improvement relative both to nongrafted ischaemic controls and rats with CA3 grafts, but performed significantly worse than sham operated controls. Thus, rapid 'all or none' learning in the water maze failed to detect a residual impairment in the CA1 grafted animals, that was seen in the runway task, although it picked up the substantial deficit in the CA3 and ischaemic control groups (Hodges et al., 1996). Procedural and rapid acquisition working memory tasks, therefore, differ in the extent to which they are sensitive to impairments in spatial information processing, and in the extent to which they can be used for steady state and repeated assessments of effects of brain damage or treatments.

Interestingly although runway error rates were higher in ischaemic groups than controls. Figure 6.9c shows that all the groups showed a similar pattern of errors at the four choice points (barriers), with fewer error at the first and last barrier, than at the two intermediate ones. Since errors during the first discrimination trial did not show this pattern, attentional or

---

FIG 6.9 Effects of ischaemia (15 minutes of 4 VO) and transplants on spatial working memory assessed in the water maze (a) and three-door runway (b, c). Ischaemic controls (ISC) and ischaemic groups with fetal grafts from the CA3 (ISC+CA3) and dentate gyrus (ISC+ DG) hippocampal fields showed marked impairment in matching to position in the water maze, and pressing at opening doors in the runway on trials 2–6 when all doors were black, relative to controls. On Trial 1 in the runway (see Fig. 6.1h) all groups made a similar number of errors in learning to discriminate black (closed) from white (opening) doors. Ischemic rats with CA1 grafts (ISC+CA1) were significantly better than ischaemic controls on both water maze and runway tasks. However, working memory in the water maze improved to control (CON) level, whereas in the runway the ISC+CA1 group still showed substantial impairment relative to controls. Despite the higher error rates in the ischaemic groups relative to controls, the pattern of errors at the four barriers in the runway (c) was almost identical in all groups, with lower error rates at the first. and particularly the fourth barrier than at the intermediate ones. If this pattern reflects a serial position effect, these data indicate that storage of information did not differ in ischaemic and control groups, although ischaemic rats were clearly impaired in the use of spatial (door position) information. (Reproduced from Hodges et al., 1996, with kind permission from Elsevier Science Ltd., The Boulevard, Langford Lane, Kidlington, OX5 1GB, UK.)

a: Working memory in the water maze

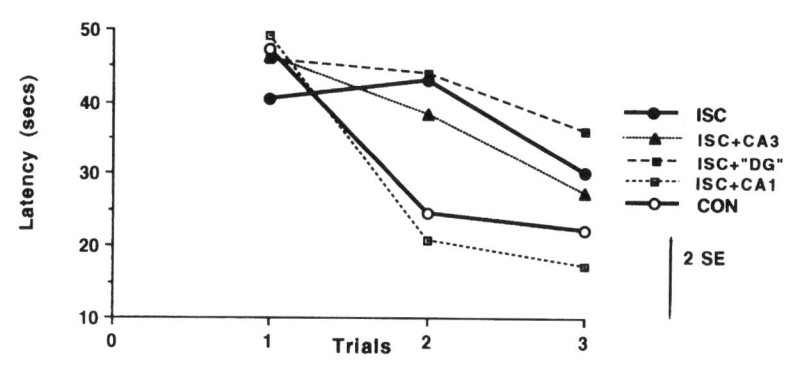

b: Working memory in the three-door runway (errors over 6 trials)

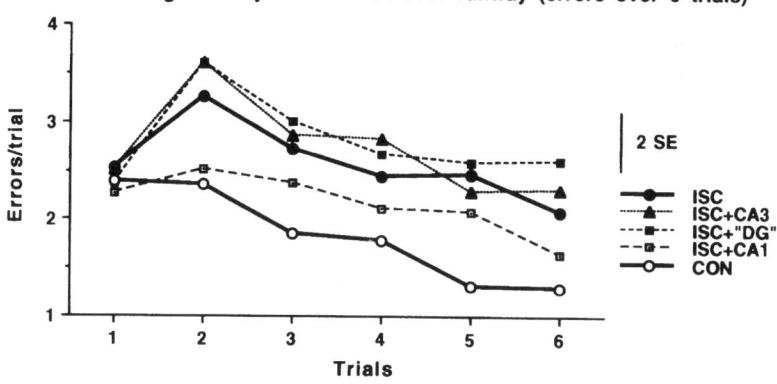

c: Error distribution at barriers in the three-door runway

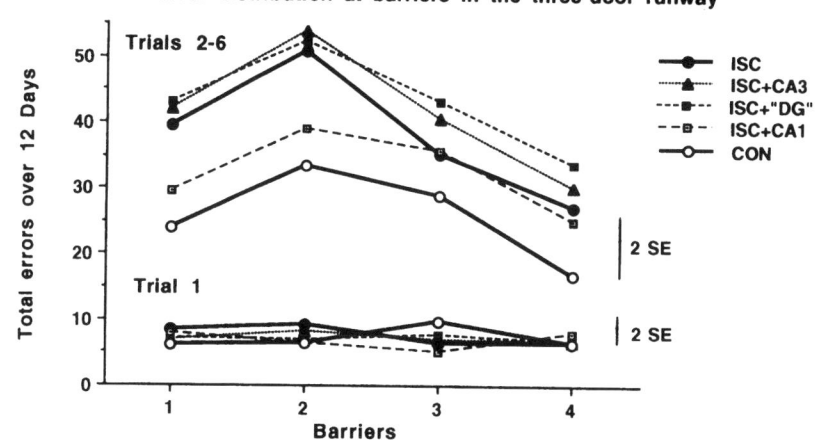

motivational factors (e.g., release from the start box, food behind the last barrier) are not likely to account for the curvilinear distribution of errors in the working memory phase (Trials 2–6). The pattern may therefore reflect a 'serial position' effect, whereby information from the first and last items of a list is retained better than from intermediate positions. Serial position curves are robust in human memory research (Glanzer & Cunitz, 1966) and thought to reflect retrieval from long-term (the primacy effect) and short-term memory (the recency effect). However, the extent to which they occur in animal studies is controversial. Gaffan (1983) and Gaffan and Gaffan (1992) suggest that they arise through statistical and procedural artefacts, for example, use of an initiating response or start box may alert the animal to the first item(s) in the list. Nevertheless, serial position effects have been reported occur in several studies, have generally been found to be more reliable with spatial than nonspatial tasks, to occur in the absence of an attention-enhancing initiating response (Castro & Larsen, 1992), and to show a stronger recency than primacy effect, as in our three-door data (DiMattia & Kesner, 1984). The recency effect has also been shown to be more sensitive to distraction and delay than the primacy effect (Harper et al., 1993), as in human studies. Use of order of spatial recognition, as opposed to familiar versus novel places, to study spatial memory impairment in brain damaged animals is relatively rare. However, Beracochea et al. (1989) showed that mice with mamillary body lesions (a region that is susceptible to alcohol-induced brain damage in humans) were not impaired if recent arms in the radial maze 'list' were presented, but showed chance levels of recognition if the first arm was presented, suggesting a failure to store spatial information in excess of short-term memory capacity. This finding contrasts with our three-door runway results, which (if they do reflect a serial position phenomenon) suggested that ischaemic rats and controls showed a similar profile of spatial information storage, even though the ischaemic groups were far worse at assimilating or using spatial information and made more mistakes.

WORKING MEMORY AND THE HIPPOCAMPUS

Because impairment in spatial working memory tasks is often seen after damage to the hippocampus in the rat, these tasks have played a dominant role in the development of theories of hippocampal function. A basic controversy has centred around whether the short-term storage (Olton et al., 1979, 1980; Rawlins, 1985) or spatial components of working memory tasks are sensitive to hippocampal damage. More recently Eichenbaum et al. (1994) have argued that these functions are dissociable: the hippocampus processes relational (including spatial) information, whilst the parahippocampal region stores information in working memory. The issue is difficult to resolve since many different methods (e.g., fornix-fimbria, electrolytic, aspiration or excitotoxic lesions) have been used to damage the hippocampus, and different tasks with both spatial and non spatial components (e.g., objects or different goal boxes in a Y-maze) have been used to assess working memory, but it is important since it raises the question of how far impairment in spatial tasks can be assumed to reflect a general working memory deficit. In the rat, evidence for hippocampal involvement in nonspatial working memory tasks is mixed; Raffaele and Olton (1988), Mumby et al., (1992) and Wood et al. (1993a,b) report impairments in delayed object recognition after hippocampal damage, whereas Aggleton et al. (1986), Rothblat and Kromer (1991) and Kelsey and Vargas (1993) found that rats with damage to the hippocampus or its inputs discriminated successfully between objects, but performed poorly in T-maze alternation, or spatial variants of the discrimination tasks. Factors, such as repeated within-session use of the same stimuli, use of simple, as opposed to complex stimuli and of discrete rather than continuous trials, have recently suggested that interference may play a key role in the emergence of deficits in nonspatial working memory tasks in rats with hippocampal aspiration or fornix lesions, consistent with the proposal of Eichenbaum et al. (1994) that it is the the relational rather than the storage requirements of tasks that is sensitive to detrimental effects of hippocampal damage. However, the ample demonstrations cited above, that rodents with hippocampal damage can show working memory deficits in spatial but not in nonspatial or less spatially demanding tasks (Nunn & Hodges, 1994; Nelson et al., 1997b; Hodges, 1996) indicate that findings from spatial tests of working memory in rodents do not necessarily apply to nonspatial tasks. Thus, in rodent models of brain damage, such as global ischaemia or medial septal lesions, in which hippocampal damage is anticipated, it is particularly important to assess animals in both spatial and nonspatial working memory procedure, before concluding that they show a generalised working memory impairment. The often selective deficit in spatial tasks shown by rodents with hippocampal damage contrasts with the global anterograde amnesia seen in humans that has been linked to hippocampal damage by post-mortem and neuroimaging studies (Squire et al., 1990). In nonhuman primates deficits induced by hippocampal damage are also not predominantly seen in spatial tasks and recent findings suggest that working memory impairment is related to parahippocampal rather than hippocampal damage (Suzuki et al., 1993). Evidence from Ridley's laboratory (Ridley et al., 1995) suggests that marmosets with intra-hippocampal CA1 lesions show marked deficits in learning conditional tasks, whether spatial or nonspatial, which involve learning a relationship between a rule and its exemplars, and where more than one response is associated with reward. Impairment is particularly marked in visuospatial conditional discriminations, where one pair of objects requires selection of the left food well, and another pair requires selection of the right food well. Hence two different responses are equally rewarded. Where deficits in spatial tasks are seen following hippocampal damage in humans and non-human primates, these have been suggested to involve broader deficits in learning relationships between objects (Cave & Squire, 1991; Gaffan & Harrison, 1989), and the memory for these relationships in primates has been likened to human episodic memory (Gaffan, 1994). It may be, therefore, that the sensitivity of hippocampal damage in rats to spatial tasks is also an example of a broader relational deficit, as Eichenbaum et al. (1994) suggest, which is more easy to detect in spatial than in nonspatial tasks, because of rodent spatial abilities.

ASSESSMENT OF RECOVERY OF FUNCTION IN SPATIAL TASKS

Two basic, but not mutually exclusive, processes have been put forward to account for improvements in task performance after brain damage: recovery of the damaged systems through sprouting, trophic mechanisms or enhanced efficiency of residual neurones, or compensation achieved either by adaptation of undamaged regions to take on additional functions, or by the use of different strategies to solve problems by methods accessible to intact systems. Use of functional neuroimaging, such as positron emission tomography (PET) scans, or more recently nuclear magnetic resonance imaging (NMRI) has highlighted the astonishing capacity of the brain for functional reorganisation. Weiller et al. (1993) found large ventral extensions of the hand field into the sensorimotor area (SMA) normally allocated to the face in patients with recovered finger-thumb apposition following stroke damage to the posterior limb of the internal capsule. Chollet and Weiller (1994) suggest that motor circuits (primary motor cortex, premotor cortex and the SMA) may operate in parallel, rather than hierarchically, and are thus able to substitute or compensate for each other. However, in systems which utilise fast highly specific transmission of information, as in the visual cortex, damage appears to be irreversible and resistant to neuronal remodelling, so that compensatory changes are more likely to involve the use of alternative sensory systems, such as touch or sound to substitute for loss of vision, exemplifying the 'general adaptive capacity' of the brain (Finger, 1978).

Many different approaches to rehabilitation are employed in clinical practice, but methods aiming to promote functional recovery of cognitive abilities in animals have focused on three main strategies: (1) use of environmental enrichment to enhance sensory information processing, (2) cerebroprotective drugs given shortly after trauma to reduce the extent of brain damage, and (3) intracerebral transplants of fetal or cultured and genetically engineered cells aimed to promote recovery by a variety of mechanisms ranging from the supply of neurotransmitters, enzymes or trophic factors to the repair of neural circuits in a point-to-point manner (see Sinden et al., 1995, for a review). Broadly speaking, enriched conditions (EC) aim to support compensatory mechanisms, cerebroprotective methods aim to prevent brain damage, whereas graft methods aim for brain repair. However, these distinctions are not absolute, since all three approaches are accompanied by behavioural and morphological changes, indicating both functional and structural correlates. Moreover, interventions are carried out in the context of the substantial reorganisation occurring in the brain following injury, so that is important to know whether the compensatory or recovery processes instigated by treatments complement or conflict with spontaneous processes of recovery and remodelling. Thus, attempts to isolate factors that might contribute to improvement in task performance in studies of functional recovery is just as important as in investigations of factors that might contribute to impairments following brain damage. Moreover, evidence for improvement in some tasks, but not in others, may point to mechanisms or processes that are amenable or resistant to effects of EC, cerebroprotective drugs or grafts, and provide further information about the flexibility of spatial information processing. Recovery of function will be discussed with reference to EC and transplants, since there has been relatively little investigation of the functional effects of cerebroprotective drugs (see Nunn & Hodges, 1994, and Hodges et al., 1994, for reviews).

ENVIRONMENTAL ENRICHMENT

Effects of enriched conditions in both normal and brain-damaged animals are not always apparent or easy to interpret, so that the extent of behavioural improvement may depend on the type of task, the type of brain damage, and the age of the animal. Whishaw et al. (1984), for example, found no evidence for an effect of EC following neonatal hemidecortication, which had less severe effects on water maze learning than adult hemidecortication. However EC substantially improved water maze learning in an adult lesioned group (Whishaw et al., 1984; Rose et al., 1989). Hence, early plasticity may improve performance to the extent that subsequent environmental manipulations exert no further enhancement. Rose et al. (1987, 1988, 1993) found no evidence for EC effects in the improvement that occurred in sensorimotor neglect, assessed by the ability to remove a sticky tape bracelet, or the ability to make go-no go conditional discrimination following unilateral cortical lesions. There was, however, evidence for an EC-induced acceleration in place learning in the water maze after impairments induced by bilateral occipital lesions, in line with Whishaw's findings in adult hemidecorticate rats. However, Rose et al. (1993) also found that EC failed to reduce the WM deficit in animals trained pre-operatively, although they assisted acquisition in rats with bilateral occipital lesions (Rose et al., 1990), suggesting that EC effects on spatial information processing were on compensation rather than on functional recovery, because rats that learned with the assistance of extra-maze patterned cues were not benefited by EC, when the use of these cues was impaired by the lesion. Rats that learned under conditions of visual impairment did show a beneficial effect of EC, arguably because they were more ready to switch to another strategy, or sensory modality. However, the nature of this alternative strategy is not known. For example, lesioned EC rats did not show an advantage when tested in the dark, but provided with spatially separated and distinctive auditory cues (Rose et al., 1993). There is some evidence to suggest that effects of EC may be highly specific, rather than fostering a general tendency to switch response strategies, as Finger (1978) suggested. Pacteau et al. (1989) found that in rats lesioned as weanlings in the dorsal hippocampus, and then given a month's experience in EC, learning and performance varied according to task. EC rats performed worse than lesion controls (which showed substantial impairment) in the Olton radial maze working memory task, with all arms rewarded, but were significantly better in learning which arm in a plus-maze contained food, when supplied with plentiful intra-maze and extra-maze cues. Similarly spontaneous alternation was normalised in a 'maximum cue' T-maze, but not in a 'minimum cue' version. Thus although effects of early enrichment persisted into adulthood, the opportunity to play with and manipulate objects appears to have facilitated only cue-place associative learning, and may even have further interfered with lesion-induced deficits in the ability to use allocentric extra-maze cues. Despite difficulties in teasing apart the effects of EC, this approach may provide methods for examining interactions between experience and brain damage in the formation of alternative strategies in spatial learning.

INTRA-CEREBRAL TRANSPLANTATION

Effects of transplants on recovery of function have been robust, in terms of behavioural change, but difficult to understand in terms of mechanisms. Mechanisms of graft action appear to differ profoundly according to the source of donor tissue and the model of host brain damage. Cholinergic-rich fetal basal forebrain grafts, for example, are typically placed at the terminal regions of the widely projecting forebrain cholinergic projection system (FCPS), on the assumption that they will release ACh *in situ*, to replace the host's deficient levels of transmitter (Gray et al., 1990). Early studies suggested that release of ACh was, indeed, an important mechanism, since noncholinergic grafts were ineffective, *in vivo* dialysis studies showed that grafts released ACh under partial control of the host, and immunohistochemical studies showed restoration of hippocampal and cortical choline acetyltransferase (ChAT) levels, although the ectopic placement of cholinergic cell bodies from the fetal basal forebrain precluded close integration of grafts into host circuits, and produced some distortion of host cytoarchitecture (see Sinden et al., 1995; Dunnett, 1990, for reviews). However, recent studies comparing effects of primary cultures from fetal basal forebrain, in which neuronal and glial components have been separated, have shown that neuronal grafts (which increase ChAT levels) are ineffective in promoting functional recovery, whereas glial grafts are as successful as the parent fetal basal forebrain suspensions, calling in question the ACh replacement hypothesis (Bradbury et al., 1995). There are striking points of contrast between the actions of grafts in animals with lesions to the FCPS, and those of intrahippocampal grafts in animals with loss of CA1 cells induced by 4 VO (Sinden et al., 1995; Hodges et al., 1994). In animals with cholinergic system damage grafts are placed in the terminal regions, and not within the lesion sites, where they have been shown to be ineffective (Hodges et al., 1991b), whereas in ischaemic rats grafts are placed close to the region of CA1 cell loss. Effective grafts in animals with cholinergic system damage, as noted, have been obtained from a variety of sources, including fetal basal forebrain and ventral mesencephalon, and primary glial cultures. In rats with ischaemic damage, only grafts containing pyramidal CA1 cells appear to be functionally effective. Cells from the CA3 region have not been shown to improve spatial learning, even though they are pharmacologically and structurally similar to CA1 cells (Netto et al., 1993; Hodges et al., 1996). Thus, graft requirements appear to be more stringent for repair of damaged hippocampal circuits as opposed to restoration of cholinergic tone. Graft methods are increasing at an exponential rate, including conditionally immortalised clonal cell lines, polymer encapsulated cells or other types of transfected carrier cells aimed to deliver transmitters, synthesising enzymes or growth factors, so the potential number of mechanism involved is vast. For this reason it is vital to employ a variety of tests of different aspects of spatial and nonspatial function, and to standardise cognitive assessments across laboratories, in order to obtain a sensitive comparative index of the effects of different grafts, and the cognitive functions that are and are not influenced by graft action.

Grafts differ in the extent and nature of their functional effects. Beneficial effects of cholinergic-rich grafts have been seen following a wide variety of lesion-induced deficits, including both working and reference memory components of Jarrard's tasks in the radial maze, serial reaction time, and passive avoidance (Sinden et al., 1992, 1995). Cognitive effects appear to involve information processing, rather than response strategy, since grafted animals, like controls, were found to be impaired in the place, but not the cue task when tested in dim light, suggesting that they were under stimulus control (Hodges et al., 1991b: see Fig. 6.3). However, in addition to the normalisation of lesion-induced deficits, several studies have reported either no effect, or increased impairments following cholinergic-rich grafts. Sinden et al. (1990), for example found that although grafts reduced passive avoidance deficits, they failed to improve conditional discrimination performance in the same basal forebrain lesioned animals. Within the narrower range of largely spatial tasks in which deficits in rats with CA1 hippocampal lesions induced by 4 VO have been detected (Nunn & Hodges, 1994), grafts of CA1 fetal cells have also had mixed effects. Acquisition in the water maze has shown substantial recovery to control levels in terms of latency to find the platform (see Fig. 6.8b), and search in the training quadrant, but rats with CA1 grafts still exhibited deficits in accuracy of recall of the platform position on probe trials, and impairment in working memory in the three-door runway (see Fig.6.9b,c) relative to controls (Netto et al., 1993; Hodges et al., 1996). Similar mixed effects of grafts have been reported in models of motor impairment following nigrostriatal lesions, with amelioration of rotational deficits, but not skilled paw reaching following dopaminergic grafts in rats (Bjorklund et al., 1994b) and monkeys (Annett, 1994). Models of gross brain damage, such as cortical and striatal infarction following MCAO, provide a challenging test of graft efficacy, but have received scant investigation to date. Recent findings suggest that grafts of striatal tissue placed in the relevant regions of infarction have improved water maze learning following MCAO (Nishino et al. 1993; Aihara et al., 1994). However, cortical grafts have not reduced deficits in paw reaching tasks (Grabowski et al., 1995), so that it is important to see whether egocentric spatial learning remains more impaired than allocentric spatial learning in stroke models.

COMBINED EFFECTS OF ENRICHMENT AND TRANSPLANTATION

Combined effects of transplants and enriched conditions (EC) offer an interesting and relatively unexplored approach to recovery from brain damage, which may harness processes of both recovery and compensation. Mohammed et al. (1990) found some evidence that EC increased the release of nerve growth factor which may contribute to increased arborisation and dendritic spine density seen in hippocampus and cortex after EC (Renner & Rosenzweig, 1987). Thus, EC may provide a more hospitable host environment for graft growth, in line with findings of improved graft survival following exogenous administration of trophic factors (Mayer et al., 1993). Consistent with this proposal Dunnett et al. (1986) found that survival and growth of cholinergic-rich grafts was more marked at 4 weeks after grafting in rats provided with EC, than controls in standard housing, although this early advantage was not seen 10 weeks after transplantation. Behavioural assessment of combined effects of grafts and EC may therefore be expected to show additive effects. In the few studies to date this has not been unequivocally demonstrated. Kelche et al. (1988) found that performance in a Hebb–Williams maze was significantly improved in EC rats grafted with solid septal transplants in the lesion cavity following fimbria-fornix aspiration lesions, relative to rats receiving EC or grafts alone, which remained as impaired as the lesion control group. These findings suggested that both grafts and transplants were required for improved spatial learning, indicating an additive effect. However, there was no evidence for an increase in AChE staining, so the effect could not be attributable to gross enhancement of graft-derived reinnervation of the host, although a more fine-grained study of synaptic connections may be required to look for enriched connectivity following EC. In the case of combined EC and cortical grafts following infarction induced by MCAO, Grabowski et al. (1995) found that EC substantially improved balance on a rotating rod and inclined plane, but rats with both EC and grafts were not significantly superior to those receiving EC alone, so there was no apparent additive effect. Deficits in skilled paw reaching were not reduced by either EC or grafts, and the possibility that this persistent motor asymmetry might disrupt performance on some spatial tasks, as shown in other studies of effects of MCAO, was not investigated. The results to date suggest that combined EC and graft treatments may be more advantageous that either treatment alone, depending on the task, but some behavioural effects of brain damage are resistant to both methods.

Studies of functional recovery following environmental enrichment and transplantation potentially have much to contribute to the study of spatial learning and memory. They can show the types of spatial behaviour that are influenced by experience and interactions between experience and early or late brain damage. Transplant studies provide a useful complement to lesion approaches, in showing the neuronal systems involved in spatial learning, and the extent to which these can be repaired by precise cell replacement, trophic mechanisms or cells that release transmitters or enzymes. However, as with lesion-only or lesion-plus EC studies, it is important to employ a variety of tests that make different demands on spatial information processing, including tasks that permit allocentric and egocentric spatial learning, associative learning, working memory in spatial and nonspatial tasks, and sensorimotor function. Only a wide-ranging profile of effects will permit inroads to be made into processes of recovery and compensation.

CONCLUSIONS

Spatial procedures offer a variety of ways of assessing learning and memory in animals, of looking at effects of drugs and of different types of brain damage. However, it cannot be assumed that findings in one type of spatial task will generalise to another, because these tasks tap a variety of processes, other than the ability to use visuospatial cues to form cognitive maps of the environment, including associative and procedural learning. To date, the formation of cognitive maps has chiefly been manipulated by use of allocentric visuospatial cues. However, rats may also use other sensory modalities (tactile, olfactory) to construct maps, together with associative and egocentric information. Tasks differ in the extent to which they permit the use of these diverse sources of spatial information, and in the extent to which they also require processing of nonspatial information. Constrained pathways and open arenas, exemplified by radial arm and water mazes, illustrate spatial tasks that differ in important ways, notably free, as opposed to restricted search, requirement for precise localisation, and motivation. The water maze is adapted to rapid allocentric spatial learning, whereas the radial maze measures relatively stable asymptotic reference memory performance, and/or working memory. Two main considerations must be borne in mind when assessing spatial dysfunction in animals models: first, that apparently similar maze tasks may be looking at differing processes, and second, that memory processes measured in spatial tasks may not generalise to performance in tasks that make fewer demands on spatial information processing. It is important, therefore, to use different spatial tasks, and to assess animals in nonspatial learning and memory, to look for consistency of effects. However, if discrepancies are found, as in the impairment of ischaemic rats in the water maze, but not the radial maze reported by Nunn et al. (1991), these can be used to home-in on selective deficits associated with different types of brain damage. Such dissociations in the performance of maze tasks therefore provide an important starting point for investigating structure-function relationships in impaired and intact animals.

ACKNOWLEDGEMENTS

Our work cited in this chapter was supported by the Wellcome Trust and the British Heart Foundation. I would like to thank J.A. Nunn, A. Nelson, A. Lebessi and T-A Perry for permission to cite their as yet unpublished findings, and D. Virley for assistance with preparing the manuscript.

REFERENCES

Abdulla, F.A., Calaminici, M-R., Stephenson, J.D. & Sinden, J.D. (1994). Unilateral AMPA lesions of nucleus basalis magnocellularis induce a sensorimotor deficit which is differentially altered by arecoline and nicotine., *Behavioural Brain Research*, 60, 161–169.

Aggleton, J.P., Hunt, P.R. & Rawlins, J.N.P. (1986). The effect of hippocampal lesions upon spatial and nonspatial tests of working memory. *Behavioural Brain Research*, 19, 133–146.

Aggleton, J.P., Hunt, P.R. & Shaw, C. (1990). The effects of mamillary body and combined amygdalar-fornix lesions on tests of delayed non-matching-to-sample in the rat. *Behavioural Brain Research*, 40, 145–157.

Aggleton, J.P., Keith, A.B. & Sahgal, A. (1991). Fimbria fornix and anterior thalamic, but not mamillary, lesions disrupt delayed non-matching-to-position memory in rats. *Behavioural Brain Research*, 44, 151–161.

Aihara, N., Mizukawa, K., Koide, H.M. & Nishino, H. (1994). Striatal grafts in infarct striatopallidum increase GABA release, reorganise $GABA_A$ receptor and improve water-maze learning in the rat. *Brain Research Bulletin*, 33, 483–488.

Annett, L.E. (1994). Functional studies of neural grafts in Parkinsonian primates. In S.B. Dunnett & A. Bjorklund (Eds.), *Functional neural transplantation* (pp.71–102). New York: Raven.

Arendt, T., Allen, Y., Marchbanks, R., Schugens, M.M., Sinden, J., Lantos, P.L. & Gray, J.A. (1989). Cholinergic system and memory in the rat: effects of chronic ethanol, embryonic basal forebrain transplants and excitotoxic lesions of the forebrain cholinergic projection system. *Neuroscience*, 33, 435–462.

Bartus, R.T., Dean, R.L., Beer, B. & Lippa A.S. (1982). The cholinergic hypothesis of geriatric memory dysfunction. *Science*, 217, 408–417.

Bättig, K. (1983). Spontaneous tunnel maze locomotion in rats. In G. Zbinden, V. Cuomo, G. Racagni, & B. Weiss (Eds.), *Application of behavioral pharmacology in toxicology* (pp.15–26). New York: Raven.

Beracochea, D.J., Alaoui-Bouarraqui, F. & Jaffard R. (1989). Impairment of memory in a delayed non-matching to place task following mamillary body lesions in mice. *Behavioural Brain Research*, 34, 147–154.

Beracochea, D., Lescaudron, L., Tako, A., Verna, A. & Jaffard, R. (1987). Build-up and release from proactive interference during chronic ethanol consumption in mice: a behavioral and neuroanatomical study. *Behavioural Brain Research*, 25, 63–74.

Biegler, R., & Morris, R.G.M. (1993). Landmark stability is a prerequisite for spatial but not discrimination learning. *Nature*, 361, 631–633.

Birch, D. & Jacobs, J.H. (1979). Spatial contrast sensitivity in albino and pigmented rats. *Vision Research*, 19, 933–937.

Bjorklund, A., Campbell K., Sirinathsinghji, D.J., Fricker, R.A. & Dunnett, S.B. (1994a). Functional capacity of striatal transplants in the rat Huntington model. In S.B. Dunnett, & A. Bjorklund (Eds.), *Functional neural transplantation* (pp.157–195). New York: Raven.

Bjorklund, A., Dunnett, S.B. & Nikkah, G. (1994b). Nigral transplants in the rat Parkinson model. In S.B. Dunnett, & A. Bjorklund (Eds.), *Functional neural transplantation* (pp.47–69). New York: Raven.

Bradbury, E., Kershaw, T., Marchbanks, R., & Sinden J. (1995). Foetal basal forebrain cell types and behavioural recovery. *Neuroscience*, 65, 955–972.

Brown, M.F. (1992). Does a cognitive map guide choices in the radial-arm maze? *Journal of Experimental Psychology*, 18, 56–66.

Brown, M.F., Rish, P.A., Von Culin, J.E. & Edberg J.A. (1993). Spatial guidance of choice behavior in the radial-arm maze. *Journal of Experimental Psychology*, 19, 195–214.

Bures, J., Buresova, O. & Nerad, L. (1992). Can rats solve a simple version of the traveling salesman problem? *Behavioural Brain Research, 52*, 133–142.

Buresova, O., Bures, J., Oitzl, M.S. & Zahalka, A. (1985). Radial maze in the water tank: an aversively motivated spatial working memory task. *Physiology and Behavior, 34*, 1003–1005.

Carvell, S.E. & Simons, D.J. (1990). Biometric analyses of vibrissal tactile discrimination in the rat. *Journal of Neuroscience, 10*, 2638–2648.

Castro, C.A. & Larsen, T. (1992). Primacy and recency effects in nonhuman primates. *Journal of Experimental Psychology, 18*, 335–430.

Cave, C.B., & Squire, L.R. (1991). Equivalent impairment of spatial and nonspatial memory following damage to the human hippocampus. *Hippocampus, 1*, 329–340.

Chollet, F., & Weiller, C. (1994). Imaging recovery of function following brain injury. *Current Opinion in Neurobiology, 4*, 226–230.

Cook, D., & Kesner R.P. (1988). The caudate nucleus and memory for egocentric localization. *Behavioral and Neural Biology, 49*, 332–343.

Corwin, J.V., Fussinger, M., Meyer, R.C., King, V.R., & Reep, R.L. (1994). Bilateral destruction of ventrolateral orbital cortex produces allocentric but not egocentric spatial deficits. *Behavioural Brain Research, 61*, 79–83.

Crowne D.P., Tokrud, P.A. & Brown P. (1992). The relation of rotation to egocentric and allocentric spatial learning in the rat. *Pharmacology, Biochemistry and Behavior, 43*, 1151–1153.

Davis, H.P. & Volpe, B.T. (1990) Memory performance after ischemic or neurotoxin damage of the hippocampus. In L.R. Squire, & E. Lindenlaub (Eds.), *The biology of memory* (pp.477–507). Stuttgart & New York: Symposia Medica Hoechst 23, F.K. Schattauer Verlag.

Dalrymple-Alford, J.C., Kelche, C. & Will, B. (1985). Is place learning impaired after lesions of the dorsal hippocampus? *Behavioural Brain Research, 20*, EBBS Meeting, Oxford, UK.

Dean, P. (1978). Visual acuity in hooded rats: effects of superior collicular or posterior neocortical lesions. *Experimental Brain Research, 18*, 433–445.

DiMattia, B.V. & Kesner, R.P. (1984). Serial position curves in rats: automatic versus effortful information processing. *Journal of Experimental Psychology, Animal Behaviour Processes, 10*, 557–563.

Dunnett, S.B. (1990). Neural transplantation in animal models of dementia. *European Journal of Neuroscience, 2*, 567–587.

Dunnett, S.B., Everitt, B.J. & Robbins, T.W. (1991). Interpreting the cognitive effects of excitotoxic basal forebrain lesions. *Trends in the Neurosciences, 14*, 494–501.

Dunnett, S.B., Whishaw, I.Q., Bunch, S.T. & Fine, A. (1986). Acetylcholine-rich neuronal grafts in the forebrain of rats: effects of environmental enrichment, neonatal noradrenaline depletion, host transplantation site, and acetylcholinesterase-positive fibre outgrowth. *Brain Research, 378*, 357–373.

Eichenbaum, H., Otto, T. & Cohen, N.J. (1994). Two functional components of the hippocampal memory system. *Behavioral and Brain Sciences, 17*, 449–518.

Eichenbaum, H., Stewart, C. & Morris, R.G.M. (1991). Hippocampal representation in place learning. *Journal of Neuroscience, 10*, 3531–3542.

File, S.E. (1993). The interplay of learning and anxiety in the elevated plus-maze. *Behavioural Brain Research, 58*, 199–202.

Finger, S. (1978). Environmental attenuation of brain lesion symptoms. In S. Finger (Ed.), *Recovery from brain damage: Research and theory* (pp.297–329). New York: Plenum.

Furuya, Y., Yamamoto, T., Yatsugi, S. & Ueki, S. (1988). A new method for studying working memory by using the three-panel runway apparatus in rats. *Japanese Journal of Pharmacology, 4*, 183–188.

Gaffan, D. (1983). A comment on primacy effects in monkey's memory for lists. *Animal Learning and Behavior, 11,* 144–145.

Gaffan, D. (1994) Scene-specific memory for objects: a model of episodic impairment in monkeys with fornix transection. *Journal of Cognitive Neuroscience, 6,* 305–320.

Gaffan E.A. & Gaffan D. (1992). Less than expected variability in evidence for primacy and von Restorff effects in rats' nonspatial memory. *Journal of Experimental Psychology, Animal Behaviour Processes, 18,* 298–301.

Gaffan, D. & Harrison, S. (1989). Place memory and scene memory: effects of fornix transection in the monkey. *Experimental Brain Research, 74,* 202–212.

Gallagher, M., Burwell, R. & Burchinall, M. (1993). Severity of spatial learning impairments in aging: development of a learning index for performance in the Morris water maze. *Behavioral Neuroscience, 107,* 618–636.

Glanzer, M. & Cunitz, A.R. (1966). Two storage mechanisms in free recall. *Journal of Verbal Learning and Verbal Behavior, 5,* 351–360.

Grabowski, M., Brundin, P. & Johansson, B.B. (1993). Paw-reaching, sensorimotor, and rotational deficits after brain infarction in rats. *Stroke, 24,* 889–895.

Grabowski, M., Sorenson, J.C., Mattson, N.B., Zimmer, J. & Johansson, B.B. (1995). Influence of an enriched environment and cortical grafting on functional outcome in brain infarcts of adult rats. *Experimental Neurology, 133,* 96–102.

Gray, J.A. (1971). Medial septal lesions, hippocampal theta rhythm and the control of vibrissal movement in the freely moving rat. *Electroencephalography and Clinical Neurophysiology, 30,* 189–197.

Gray, J.A., Sinden J.D. & Hodges H. (1990). Cognitive function, neural degeneration and transplantation. In B. Anderton (Ed.), *Neurodegenerative disease, seminars in the neurosciences* (Vol. 2, pp.122–142). London: Saunders.

Grigoryan, G.A., Hodges, H., Mitchell, S.N. & Gray, J.A. (1994). Interactions between the effects of propranolol and nicotine on radial maze performance of rats with lesions of the cholinergic forebrain projection system. *Behavioural Pharmacology, 5,* 265–280

Guic-Robles, E., Jenkins, W.M. & Bravo, H. (1992). Vibrissal roughness discrimination is barrel cortex-dependent. *Behavioural Brain Research, 48,* 145–152.

Guic-Robles, E., Valdiviesco, C. & Guajardo, G. (1989). Rats can learn a roughness discrimination using only their vibrissal system. *Behavioural Brain Research, 31,* 285–289.

Harper, D.N., McLean, A.P. & Dalrymple-Alford, J.C. (1993). List item memory in rats: effects of delay and delay task. *Journal of Experimental Psychology, 19,* 307–316.

Hodges, H. (1996). Maze procedures: the radial-arm and water maze compared. *Cognitive Brain Research, 3,* 167–181.

Hodges, H., Allen, Y., Sinden, J., Mitchell, S.N., Lantos, P.L. & Gray, J.A. (1991a). The effects of cholinergic drugs and cholinergic-rich neural transplants on alcohol-induced deficits in radial maze performance in rats. *Behavioural Brain Research, 43,* 7–28.

Hodges, H., Allen, Y., Kershaw, T., Lantos, P.L., Gray, J.A. & Sinden, J. (1991b). Effects of cholinergic-rich neural grafts on radial maze performance after excitotoxic lesions of the forebrain cholinergic projection system: 1. amelioration of cognitive deficits by transplants into cortex and hippocampus, but not basal forebrain. *Neuroscience, 45,* 587–607.

Hodges, H., Allen, Y., Sinden, J., Lantos, P.L. & Gray, J.A. (1991c). Effects of cholinergic-rich neural grafts on radial maze performance of rats after excitotoxic lesions of the forebrain cholinergic projection system-II: cholinergic drugs as probes to investigate lesion-induced deficits and transplant-induced functional recovery. *Neuroscience, 45,* 609–623.

Hodges, H., Gray, J.A., Allen, Y., & Sinden, J. (1991d). The role of the forebrain cholinergic projection system in performance in the radial-arm maze in memory-impaired rats. In

A. Adlkofer, & K.Thurau (Eds.), *Effects of nicotine on biological systems* (pp.389–399). Basel: Birkhauser.

Hodges, H. & Green S. (1986). Chlordiazepoxide-induced disruption of radial maze exploration in rats. *Psychopharmacology, 88*, 460–466.

Hodges, H., Sinden, J., Turner, J.D., Netto, C.A., Sowinski, P. & Gray, J.A. (1992). Nicotine as a tool to characterise the role of the forebrain cholinergic projection system in cognition. In P.M. Lipiello, A.C. Collins, J.A. Gray, & J.H. Robinson (Eds.), *The biology of nicotine* (pp.157–180). New York: Raven.

Hodges H., Sinden J., Meldrum, B.S. & Gray, J.A. (1994). Cerebral transplantation in animal models of ischaemia. In S.B. Dunnett, & A. Bjorklund (Eds.), *Functional neural transplantation* (pp.347–385). New York: Raven.

Hodges, H., Sowinski, P., Fleming, P., Kershaw, T.R., Sinden, J.D., Meldrum, B.S. & Gray, J.A. (1996). Contrasting effects of foetal CA1 and CA3 hippocampal grafts on deficits in spatial learning and working memory induced by global cerebral ischaemia in rats. *Neuroscience, 72*, 959–988.

Hodges, H., Sowinski, P., Sinden, J.D., Netto, C.A. & Fletcher, A. (1995). The selective 5-HT_3 antagonist, WAY100289, enhances spatial memory in rats with ibotenate lesions of the forebrain cholinergic projection system. *Psychopharmacology, 117*, 318–332.

Hodges H., Thrasher, S. & Gray, J.A. (1989). Improved radial maze performance induced by the benzodiazepine receptor antagonist ZK 93 426 in lesioned and alcohol-treated rats. *Behavioural Pharmacology, 1*, 44–54.

Jarrard, L.E. (1986). Selective hippocampal lesions and behavior: implications for current research and theorizing. In R.L. Iversen, & K.H. Pribram (Eds.), *The hippocampus* (Vol. 4, pp.93–126). New York: Plenum.

Jarrard, L.E. (1993). Review: on the role of the hippocampus in learning and memory in the rat. *Behavioral and Neural Biology, 60*, 9–26.

Kelche, C., Dalrymple-Alford, J.C. & Will, B. (1988). Housing conditions modulate the effects of intracerebral grafts in rats with brain lesions. *Behavioural Brain Research, 28*, 287–295.

Kelsey, J.E. & Vargas, H. (1993). Medial septal lesions disrupt spatial, but not nonspatial working memory in rats. *Behavioral Neuroscience, 107*, 565–574.

Kesner, R.P., Farnsworth, G. & DiMattia, B.V. (1989). Double dissociation of egocentric and allocentric space following prefrontal and parietal cortex lesions in the rat. *Behavioural Neuroscience, 103*, 956–961.

King, V.R. & Corwin, J.V. (1992). Spatial deficits and hemispheric asymmetries in the rat following unilateral and bilateral lesions of the posterior parietal or medial agranular cortex. *Behavioural Brain Research, 50*, 53–68.

Kiyota, Y., Miyamoto, M. & Nagaoka, A. (1991). Relationship between brain damage and memory impairment in rats exposed to transient forebrain ischemia. *Brain Research, 538*, 295–302.

Komisaruk, B.R. (1970). Synchrony between limbic system theta activity and rthymical behavior in rats. *Journal of Comparative and Physiological Psychology, 70*, 482–492.

Mallet, P.E. & Beninger, R.J. (1993). The double Y-maze as a tool for assessing memory in rats. *Neuroscience Protocols, 10*, 2–11

Mandel, R.J., Gage, F.H. & Thal, L.J. (1989). Enhanced detection of nucleus basalis magnocellularis lesion-induced spatial learning deficit in rats by modification of training regimen. *Behavioural Brain Research, 31*, 221–229.

Markowska, A.L., Long, J.M., Johnson, C.T. & Olton, D.S. (1993). Variable-interval probe tests as a tool for repeated measurements of spatial memory in the water maze. *Behavioral Neuroscience, 107*, 627–632.

Mayer E., Fawcett, J.W. & Dunnett, S.B. (1993). Fibroblast growth factor promotes the survival of embryonic ventral mesencephalic dopaminergic neurons: II. Effects on nigral transplants in vivo. *Neuroscience, 56,* 389–398.

McDonald, R.J. & White, N.M. (1994). Parallel information processing in the water maze: Evidence for independent memory systems involving dorsal striatum and hippocampus. *Behavioral and Neural Biology, 61,* 260–270.

McNamara, R.K. & Skelton, R.W. (1993). The neuropharmacological and neurochemical basis of place learning in the Morris water maze. *Brain Research Reviews, 18,* 33–49.

Moghaddam, M. & Bures, J. (1996). Contribution of egocentric spatial memory to place navigation of rats in the Morris water maze. *Behavioural Brain Research, 78,* 121–129.

Mohammed, A.K., Winblad, B., Ebendahl, T. & Larkfors, L. (1990). Environmental influence on behaviour and nerve growth factor in the brain. *Brain Research, 528,* 69–72.

Morris, R.G.M., Garrud, P., Rawlins, J.N.P. & O'Keefe, J. (1982). Place navigation is impaired in rats with hippocampal lesions. *Nature, 297,* 681–683.

Morris, R.G.M., Hagan, J.J. & Rawlins, J.N.P. (1986). Allocentric spatial learning by hippocampectomised rats: a further test of the 'spatial mapping' and 'working memory' theories of hippocampal function. *Quarterly Journal of Experimental Psychology, 35B,* 365–395.

Mumby, D.G., Wood, E.R. & Pinel, J.P.J. (1992). Object recognition memory is only mildly impaired in rats with lesions of the hippocampus and amygdala. *Psychobiology, 20,* 18–27.

Nelson, A., Lebessi, A., Sowinski, P. & Hodges, H. (1997a). Comparison of effects of global cerebral ischaemia on spatial learning in the standard and radial water maze: relationship of hippocampal damage to performance. *Behavioural Brain Research, 85,* 93–115.

Nelson, A., Sowinski, P. & Hodges, H. (1997b). Differential effects of global ischaemia on delayed matching- and non-matching-to-position tasks in the water maze and Skinner box. *Neurobiology of Learning and Memory, 67,* 228–247.

Netto, C.A., Hodges, H., Sinden, J.D., Le Peillet, E., Kershaw, T., Sowinski, P., Meldrum, B.S. & Gray, J.A. (1993). Effects of foetal hippocampal grafts on ischemic-induced deficits in spatial navigation in the water maze. *Neuroscience, 54,* 69–92.

Nishino, H., Koide, K., Aihara, N., Kumazaki, M., Sakurai, T., & Nagai, H. (1993). Striatal grafts in the ischemic striatum improve pallidal GABA release and passive avoidance. *Brain Research Bulletin, 32,* 517–520.

Nunn, J.A. & Hodges, H. (1994). Cognitive deficits induced by global cerebral ischaemia: relationship to brain damage and reversal by transplants. *Behavioural Brain Research, 65,* 1–31.

Nunn, J.A., Le Peillet, E., Netto, C.A., Sowinski, P., Hodges, H., Meldrum, B.S. & Gray, J.A. (1991). CA1 cell loss produces deficits in the water maze but not in the radial maze. *Society for Neuroscience Abstracts, 17,* 108.

Nunn, J.A., Le Peillet, E., Netto, C.A., Sowinski, P., Hodges, H., Gray, J.A. & Meldrum, B.S. (1994). Global ischaemia: hippocampal pathology and spatial deficits in the water maze. *Behavioural Brain Research, 62,* 41–54.

O'Keefe, J. (1983). Spatial memory within and without the hippocampal system. In W. Seifert (Ed.), *Neurobiology of the hippocampus* (pp.375–403). London: Academic Press.

O'Keefe, J. & Conway, D.H. (1980). On the trail of the hippocampal engram. *Physiological Psychology, 8,* 229–238.

O'Keefe, J., & Nadel, L. (1978). *The hippocampus as a cognitive map.* Oxford: Clarendon.

O'Keefe, J. & Speakman, A. (1987). Single unit activity in the rat hippocampus during a spatial memory task. *Experimental Brain Research, 68,* 1–27.

Olton, D.S., Becker, J.T. & Handelman, G.E. (1979). Hippocampus, space and memory. *Behavioral and Brain Sciences, 2,* 313–365.

Olton, D.S., Becker, J.T. & Handelman, G.E. (1980). Hippocampal function: working memory or cognitive mapping? *Physiological Psychology, 8*, 239–246.

Olton, D.S., Collison, C. & Werz, M. (1977). Spatial memory and radial arm maze performance in rats. *Learning and Motivation, 8*, 289–314.

Olton, D.S. & Collison, C. (1979). Intramaze cues and 'odor trails' fail to direct choice behavior on an elevated maze. *Animal Learning and Behavior, 7*, 221–223.

Olton, D.S. & Pappas, B.C. (1979). Spatial memory and hippocampal function. *Neuropsychologia, 17*, 669–682.

Otto, T. & Eichenbaum, H. (1992). Complementary roles of the orbital prefrontal cortex and the perirhinal-entorhinal cortices in an odor guided delayed-nonmatching-to-sample task. *Behavioral Neuroscience, 106*, 762–775.

Pacteau, C., Einon, D. & Sinden, J.D. (1989). Early rearing environment and dorsal hippocampal ibotenic acid lesions: long-term influences on spatial learning and alternation in the rat. *Behavioural Brain Research, 34*, 79–96.

Peele, D.B. & Baron, S.P. (1988). Effects of scopolamine on repeated acquisition of radial-arm maze performance by rats. *Journal of Experimental Analysis of Behavior, 49*, 275–290.

Raffaele, K.C. & Olton, D.S. (1988). Hippocampal and amygdaloid involvement in working memory for nonspatial stimuli. *Behavioral Neuroscience, 102*, 349–355.

Rawlins, J.N.P. (1985). Associations across time: the hippocampus as a temporary memory store. *Behavioral and Brain Sciences, 8*, 479–496.

Rawlins, J.N.P., Lyford, G.L., Seferiades, A., Deacon, R.M.J. & Cassaday, H.J. (1993). Critical determinants of nonspatial working memory deficits in rats with conventional lesions of the hippocampus or fornix. *Behavioral Neuroscience, 107*, 420–433.

Renner, M.J. & Rosenzweig, M.R. (1987). *Enriched and impoverished environments: Effects on brain and behaviour.* New York: Springer.

Ridley, R.M., Timothy, C.J., Maclean, C.J. & Baker, H.F. (1995). Conditional learning and memory impairments following neurotoxic lesion of the CA1 field of the hippocampus. *Neuroscience, 67*, 263–275.

Rose, F.D., Al-Khamees, K., Davey, M.J., & Attree, E.A. (1993). Environmental enrichment following brain damage: An aid to recovery or compensation? *Behavioural Brain Research, 5*, 93–100.

Rose, F.D., Davey, M.J. & Attree, E.A. (1993). How does environmental enrichment aid performance following cortical injury in the rat? *NeuroReport, 4*, 163–166.

Rose, F.D., Davey, M.J., Love, S. & Attree, E.A. (1990). Water maze performance in rats with bilateral occipital lesions: a model for use in recovery of function research. *Medical Science Research, 18*, 167–169.

Rose, F.D., Davey, M.J., Love, S. & Dell, P.A. (1987). Environmental enrichment and recovery from contralateral sensory neglect in rats with large unilateral neocortical lesions. *Behavioural Brain Research, 24*, 195–202.

Rose, F.D., Dell, P.A. & Davey, M.J. (1989). Post-surgical environmental enrichment and functional recovery in the hemidecorticate rat: alternative interpretations. *Medical Science Reviews, 17*, 481–483.

Rose, F.D., Dell, P.A., Love, S. & Davey, M.J. (1988). Environmental enrichment and recovery from a complex Go/Nogo reversal deficit in rats following large unilateral neocortical lesions. *Behavioural Brain Research, 31*, 37–45.

Rothblat, L.A. & Kromer, L.F. (1991). Object recognition memory in the rat: the role of the hippocampus. *Behavioural Brain Research, 42*, 25–32.

Schwarting, R.K.W., Pei, G., Soderstrom, S., Ebendal, T. & Huston, J.P. (1994). Unilateral stimulation or removal of rat vibrissae: analysis of nerve growth factor and tyrosine hydroxylase mRNA in the brain. *Behavioural Brain Research, 60*, 63–71.

Selden, N.R.W., Cole, B.J., Everitt, B.J. & Robbins, T.W. (1990). Damage to ceruleo-cortical noradrenergic projections impairs locally cued but enhances spatially cued water maze acquisition. *Behavioural Brain Research, 39*, 29–51.

Sinden, J.D., Allen, Y.S., Rawlins, J.N.P. & Gray, J.A. (1990). The effects of ibotenic acid lesions of the nucleus basalis and cholinergic-rich neural transplantation on win-stay/lose-shift and win-shift/lose-stay performance in the rat. *Behavioural Brain Research, 36*, 229–249.

Sinden, J.D., Hodges, H. & Gray, J.A. (1995). Neural transplantation and recovery of cognitive function. *Behavioral and Brain Sciences, 18*, 10–35.

Sinden J.D., Patel, S.N. & Hodges, H. (1992). Neural Transplantation: problems and prospects for therapeutic application. *Current Opinion in Neuropathology, 5*, 902–908.

Squire, L.R. (1986). Mechanisms of memory. *Science, 232*, 1612–1619.

Squire, L.R., Amaral, D.G. & Press, G.A. (1990). Magnetic resonance imaging of the hippocampal formation and mamillary nuclei distinguish medial temporal lobe and diencephalic amnesia. *Journal of Neuroscience, 9*, 3110–3117.

Sutherland, R.J. & Dyck, R.H. (1984). Place navigation by rats in a swimming pool. *Canadian Journal of Psychology, 382*, 322–344.

Suzuki, W.A., Zola-Morgan, S., Squire, L.R., & Amaral D.G. (1993). Lesions of the perirhinal and parahippocampal cortices in the monkey produce long-lasting memory impairment in the visual and tactile modalities. *Journal of Neuroscience, 13*, 2430–2451.

Tolman, E.C. (1948). Cognitive maps in rats and men. *Psychological Reviews, 5*, 189–208.

Tonkiss J., Feldon, J. & Rawlins, J.N.P. (1990). Section of the descending columns of the fornix produces delay- and interference-dependent working memory deficits. *Behavioural Brain Research, 36*, 113–126.

Turner, J.J., Hodges, H., Sinden, J.D. & Gray, J.A. (1992). Comparison of radial maze performance of rats after ibotenate and quisqualate lesions of the forebrain cholinergic projection system: effects of pharmacological challenge and changes in training regime. *Behavioural Pharmacology, 3*, 359–373.

van Luijtelaar, E.L.J.M., van der Staay, F.J. & Kerbusch, J.M.L. (1989). Spatial memory in rats: a cross validation study. *Quarterly Journal of Experimental Psychology, 41B*, 287–306.

Volpe, B.T., Davis, H.P., Towle, A. & Dunlap, W.P. (1992). Loss of hippocampal CA1 neurons correlates with memory impairment in rats with ischemic or neurotoxin lesions. *Behavioral Neuroscience, 106*, 457–464.

Weiller, C.R., Ramsay, S.C., Wise, R.J.S., Friston K.J. & Frackowiak, R.S.J. (1993). Individual patterns of functional reorganization in the human cerebral cortex after capsular infarction. *Annals of Neurology, 33*, 181–189.

Whishaw, I.Q. (1985). Formation of a place learning-set by the rat: a new paradigm for neurobehavioural studies. *Physiology and Behaviour, 35*, 139–143.

Whishaw, I.Q. & Tomie, J. (1987). Cholinergic receptor blockade produces impairments in a sensorimotor subsystem for place navigation in the rat: evidence from sensory, motor and acquisition tests in a swimming pool. *Behavioral Neuroscience, 101*, 603–616.

Whishaw, I.Q., Zaborowski, J.A. & Kolb, B. (1984). Postsurgical enrichment aids adult hemicorticate rats on a spatial navigation task. *Behavioral and Neural Biology, 42*, 183–190.

Will, B. & Kelche C. (1992). Environmental approaches to recovery of function from brain damage: a review of animal studies (1981 to 1991). In F.D. Rose, & D.A. Johnson (Eds.), *Recovery from brain damage: Reflections and directions. Advances in Experimental Medicine and Biology, 325* (pp.79–103). New York: Plenum.

Wood, E.R., Bussey, T.J. & Phillips A.G. (1993a). A glycine antagonist 7–chlorokynurenic acid attenuates ischemia-induced learning deficits. *NeuroReport, 4*, 151–154.

Wood, E.R., Mumby, D.G., Pinel, J.P.J. & Phillips A.G. (1993b). Impaired object recognition memory in rats following ischemia-induced damage to the hippocampus. *Behavioral Neuroscience, 107*, 51–62.

Yoerg, S.I. & Kamil, A.C. (1982). Response strategies in the radial arm maze: running round in circles? *Annals of Learning and Behaviour, 10*, 530–534.

7 Long-distance Travels and Homing: Dispersal, Migrations, Excursions

Jacques Bovet
Université Laval, Québec, Canada

A PRELIMINARY OVERVIEW AND A PREFATORY NOTE

In the real world, that is in the sky, fields, woods, creeks, lakes, oceans, etc. free-ranging animals do not typically spend their life roaming about erratically. They rather tend to display distinct bouts of *sedentariness* in relatively small stretches of space that they go all over repeatedly. Biologists call such a patch of "personal" space a *home range*. An animal is not a prisoner of its home range, however, and there are three major types of instance when it will travel outside. One is a move by which an animal leaves a home range forever to settle into a new one. Such a move is called *dispersal*, and its most common occurrence is for juveniles leaving their natal home range to settle into a distant reproductive home range. Another type consists of so-called *return migrations*, which are periodical trips back and forth between two established, but distant home ranges. Its most common occurrence is in seasonal migrations between a summer and a winter home range, as in typical migratory birds. The last major kind of off-home range travel occurs when an animal makes a relatively short-lasting foray away from, then back to its home range, without really settling down anywhere along the route of this so-called *excursion*. In a migration or an excursion, the return of the individual to a previously occupied home range is an instance of *homing*, and a demonstration of *site fidelity*.

Compared to the conditions of the spatial tasks discussed in the other parts of this Handbook, the environments, routes and goals involved in these off-home range movements are at a totally different scale relative to the size of the animal, as illustrated in Fig. 7.1 (hence the designation

240 BOVET

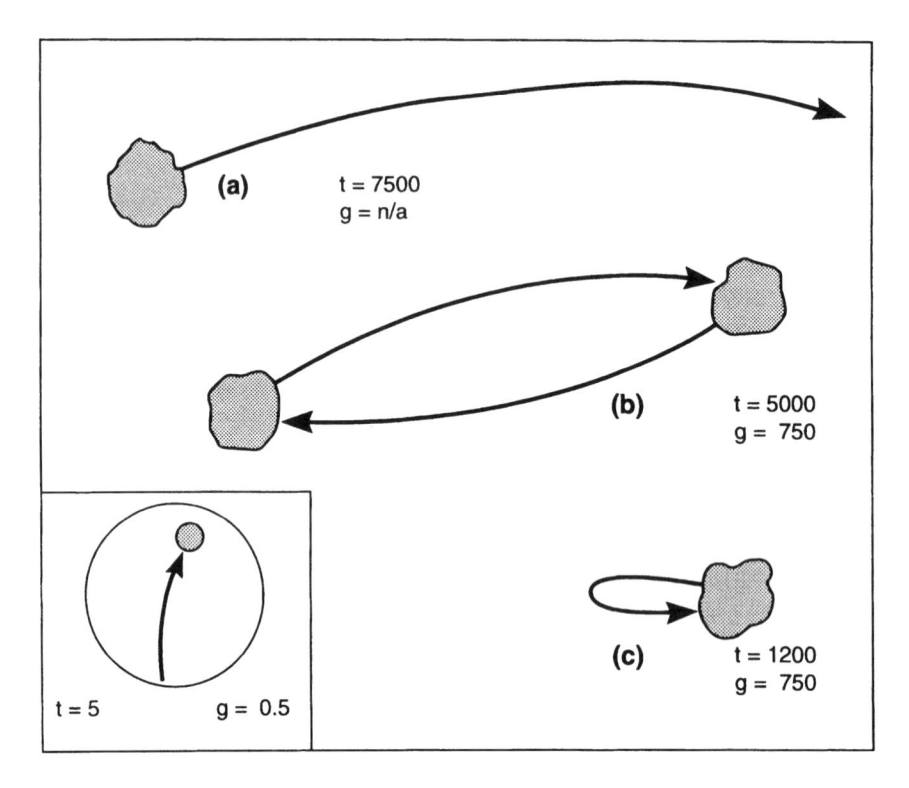

FIG. 7.1 Schematic representations of (a) dispersal, (b) return migrations, (c) an excursion by a free-ranging animal. Shaded areas are home ranges. Arrows symbolize travel trajectories, and point toward the goal. Assuming that the animal is a white-tailed deer, t is the approximate span of the trajectory and g the approximate diameter of the goal, relative to the size of the animal (= 1). Inset: Analogous representation of the trajectory of a rat in a Morris' swimming pool. The dot is the goal (e.g., a hidden platform).

long-distance travels). This has profound implications on the paradigms and methods of research on the behavioural strategies, orientational abilities and learning processes involved in these kinds of travels. Except for microscopic animals, the distances from start to goal are several orders of magnitude larger than the size of a standard research laboratory. This requires that at least part of the observations and experiments be performed in the field, with wild-born and/or free-ranging animals. This in turn has all sorts of practical implications, the discussion of which might be too long or too trivial here but which, taken together, make that these long-distance travels have been and still are an object of study by zoologists rather than by psychologists (with a few notable exceptions, of course). As a consequence and in comparison with other fields of spatial research, paradigms and methods are here full of explicit or implicit reference to natural history, ecology, adaptive value and evolution. Conversely, because of the constraints and imponderables of research in large-scale natural environments, these paradigms and methods are perhaps less finicky about standardization of experimental conditions and subjects.

In order to remain within the relatively restricted taxonomic framework of this Handbook, the present chapter focuses on mammals, particularly on terrestrial ones. Readers should be aware from the start, and keep in mind throughout the chapter, that contrary to other aspects of spatial behaviour discussed in the Handbook, long-distance travels and orientation have been far less studied in mammals than in several other taxonomic groups, notably birds. Indeed, paradigms and methods in the study of mammalian travels traditionally remain at a relatively unsophisticated level by comparison with what is found for other groups, and have often lagged behind more advanced research in these other groups. Interested readers are referred to Papi (1992a) for an overview of long-distance travel strategies and orientation mechanisms in the animal kingdom. Able (1995) provides a concise account of the history of long-distance navigation research in birds, with stress on paradigms. A detailed review of experimental and analytical techniques of bird orientation research is found in Helbig (1991).

OBSERVING AND RECORDING TRAVELS IN THE FIELD

The actual trajectory covered by an animal while performing spatial behaviour is a powerful indicator of the processes involved. Unfortunately, contrary to what can be done in the laboratory (as shown in other chapters of this Handbook), and except for a few special cases (e.g., ants: see Wehner, 1992), it has long been and still is technical'y difficult, if not impossible, to map the local or the long-distance movements of free-ranging animals on the basis of a direct recording of the continuous paths that they run, fly or swim along while travelling spontaneously or as the result of an experimental manipulation (e.g., in a displacement experiment: see "Displacement experiments", p.259). Our knowledge of their spatial behaviour is therefore often based on the (sensible) reasoning that discrete consecutive recordings of the presence of a given individual at discrete locations are evidence that this animal travelled between these locations. They are, however, only that; in particular, they say nothing about the straightness or sinuosity of the route taken between the two locations.

Capture-Marking-Recapture

The so-called Capture-Marking-Recapture method (CMR) has long been (and still is) a common method for detecting the presences of a known individual at discrete locations. In general terms and as the name indicates, animals have to be captured alive by whatever means, and marked in such a way that when eventually recaptured at a new location, they can be recognized as "previously captured at another location", and inferred to have moved from the former to the new site of capture. Capture devices (traps, nets, etc.) are almost as varied as the species studied. Many are available on a commercial basis, and are usually identified by brand name in the primary literature. Marking devices or procedures are also as diverse. Individually numbered or coloured leg-rings are classics of their kind for birds, as are individual combinations of colour-spots on the abdomen for bees or ants. The most common devices for marking mammals are individually numbered ear-tags. Tags originally manufactured and commercialized for marking farm animals or fish can be easily clamped on the ear (pinna) of many large or small mammals. Marking by means of ear- or toe-clipping is no longer performed, for ethical and practical reasons.

Two major limitations must be taken into account for a proper interpretation of CMR data. One is that the patterns of travels that can be potentially recorded are totally dependent on, and limited to the pre-existing pattern of spatial distribution of the capture devices. The other is a matter of *trappability*, that is the readiness of animals to enter capture devices. Trappability is usually assumed not to vary, neither among conspecifics nor, for a given individual, with space or time. This assumption remains largely untested, despite the evidence that it should not (e.g., the cases of trap-addiction or of trap-shyness that are part of the anecdotal repertoire of every practitioner in the field; see also Khan, 1992; Drickamer et al., 1995; Gehrt & Fritzell, 1996).

Radio-tracking

Since the 1960s, CMR methodologies have been progressively complemented if not replaced by telemetric procedures called *radio-tracking*. An animal is captured alive by whatever means, and fitted with a radio transmitter tuned to an individual radio frequency. The animal is then let free, and its eventual spatial positions (= *fixes*) are detected and identified from a distance whenever needed, by means of an appropriate radio-receiving system. Potential advantages over CMR are obvious. Since recaptures are no longer necessary, problems of trappability and of spatial distribution of capture devices are solved, so to speak, by default. Basically, consecutive fixes provide the same kind of information as consecutive places of recaptures. Practically, however, since fixes are not limited to the positions of pre-established traps and can be taken at intervals much shorter than trapping data, radio-tracking provides a more complete and finely grained picture.

Radio-tracking devices are manufactured throughout the world, in configurations adapted to all sorts of animal species and habitat or climatic conditions. They are usually identified by model and brand name in the primary literature. In mammals, transmitters are commonly fastened on the animal as a collar or a harness around the neck or shoulders, or implanted intraperitoneally. Depending on the species' morphology, other placements are feasible (e.g., on a beaver's tail: Sokolov et al., 1977; or through a polar bear's pinna: Hansen, 1995). The choice of an adequate transmitter results from a compromise between its mass, the carrying distance of its signal, and the life span of its power supply, the latter two being limited by the former, and in inverse relation to each other. By rule of thumb, the mass of the device (including battery, antenna and fastener) should be <5% of the animal's body mass. A survey of the recent mammalogical literature indicates that radio-tracking devices are now available for animals weighing ≥20grams.

Techniques of radio-tracking data collection and analysis are the topic of a number of handbooks (e.g., Mech, 1983; White & Garrott, 1990). Samuel and Fuller's (1994) account is commendable as an up-to-date starter. From a practical, down-to-earth point of view, the limitations of radio-tracking revolve around whether or not a signal is received, and whether its origin can be pinpointed accurately. The carrying distance of the signal depends not only on the power of the transmitter, but also on the features of the landscape between transmitter and receiver (e.g., flat and bare vs. hilly and densely vegetated). Thus, the absence of a signal, even when not due to an inconspicuously broken wire in the system, is not foolproof evidence that the animal is very far away.

Receivers are typically equipped with a directional antenna that points toward the origin of a signal with some margin of error. In order to determine the position of the signal (not only its direction), fixes are usually taken using two spaced out receivers that work in synergy (a procedure called *triangulation*, because the straight line between the two receivers is the basis of a triangle, the third apex of which is the transmitter). However, because of the range of error inherent in the system, this procedure yields only a polygon which contains the origin of the signal (Fig. 7.2). The size and shape of the polygon depends on a number of factors (e.g., distance from transmitter to receivers, angle between the radio bearings).

There are many possible variations. Because of problems with the transmission of radio waves in deep and/or salt water, the tracking of aquatic animals is often made using acoustic devices (Winter, 1983). Once the signal (radio or sound) is detected, the observer can *home-in* on its origin, that is, come nearer and nearer to it by trial and error until it can be pinpointed accurately (or the animal itself can be seen). This is, of course, a

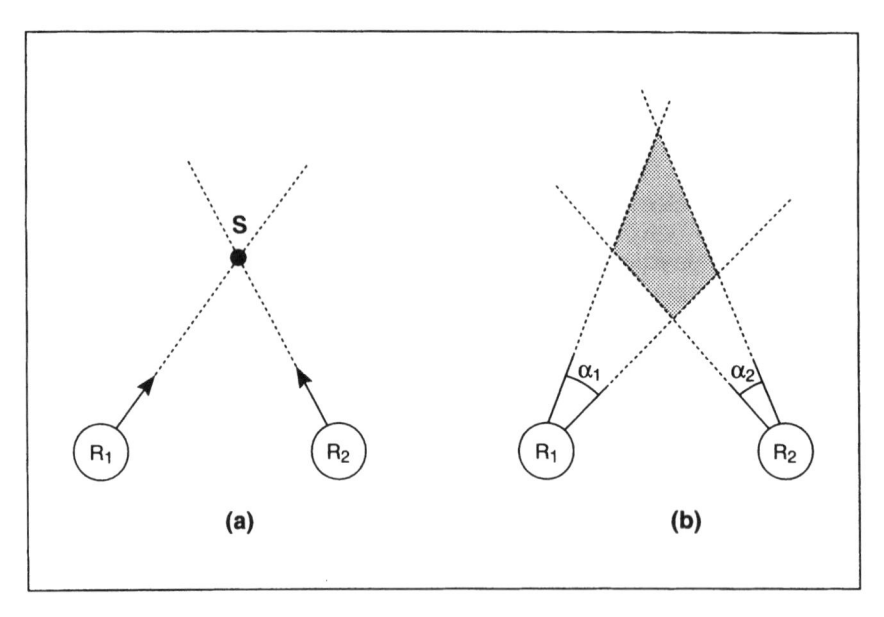

FIG. 7.2 Radio-tracking: theory and practice of triangulation. (a) In theory, the source of the radio signal (S) is at the place of convergence of the bearings yielded by each of the two recorders R_1 and R_2. (b) In practice, however, because these bearings are inaccurate within the angular ranges of error α_1 and α_2, the source of the signal can be anywhere in the shaded polygon.

time-consuming procedure that implies a close-range interference with the animal. It is often associated with radio-tracking from an aircraft. If the receiver can be moved very fast from one site to another (compared to the speed of the target animal), triangulation can still be performed with reasonable accuracy with two discrete recordings by the same receiver. The Argos system of radio-tracking makes use of an orbiting satellite as a receiver; locations are computed from the Doppler shift in the frequency of the signals as the satellite moves closer to, then away from the sources (see Fancy et al., 1989). Depending on the species studied, the difficulties of an initial capture for fastening the transmitter can be bypassed by various means such as darting it into the skin by means of an arrow and crossbow (e.g., for whales: Goodyear, 1993), or having it swallowed down together with a bait (e.g., for deep-sea fish: Bagley et al., 1994).

A recent development in radio-tracking is the use of GPS technology (for global positioning system) [see Rempel et al., 1995; also *Telonics Quarterly*™, 1996, vol. 9(1)]. It is, so to speak, the inverse of the Argos system: the "item", the position of which has to be determined, carries a receiver which detects and analyses the radio signals sent by at least 3 orbiting satellites out of a constellation of 24. In its standard applications by soldiers, sailors, hunters, farmers, etc., the system provides instantaneous positional information to the user-carrier. However, in its applications for the study of animal travel, it is the animal, not the user who carries the apparatus, and the (sequential) information must be stored by the receiving device for eventual transfer into the user's computer. If the apparatus is also equipped with an appropriate transmitter (which has, of course, an impact on the overall mass of the device), the information can be recovered from a distance (e.g., via a FM-link or an Argos satellite); if not, the animal has to be recaptured, which implies that it has to be relocated by other means than GPS.

Radio-tracking is particularly well suited for the study of movements of large animals, who can carry relatively heavy, thus powerful transmitters or receivers, and are not easily captured or recaptured. As a consequence, these animals are no longer studied by means of CMR. For smaller animals, however, CMR remains a useful, if not the only alternative.

Miscellaneous opportunistic techniques

Capture-Marking-Recapture (CMR) and radio-tracking are robust, all-purpose, transtaxonomic methodologies that have, however, definite limitations, as mentioned above. Many authors dealing with specific spatial problems in a particular species have tried to bypass these limitations in developing what could be called *opportunistic techniques* that exploit some particular behavioural or morphological aspects of that species and/or some special features of the environment. Because of space constraints, the following survey is limited to a few examples of techniques likely to provide detailed travel trajectories.

Spool and line devices. The animal carries on its back a spool of thread, the free end of which is attached to a start mark. While moving along, the animal unloads the spool; the thread clings to the bumps of the landscape (trees, logs, rocks, etc.), leaving thus an approximate picture of the animal's trajectory that is easily "rewound" (and surveyed) by the observer. The procedure seems particularly adequate for slow-moving animals which are unlikely to produce inextricable knots (e.g., toads: Sinsch, 1987; tortoises: Hailey & Coulson, 1996). Its use has also been advocated for recording the three-dimensional movements of arboreal mammals (e.g., Miles et al., 1981; Anderson et al., 1988; Key & Woods, 1996).

Fluorescent powders. The animal is dusted with a mix of fluorescent pigments, the composition of which can be modified for each individual studied. Once released, the animal leaves behind an accurate track of its travels, that can be eventually detected and recorded, using an ultraviolet flashlight (Duplantier et al., 1984; Lemen & Freeman, 1985). The technique seems safe in terms of toxicity (Stapp et al., 1994). Recording the sequence of events can become a problem if the track tangles.

Footprints in the substrate. When conditions are favorable (e.g., no wind, no precipitation), animals moving on a soft substrate like sand or snow leave a track of their travels that can be eventually recorded with great accuracy, literally step by step. Problems of interpretation inevitably arise with the tangling of tracks left by one or several individuals. However, when a track with little tangling can be positively ascribed to a single individual, the information it provides is particularly enlightening (Bovet, 1968; Ramsey & Andriashek, 1986; Fielden, 1991).

Radar. Provided they are large enough, flying animals (either individually or as a group) can be spotted by radar and tracked on the radar screen like an ordinary airplane. Radar-tracking is a powerful tool for the study of bird migrations (see Alerstam, 1990). A technique based on harmonic radar (as opposed to more traditional pulse radar) is currently being developed for tracking individual bees flying at low altitude (Riley et al., 1996).

Luminescent markers. Nocturnal animals can be equipped with a luminescent device which is detectable visually from a distance (Buchler, 1976). If these animals move on the ground in a flat and fairly bare environment, their travels can be monitored from the top of a tower by means of an optical apparatus continuously aimed at the luminescent marker. The positions of the animal are computed, using the azimuth of the line of sight, its angular deviation from the vertical, and the height of the tower (Benhamou, 1990).

Surveillance. Many free-ranging animals habituate readily to the close presence of a human observer (even black bears do: L.L. Rogers, personal communication; see also Anonymous, 1994). If such an habituated animal moves, the observer can "shadow" it by keeping visual contact and record its travels, by reference to a printed map, to a grid of pre-established survey marks, or to a sequence of marks left behind while moving along with the animal. The tracking of homing pigeons by helicopter is an example (e.g., Wagner, 1970). The effectiveness of the surveillance will be enhanced if it is *radio-assisted*, using the home-in procedure mentioned in the radio-tracking section above (p.243): if the shadowed animal gets out of sight for whatever reason (a common occurrence in a forest environment), the observer can relocate it by radio and resume visual surveillance very quickly (Bovet, 1984; Lair, 1987). For obvious reasons, the technique is essentially limited to conspicuous, diurnal animals, that move at a suitable speed (from a human point of view).

High-tech recorders. Recent advances in microelectronics have made it possible to equip free-ranging animals with devices that continuously record and store various indices which, once analysed, can be translated into spatial localizations. For instance, birds can be fitted with a device that records the duration of bouts of flight in any one compass direction (Bramanti et al., 1988), and elephant seals with a device that records light intensities at sea-surface with reference to an internal clock (DeLong et al., 1992). As with basic global positioning systems (GPS: see above p.244), the data are lost if the devices cannot be recovered.

These are but a few examples of opportunistic techniques. Anyone wishing to know what techniques have been developed for a given species is advised to search for the pertinent literature through the *Zoological Record* (a most comprehensive and uncommonly well cross-indexed database for zoological references, available on-line, as well as in CD-ROM or paper versions).

THE HOME RANGE

The concept of "home range" should be basic in the study of long-distance travels, since it defines what the "goal" is going to be in most cases (Fig. 7.1). The idea that free-ranging animals tend to restrict their daily movements to a limited area for lapses of time > 24 hours (e.g., weeks, or a season, if not a lifetime) was already present in the works published on long-distance orientation near the end of the 19th century (e.g., Darwin, Fabre: see p.251). It is, however, only since around the 1930s that formal attempts have been made to measure the size and shape of that area, and to define it operationally under the name of "home range". It might be pointed out here that, contrary to matters of long-distance travels and orientation (see the introductory remarks above, p.241), the notion of home range has been and still is more studied in mammals than in any other taxonomic group.

Measuring home range size and shape

Let us first examine a fictitious example of an application of CMR for measuring home range sizes and shapes (Fig. 7.3). We study a species of woodmouse, and we set a number of mouse-traps in the woods that are our study area, along a standardized pattern of straight columns and rows that form a 1-hectare grid of 10-metre wide square mesh, with a trapping station at each knot. We use live-traps (of course), that we tend every evening and check every morning over a period of one month. On the first day, a mouse is captured at station D4. Besides taking standard data on this animal (mass and/or length, sex, reproductive condition, parasites, etc.), we tag it with an individually numbered ear-tag. We then release the mouse on the spot. Eventual recaptures/releases of this mouse are recorded and it soon appears that: (1) they all occur within a restricted area; (2) the size of this area does not increase with additional recaptures. On a map of the grid, we then graphically enclose all sites of capture of this mouse in a *minimum convex polygon*, the size and shape of which is deemed as representative of those of the mouse's home range, which we define (following Jewell, 1966, for instance) as "the area over which (this) animal normally travels in pursuit of its routine activities".

It is obvious that a meaningful application of CMR implies that the size of the grid and the inter-trap distances should be such that the former is larger than an individual home range, and the latter are smaller than a home range length or width. Since home range size and shape are not known a priori, a problem of logic arises, akin to the classic riddle: "What came first: the chicken or the egg?" It has to be solved by trial and error. Appropriate settings of these parameters will vary not only with the species studied but also, for a given species, with habitat type (e.g., Gorman & Zubaid, 1993).

When feasible, the taking of radio fixes bypasses these problems of grid and mesh sizes. Furthermore, it provides a more finely grained picture, which often reveals that the number of localizations per surface unit varies from many in some areas to few in others (Fig. 7.4). In order to take into account this variation in *utilization density*, the current trend is to characterize home range size and shape in probabilistic terms, e.g., as the smallest possible area where the probability of encountering the animal reaches a biologically meaningful level. A common practice is to set that level at 95%, and to consider fixes that are beyond the 95% isopleth as *outliers* (i.e., spots that are not normally visited when pursuing routine activities; or, complementarily, spots that are visited when pursuing non-routine activities). In some cases, it is meaningful to delimit, within the 95% isopleth, a so-called *core area* with a particularly high utilization

---

FIG. 7.3 Fictitious example of how to determine the size and shape of a mouse's home range, using the CMR method. (a) The trapping grid. Each dot is a trapping station, identified by a letter and a digit (for the column and line it belongs to). Digits within the grid identify consecutive places of capture. The full line is the contour of the minimum convex polygon that encloses all places of capture and describes home range size and shape (see text). (b) Variation of the surface area of the minimum convex polygon with an increasing number of captures: in this example, the surface stops increasing after capture number 14.

(a)

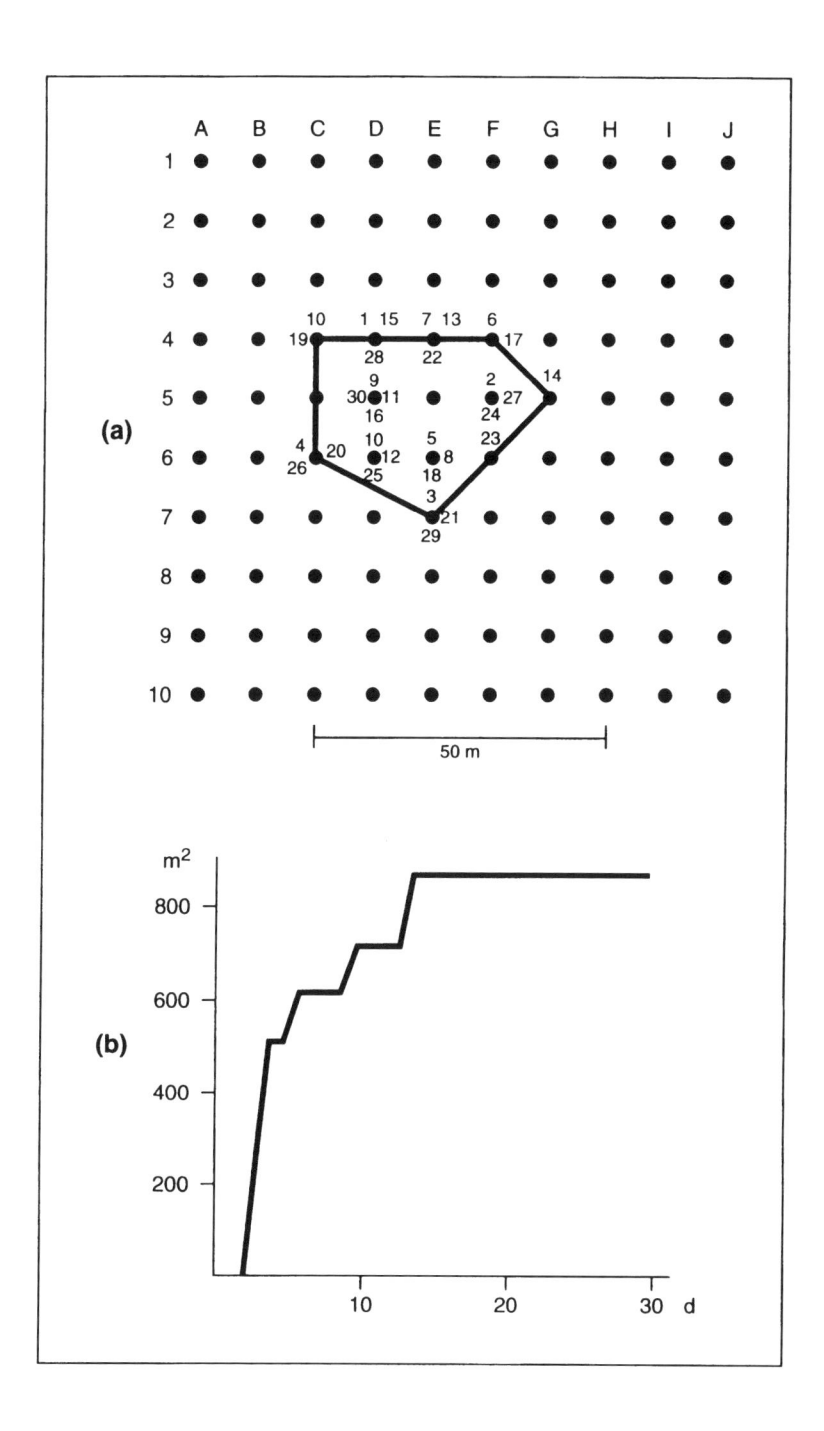

	A	B	C	D	E	F	G	H	I	J
1	•	•	•	•	•	•	•	•	•	•
2	•	•	•	•	•	•	•	•	•	•
3	•	•	•	•	•	•	•	•	•	•
4	•	•	•	•	•	•	•	•	•	•
5	•	•	•	•	•	•	•	•	•	•
6	•	•	•	•	•	•	•	•	•	•
7	•	•	•	•	•	•	•	•	•	•
8	•	•	•	•	•	•	•	•	•	•
9	•	•	•	•	•	•	•	•	•	•
10	•	•	•	•	•	•	•	•	•	•

|— 50 m —|

(b)

m^2 vs d

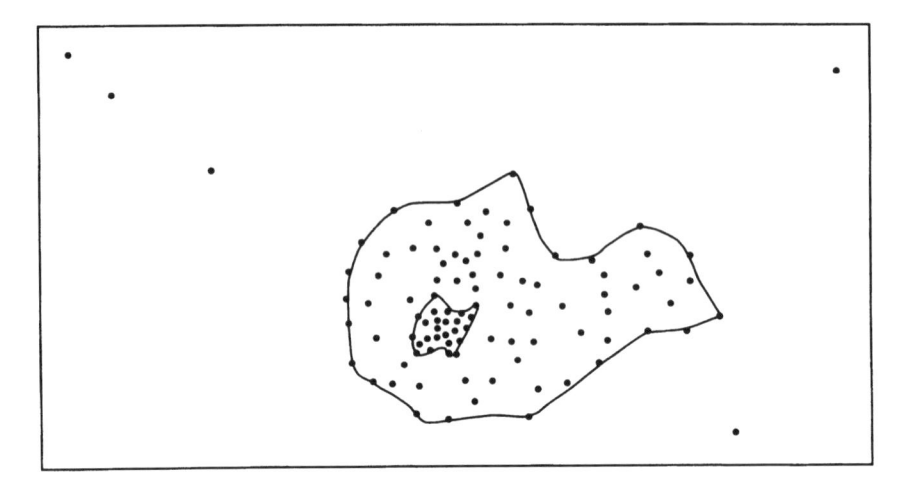

FIG. 7.4 Fictitious example of a "probabilistic" determination of home range size and shape (see text for definitions). Each dot is a radio-fix. The inner closed contour is the 25%-isopleth, which delimits the core area; the outer closed contour is the 95%-isopleth, which delimits the home range; the dots outside the 95%-isopleth are outliers.

density, defined as the smallest possible area with a 50% (or 40%, or 25% . . .) probability of encountering the animal. The pros and cons of various methods for analyzing home range data obtained by radio-tracking in mammals are throughly reviewed by Harris et al. (1990). For radio-tracking applications of the principles exposed in Figs. 7.3 and 7.4, see for instance, Clevenger (1993) and Krebs et al. (1995). Note, however, that the principle of a stabilization of the apparent home range size after a number of recaptures or fixes (Fig. 7.3b) is the topic of a current controversy (White et al., 1996, vs. Gautestad & Mysterud, 1995 and Doncaster & MacDonald, 1996).

Travels within the home range boundaries

By operational definition (see Fig. 7.1), travels within one's home range are *short-distance* travels, and are thus not the topic of this chapter (except as a matter of contrast to long-distance travels). This section is therefore limited to a very brief survey, with indications of sources of reference to recent work in this area.

Within their home ranges, animals tend to travel along an entangled network of well-defined routes or itineraries, which they use over and over again (e.g., Jamon, 1994), and that are closely linked to "structural features" in the environment, such as rocks, logs, trees, etc. (e.g., McMillan & Kaufman, 1995). In the zoological literature, it is commonly assumed that the repeated use of the same itineraries induces the development and learning of a map-like representation of one's home range, in a way likely to be analogous to the learning of routes/places in a maze, which is discussed at length elsewhere in this Handbook (see also Gallistel, 1990; Poucet, 1993). The adaptive value of the whole process would be to allow animals to know, at any place within their home range, the most "economical" route to any known potential goal (e.g., a nest, or a food cache, or a shelter; see Clarke et al., 1993), the economy being not so much a matter of distance rather than a matter of duration, safety and/or ability to "negotiate" obstacles (see Stamps, 1995). This advantage rests on a great *familiarity* with the home range and the potential landmarks it contains; it is acquired by a process of *site attachment* (Ketterson & Nolan, 1990) and kept by sedentariness.

The way by which an animal memorizes the place of "inconspicuous" resources within its home range (e.g., resources which have been concealed by the animal itself) has been the object of much experimental work which, with its emphasis on the role of distant landmarks, is definitely related to the Morris water task paradigms discussed elsewhere in this Handbook (for example, localization of food caches: e.g., Herz et al., 1994). The adaptive value of this kind of learning is that the knowledge of the clues that give access to the resources is available to the only owner, and not to its potential competitors or predators (see Jacobs, 1992). For other, more or less analogous lines of research, see Collett (1987); Etienne (1992); Etienne et al. (1995, 1996); Alyan and Jander (1994); Benhamou et al. (1995); and references therein.

Some of the paradigms that are traditionally associated with long-distance travels also warrant attention in the study of short-range orientation; for instance *compass orientation*, by which an animal determines its heading by reference to the representation it has of a "compass card" and which does not need kilometres of empty space for being expressed (see "The quest for the compass", p.262). For example, many birds can display compass-oriented behaviour in arenas or cages that are much smaller than their home range (for examples and references, see Berthold, 1991); this includes the finding of cached food (Wiltschko & Balda, 1989; Sherry & Duff, 1996). Some small mammals also show compass-oriented behaviour (for references, see Bovet, 1992).

LONG-DISTANCE TRAVELS AND HOMING

A hazy history

Considerable interest in long-distance travels developed in the second half of the 19th century (e.g., Darwin, 1873; Wallace, 1873; Fabre, 1879/1882; Morgan, 1894), with emphasis on the returns of homing pigeons displaced and released far away from their loft, and on the spatial exploits of pets (mostly dogs or cats) who, after being moved by their owners to a distant new home, manage to return by themselves to the old familiar home. From the very beginning, the question was: How do these or other animals find their way back home in this kind of artificial circumstance? Early (and not so early) answers to that question were based on fragmentary circumstantial evidence and/or on the coarse results of displacement experiments (see "Dislacement experiments" on p.259) with a variety of invertebrate and vertebrate animals, in which authors tried to replicate more or less systematically and/or critically the "pet stories" referred to earlier, with little explicit concern for their relevance to corresponding spontaneous behaviour, which remained largely undocumented until quite recently (see the sections, "Observing and recording travels in the field" on p.241, and "Spontaneous behaviour" on p.255). Because of the lack of appropriate techniques for observing and recording the trajectories of subjects while they were trying to home, these answers usually remained at a speculative level and often had to call upon the parsimony principle to decide among possible alternatives. It is certainly of "paradigmatic" importance that many, if not all major breakthroughs in this area were based on the direct observation of the trajectories of homing animals and of how these trajectories were affected by experimental manipulations [to name a few discoveries: at the beginning of the 20th century, the use of the sun as a guiding cue by ants (for references, see Wehner, 1992) and the role of mucus on trail following by limpets (e.g., Piéron, 1909); in the 1950s, the use of a time-compensated sun compass by arthropods and birds (for references, see Wehner, 1992, and Papi & Wallraff, 1992)]. Generalizing on a statement by Able (1995, p.601) about the study of long-distance orientation and navigation in birds, one can say that: "advances have tended to be more rapid and with fewer dead ends when we have followed the leads presented to us by the spontaneous behavior of (animals) and formed our hypotheses not in the vacuum of theory but in the empirically-driven milieu of our experimental results".

Homing phenomena and strategies

A *homing strategy* is "a course of action taken by an animal in order to return to its home range" from a distant place, regardless of whether this distant place has been self-selected by the animal during a spontaneous off-home range movement, or imposed by an experimenter in a displacement experiment" [see "Displacement experiments" on p.259]. "The concept . . . implies a set of alternative rules of conduct, the choice of which is decided upon depending on circumstances, as and when they appear" (Bovet, 1992, p.343). Such possible rules of conduct, or *homing phenomena*, have been classified by different authors in different ways and on different criteria, following the advancement of knowledge in the field, but resulting into a confusing terminological jungle (review in Papi, 1992b). The most recent classification is that proposed by Papi (1992b) himself. It includes all the processes that are currently under discussion and is based on the origin of the information that an animal uses to find its way back home from a distant place. What follows is an attempt at further characterizing Papi's categories with reference to levels of spatial knowledge or memory that can be assumed to be required for any such rule of conduct to be applicable. It is taken for granted that as soon as a homing individual happens to re-enter its home range, irrespective of what rule(s) of conduct it has observed, it recognizes the place as being part of the home range (thus as being its goal), and switches to short-distance travel procedures (see "Travels within the home range boundaries", p.250).

Random or systematic search. When starting its homing trip, an animal could have no knowledge nor any clue whatsoever as to its own position relative to its home. This "lost" animal, however, has a certain probability of homing within a reasonably short time by just moving along. This probability will be low if the trajectory is "Brownian", that is, in the present context, if the probability of any one "step" having such-and-such length and orientation is totally independent of the length and orientation of the previous step. But it can be substantially increased with the observance of elementary rules of systematic search (e.g., covering a centrifugal spiral trajectory).

Trail following. An animal which marks its outbound trajectory by whatever means (e.g., with mucus, an odour, or footprints, etc.) could backtrack it, thus home, by simply following the marks, like Tom Thumb and his pebbles. This rule of conduct does not involve any spatial memory or knowledge, in the sense that the animal does not remember a route, but rather an essentially single signal which stretches all along the route.

Genetically based orientation. An animal could be genetically programmed to perform long-distance movements along a certain compass axis, also possibly over a certain distance; an outbound trip would be along some compass direction, the eventual homing journey along the opposite direction. Even though the wording might seem inappropriate to some readers of this Handbook, this rule of conduct is based on the long-term genetically encoded spatial memory of two opposite compass directions and a distance, the latter being probably not encoded as such, but as travel time (for references and discussion, see Gwinner, 1996; Helbig, 1996).

Route-based orientation. An animal which, while moving along, memorizes some spatial aspects of its outward journey could use them for determining its eventual homing route. (1) If the outward route is relatively direct, thus consistently oriented in a given compass direction, the animal could simply memorize that direction and eventually home by *course reversal*, keeping its homing trajectory relatively straight and pointed at an overall compass bearing inverse of that of the outward journey. (2) *Route reversal* would occur if the animal memorizes a series of consecutive landmarks in the environment during its (possibly winding) outward journey, and retraces it in inverse order for homing. (3) If, by cumulative assessment of its turns and translations during its journey, the animal is able to evaluate continuously its position relative to its home site, it could home by *dead reckoning* or *path integration* (for discussion of various models, see Gallistel, 1990; Benhamou & Séguinot, 1995; Maurer & Séguinot, 1995; McNaughton et al., 1996). Compared to course or route reversal, path integration allows more flexible trajectory patterns, such as a winding outward route and a direct homing route. Information used for route-based orientation is often assumed to be stored in short-term memory, and to be deleted after completion of a round-trip back to home. However, it is plausible that information collected during an outward journey for eventual use for homing could also be kept in long-term memory "for further reference" [e.g., for an eventual seasonal migration; or an occasional visit to a distant resource; or the building of an extended cognitive map (Gallistel, 1990)].

Pilotage. An animal could have enough familiarity with sets of landmarks outside its home range to organize them as a cognitive map, enabling it to reach its home range by switching from one landmark to the other in whatever appropriate order, irrespective of what information it might or might not have picked up during the outward journey. This is obviously a process that rests on information stored in long-term memory.

True navigation. Independently of what information it might or might not have picked up during the outward journey, an animal going to return could find in its immediate environment the information necessary to "know" or "compute" the compass direction to its home, and have the means of selecting that direction in that environment; it could then home by travelling along that direction until it hits its home range. This is the well-known map-and-compass concept proposed by Kramer (1953) when it was becoming evident that a number of animals not only have the ability to select compass directions, but make use of it in actual homing tasks. (1) The information necessary to "know" the appropriate direction could be drawn from a "home-centred cognitive map", where a number of distant cues (visual, acoustic, olfactory...) are coded according to their compass bearing from home, as memorized by the animal while being in its home range. The resulting process would be *map-based navigation*. (2) On the other hand, this information could be obtained from the scalar values of two factors (physical, chemical...) varying along gradients that are at an angle with each other so that their isolines form a grid. The direction to home could then be "computed" by comparing the local values of the two factors with those prevailing at home. The resulting process would be *grid-based navigation*. True navigation requires the long-term memory storage of pertinent spatial information. In the case of map-based navigation, this would be in the form of an elaborate "polar" cognitive map. But in the case of grid-based navigation, the memory of the scalar values of the two critical factors at home would suffice.

Although there is a tendency for authors to stress either of those phenomena as "the" explanation for the homing performances of their study animals, it should be clear that they are not mutually exclusive; a homing strategy is not only the choice of an appropriate rule of conduct, it can also be a combination and/or a succession of several rules for the same homing trip, depending on conditions.

Spontaneous behaviour

Spontaneous long-distance travels occur in the context of dispersal, return migrations, or excursions (see the first section on p.239, and Fig. 7.1 for definitions and additional explanations) that lead an animal at least a few home range lengths away from its place of departure (and eventually back to it in the case of migrations and excursions). This section focuses on two specific aspects of spontaneous long-distance travels in mammals: the overall geometry of their trajectories, and the frequency of their occurrence, which are the crux of what it is in a free-ranging mammal's life that theory and experimental work should account for with respect to the topic of this chapter. [See Papi (1992a) for comparable aspects in other taxonomic groups. Questions related to other aspects (e.g., the adaptive value) of dispersal are discussed for instance in Anderson (1989) and Stenseth & Lidicker (1992); of return migrations in Sinclair (1983); of excursions in Baker (1982) and Cowan (1983)].

Dispersal

Much of what is known of the spatio-temporal aspects of dispersal in many species or populations is limited to places (and approximate times) of departure and/or arrival. There is, however, an increasing number of studies where trajectories are documented, in whole or in part (e.g., Harrison, 1992; Schwartz & Franzmann, 1992; Rado et al., 1992; Ross & Jalkotzy, 1992; Belden & Hagedorn, 1993; Durner & Amstrup, 1995). Overall, they suggest that dispersing individuals travel along routes that are made of a single or only a few long, essentially direct stretches (see example in Fig. 7.5a). The relative straightness of these routes does not seem to be linked to prominent linear structures of the terrain, other than accidentally. Bee-line distance and overall compass direction of dispersal can vary widely among dispersers originating from the same population. Dispersal occurs in virtually any species. Not all individuals of a given species or population disperse, however, and for those which do, dispersal is often a once-in-a-life event, that normally occurs about the time of reaching sexual maturity.

Return migrations

Not all mammals perform seasonal return migrations. For those which do (notably among bats, ungulates and marine mammals), much of the evidence available on their migratory behaviour is limited to places and times of departure and/or arrival, as for dispersal. However, since the components of a full migratory round-trip are usually once-a-year events, individuals who are sufficiently long-lived perform several migratory cycles in their lifetime. Overall, there is much evidence for a high rate of site fidelity: year after year, individuals tend to go back to, and resettle in the same seasonal home ranges (e.g., Clapham et al., 1993; Baker et al., 1995; for further examples and references, see Bovet, 1992). When documented (e.g., Brown, 1992; Danilkin et al., 1992; LeBoeuf et al., 1993; Grigg et al., 1995; for further examples and references, see Bovet, 1992), actual trajectories bear close analogy with those observed in dispersal: routes are essentially direct and their straightness cannot be systematically related to linear structures in the environment; distances and overall courses can vary widely among conspecifics, even when they share a same seasonal range. Furthermore, although both the return and the outbound trajectories of a full migratory cycle are relatively straight and originate where the other ends, the former is not necessarily a backtrack of the latter (see example in Fig. 7.5b).

Excursions

Excursions are of short duration (e.g., a few hours in rodents), they do not involve a long-term settlement in an alternative home range, and they both originate from, and end in the current home range (see Lawrence, 1928: ". . . the point of an excursion is that you come home again"). For all these reasons, excursions are not easy to detect even "after the fact" and are thus

7. LONG-DISTANCE TRAVELS AND HOMING   257

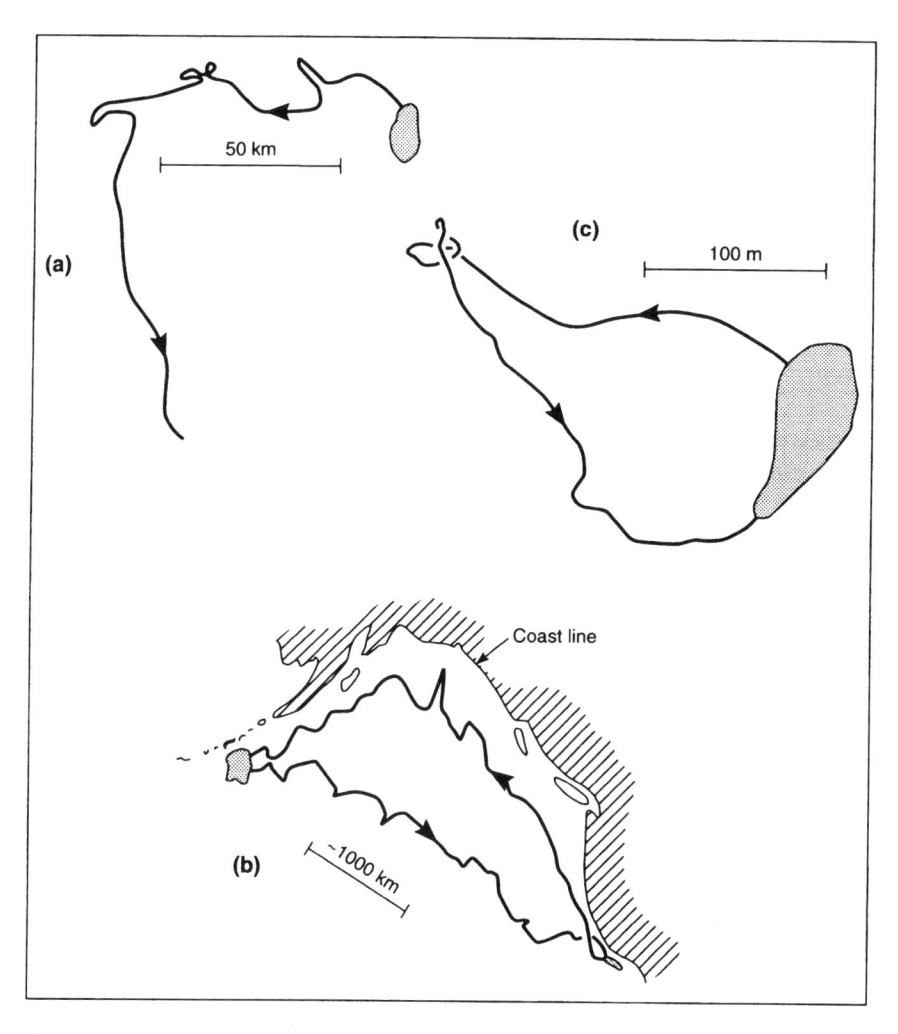

FIG. 7.5 Examples of trajectories recorded during spontaneous long-distance travels. Shaded areas are home ranges. (a) Dispersal by a cougar (adapted from Belden & Hagedorn, 1993). (b) Return migration by a northern elephant seal, from California to Alaska and back (adapted from DeLong et al., 1992). (c) Excursion by a red squirrel (original, courtesy H. Lair).

still relatively poorly documented. The slowly accumulating evidence (e.g., Gese & Mech, 1991; Thomson et al., 1992; Smith, 1993; Christian, 1994; Larsen & Boutin, 1994; Salsbury & Armitage, 1994; for further references, see Bovet, 1992) is to the effect that, as in return migrations, the trajectories of the outbound and inbound (= homing) legs of an excursion are essentially direct, are not an exact inverse copy of each other, and do not systematically follow linear structures in the environment (see example in

Fig. 7.5c). Consecutive excursions by a same individual can be in a variety of compass directions. There seems, however, to be a trend for the span of excursions not to vary much from a standard value for a given species. Excursions are performed by any class of individuals (juveniles or adults, males or females), but infrequently: they are not "routine activities" (*sensu* Jewell, 1966: see the section, "Measuring home range size and shape", p.248).

What theory and experiments should be concerned with

Recurrent features of the three categories of spontaneous long-distance travels that theory and experiments should take into account and/or explain are that: (1) they are uncommon in an individual mammal's life, leaving thus little opportunity for an elaborate training process; (2) when they are performed, their trajectories are essentially direct, but are not systematically related to linear structures in the environment; (3) the orientation of these direct trajectories can be in a variety of compass directions; (4) when homing is involved, the homing trajectory is not a backtrack of the outward route other than accidentally; and (5) the goal is not a point in space, but an area which is orders of magnitude larger than the animal itself, which leaves room for inaccuracy in orientation at no vital cost.

By contrast, travels inside one's home range follow winding trajectories that are organized within relatively narrow limits along a network of tangled routes that are run over and over again, allowing thus for the learning of routes and/or landmarks and the building of a cognitive map by virtue of repetition (e.g., by trial and error). Goals tend to be punctual, the same order of magnitude as the animal itself. In these respects, paradigms related to travels within the home range are akin to paradigms discussed in other chapters of this Handbook (e.g., mazes, the Morris water task); but those related to long-distance travels are in a class apart.

The case of invertebrates

What precedes in this section, "Spontaneous behaviour", is a view on mammals, that applies to other vertebrates with some adjustments (e.g., for return migrations in birds; see Alerstam, 1990). Invertebrates, however, differ on a number of major aspects. With only a few documented exceptions (e.g., lobsters: Pezzack & Duggan, 1986), they do not perform real return migrations and their long-distance dispersal behaviour, which entomologists often call migration, is often a complex process based on the active choice of an appropriate, passive mean of transportation (e.g., wind, current) (see reviews in Drake & Gatehouse, 1995). On the other hand, the foraging trips of some arthropods (e.g., bees, ants) and molluscs (e.g., limpets) bear some resemblance to mammalian excursions (see Wehner, 1992; Chelazzi, 1992): they are round-trips of short duration (minutes to hours), the span of which can be as much as tens or hundreds of body lengths or more; however, they are performed routinely, and the goal of their homing legs tends to be punctual (e.g., the nest entrance for an ant, or the "home scar" for a limpet).

Experiments

Displacement experiments

Virtually all field experiments on long-distance travels are based on the forced *displacement* of individuals to a distant *release site* where, as the name implies, they are liberated. Their eventual behaviour is an indicator of the way they manage to solve the spatial problems they are then facing as a result of the experimental interference.

The most common kind of displacement experiment consists of displacing a subject from its home range and releasing it at a place that is known or can be reasonably assumed to be well outside the home range. The displacement can thus be considered a mimic of the outward leg of an excursion (Baker, 1982). It is essentially comparable to the dummies (or models) used in ethology, by which the experimenter can manipulate the information that an animal needs as a releasing or orientating factor for behaving according to its current motivational state. In our context, the motivation is for homing, and the pertinent information is that which an "excursionist" can leave behind or gather during the outward journey (for eventual trail following or route-based orientation, respectively), or perceive at the release site (for eventual pilotage or true navigation). Historically, and starting with the pioneering experiments that Fabre (1879/1882) performed at Darwin's suggestion, the trend has been to try to eliminate any potential input of outward journey information, in order to check whether subjects have a so-called (and somehow mystical) "sense of direction" that would allow them to head toward home from any distant unfamiliar place. This is achieved by various means, such as displacing subjects in opaque, nonventilated containers along winding routes (this latter procedure being applied in order to make it difficult for animals to evaluate their position by some dead reckoning process). It is only recently that, with an increasing interest in route-based orientation, the trend has begun to shift from eliminating outward journey information to modifying it, in order not to deprive subjects of information, but to provide them with "misleading" cues (e.g., magnetic: Mather & Baker, 1981; olfactory: Ganzhorn & Burkhardt, 1991; visual: Bovet, 1995).

The nature and quality of data will vary, depending on the experimental design (see Fig. 7.6) and the logistics available for monitoring the animals' behaviour and whereabouts after release.

Homing success and time. Continuous monitoring of the home site area(s) by radio-fixes or Capture-Marking-Recapture (CMR) will provide evidence on whether and how long after release subjects have returned to their former home range. If several subjects are displaced in similar conditions (e.g., direction, distance, past experience, manipulation of outward journey information, etc.), the proportion which successfully returns represents the *homing success* for that set of conditions. In the mammal literature, homing success has long been the main, if not the only result considered in displacement experiments (review by Joslin, 1977). With practice, however, it appears that interpreting homing success figures in terms of how the spatial task is performed remains inconclusive because irrespective of displacement procedures, actual homing can be achieved by many different processes, and failure to home can be due to a number of circumstances that are unrelated to orientation (e.g., death through predation, loss of motivation; see Bovet, 1992). This is why, as stressed by Wilkinson (1952) on the basis of simulations, adequate data on the orientation of travel are crucial for identifying what processes are at work. Indeed, far more progress has been accomplished in the study of long-distance travels in homing pigeons on the basis of direct evidence on orientation (discussed below) than on homing success figures (see Papi & Wallraff, 1992; Able, 1995).

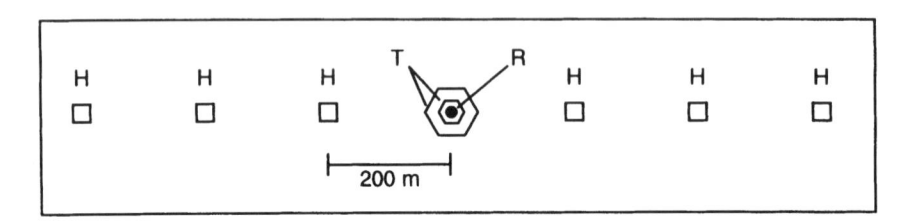

FIG. 7.6 Example of an elaborate CMR design for a displacement experiment with red-backed voles (adapted from Bovet, 1980). A grid of traps is set in each of six home areas (H) that are established along a 1200-metre straight line. Voles captured there early in the night are displaced 200, 400 or 600 metres direct to a common release site (R), where they are released at about midnight. In order to get away from the release site area, they must pass the two closed lines of traps (T) set at 20 and 40 metres around R, with one trap every 2.5 metres. Trapping operations are continued for several days after the last displacement has been performed. The places of recaptures made along T in the few hours after release are assumed to be indicative of intermediate orientation. Recaptures made in the original home area indicate successful homing. Repeated recaptures along T or in a wrong home area indicate the proportions of lost animals that have settled near the release site, or at a distance in the "good" or "bad" direction.

Orientation of travel. If a subject's trajectory can be recorded continuously by appropriate means (see, "Observing and recording travels in the field", p.241), the orientation of travel (and changes thereof) with respect to any meaningful direction of reference (e.g., home range, some environmental factor, north, or predicted as a result of experimental manipulation, etc.) can be determined at any stage of the process and from any place along the route (e.g., Bovet, 1995; for further examples and references on mammals, see Bovet, 1992). When continuous tracking is not possible, information on the *initial orientation* of travel can be obtained by recording the so-called *vanishing bearing*, that is, the azimuth of the place where the subject disappears from the observer's sight with respect to the direction of reference as seen from the release site. This datum is readily and routinely recorded in work with homing pigeons (for references, see Papi & Wallraff, 1992). With mammals, however, it is of little use, because the subjects usually head for, and hide under the closest cover within seconds after release. Attempts to bypass this problem by releasing displaced animals in artificial arenas with no place to hide have produced controversial results (for references and discussion, see Bovet, 1992). However, pertinent information can be obtained on the *intermediate orientation* of travel at a later stage by an appropriate application of the CMR method (e.g., lines of traps set at some distance from and around the release site: Fig. 7.6) or by radio-fixes. The datum is then the bearing of the place of capture as seen from the release site, with respect to a direction of reference (e.g., that from release site to home). Its value is subject to the general limitations that affect CMR or radio-fixing data, among others that it is not indicative of the sinuosity of the trajectory between release site and place of capture (see the section, "Observing and recording travels in the field", p.241).

Final fate of nonhomers. If the whole study area is searched systematically and long enough by radio-fixes or CMR, potentially useful information can also be obtained on the final whereabouts of nonhomers (see Fig. 7.6; also: Wallraff, 1993).

Homing success data are proportions, and usually analysed by standard nonparametric statistical methods (e.g., Chi-square tests). Orientational data are expressed as bearings with respect to a direction of reference. The questions that usually arise revolve around whether these bearings are significantly grouped around an average bearing, and whether this average bearing is significantly different from a predicted value or from another average bearing. Since the data are distributed along a circumference, conventional "axial" tests (e.g., t-tests) are not applicable and must be replaced by corresponding tests belonging to the so-called *circular statistics* (Fig. 7.7). Batschelet's (1981) monograph is the most widely used practical reference for standard applications. Mardia (1972) goes deeper into the theory of the tests; Cabrera et al. (1991) discuss randomization techniques; and Schnute and Groot (1992) propose methods for analysing multimodal distributions.

A less common kind of displacement experiment includes the capture, translocation and release of subjects while they are engaged in long-distance travelling, and the study of how their eventual behaviour is affected by the manipulation. Classic examples are a study of seasonal migration in birds (Perdeck, 1958) and several studies of foraging behaviour in arthropods (for references, see Wehner, 1992). No such experiments have been performed with mammals. However, the observation by Nelson (1994) of the migratory behaviour of white-tailed deer that had previously failed to home in a standard displacement experiment provides essentially the same kind of data.

The quest for the compass

Some of the homing phenomena listed in the section, "Homing phenomena and strategies" (p.252), involve an orientation based on the use of compass directions, others do not. It is, therefore, important to determine whether the individuals of the species under study have an ability for *compass orientation* and, if the case arises, whether they actually use it for establishing the trajectories of their long-distance travels.

A common experimental procedure for revealing an ability for compass orientation consists in: (1) having subjects that display within a radially symmetrical arena a behaviour which is consistently oriented in any one compass direction, either spontaneously or as a result of training; and (2) verifying that an angular shift of the environmental factor that is tested as a potential basis for compass orientation induces a corresponding shift in the orientation of this behaviour. So for instance, testing for a *magnetic compass* can be done using an arena enclosed within a set of Helmholtz coils, the switching on or off of which induces a predetermined shift in orientation of the magnetic field inside the arena; testing for a *star compass* can be done under the revolvable vault of a planetarium; and testing for a *sun compass* can be done by moving an artificial sun around the arena. In the case of the sun compass, the demonstration is all the more convincing if it can be shown that the process is *time-compensated*, that is, that the subject adjusts its bearing with respect to the sun according to the time of day.

These procedures have revealed an ability for compass orientation in a number of taxa, notably in birds (see reviews in Schmidt-Koenig et al., 1991, and Wiltschko & Wiltschko, 1991, 1995). Few studies of this kind

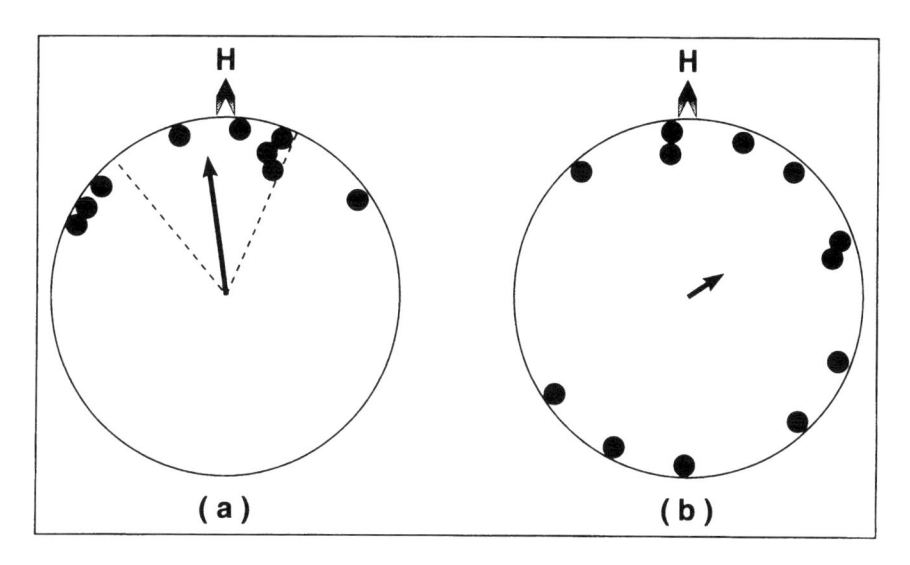

FIG. 7.7 Example of elementary applications of circular statistics, following Batschelet (1981), on data drawn from Bovet (1995). Red squirrels were displaced individually from their home range to a distant release site, >400 metres away. Soon after release, the squirrels (under radio-assisted surveillance) left the release site area along a fairly straight course but, at a median distance of about 150 metres, typically made a "U-turn" and came back to the release site. The bearings of the places where the squirrels made these U-turns are computed with respect to a direction of reference (H = 0°) which is the direction from the release site to home, and are plotted as dots on the contour of a circular diagram, the centre of which is the release site. Diagram (a) refers to 9 squirrels that were displaced along a direct route, diagram (b) refers to 12 other squirrels that were displaced along a standardized detour. For statistical treatment, each dot is considered the tip of a vector, of which the base is the centre of the circle and the length is 1. By elementary trigonometric operations (see Batschelet, 1981, for details), a mean vector is obtained (arrows in the diagrams), of which the orientation corresponds to the mean bearing of the sample [−8° in (a), +58° in (b)], and the length [0.782 in (a), 0.232 in (b)] can vary between 0 (when the distribution of individual bearings is uniform over 360°) and 1 (when all individual bearings are identical). The Rayleigh test is based on the length of the mean vector and it tests whether there is statistical evidence for a concentration of the individual bearings around the mean bearing [there is in (a) ($P < 0.002$), there is not in (b) ($P > 0.5$)]. From the appropriate charts in Batschelet (1981, p. 86), we determine that the 95% confidence interval for the −8° mean bearing in (a) is ± 32° (dashed lines), which includes the home direction: thus, the mean bearing in (a) does not differ significantly from the home direction. On the other hand, it is meaningless to determine a confidence interval for the mean bearing in (b), since there is no significant concentration of individual bearings around it. The Mardia–Watson–Wheeler test indicates whether two samples differ significantly from each other, irrespective of whether the difference is in the mean angles, in the angular variance around the mean angles, or in both. In the present example, the test indicates a "nearly significant" difference between (a) and (b), according to commonly accepted standards ($0.10 > P > 0.05$). A visual inspection of the diagrams suggests that angular variance around the mean angles is the main reason for the difference.

have been published with respect to mammals. The ones that tested for sun compass abilities produced positive evidence; but overall, the results of those testing for magnetic compass abilities are ambiguous and/or contradictory (for a review and discussion, see Bovet, 1992).

When the point is to demonstrate the actual use of compass orientation in long-distance travels, it is obviously impractical to manipulate the orientation of the natural magnetic field, the stellar configuration of the sky or the course of the sun. However, it is possible to affect predictably a subject's use of a sun compass in the field by shifting its physiological clock according to simple chronobiological procedures (see Bünning, 1973). Since the pioneering work of Schmidt-Koenig (1961), this has become a standard tool in the study of homing pigeons. Under appropriate circumstances, it can also be used with free-ranging wild animals (e.g., lizards: Ellis-Quinn & Simon, 1991), but it has not been so far with mammals.

PERSPECTIVES FOR THE STUDY OF MAMMALS

As mentioned in the first section of this chapter, research on long-distance travels in mammals is still poorly developed when compared to work carried out with birds or arthropods, for instance. Most of the research with mammals has been essentially descriptive, even though it was performed as displacement experiments. Indeed, these "experiments" have been usually designed from a perspective with no clearly stated predictions. However, they have produced a wealth of data that can be usefully correlated with well-defined sets of circumstances, or tested *post hoc* for goodness-of-fit with theoretical models (e.g., Bovet & Bovet, 1993). On the other hand, the techniques now available for continuous tracking enable us to characterize actual trajectories in both spontaneous and experimentally induced long-distance travels. The close similarity of these trajectories in both conditions, especially their common straightness (for examples and references, see Bovet, 1992), is likely to be the touchstone that has been missing for testing sharp operational predictions.

REFERENCES

Able, K.P. (1995). Orientation and navigation: A perspective of fifty years of research. *Condor*, 97, 592–604.

Alerstam, T. (1990). *Bird migration*. Cambridge: Cambridge University Press.

Alyan, S., & Jander, R. (1994). Short-range homing in the house mouse, *Mus musculus*: Stages in the learning of direction. *Animal Behaviour*, 48, 285–298.

Anderson, J.C., Berry, A.J., Amos, J.N., & Cook, J.M. (1988). Spool-and-line tracking of the New Guinea spiny bandicoot, *Echymipera kalubu* (Marsupialia, Peramelidae). *Journal of Mammalogy*, 69, 114–120.

Anderson, P.K. (1989). *Dispersal in rodents: A resident fitness hypothesis.* Provo, UT: The American Society of Mammalogists.
Anonymous (1994). Rogers, Lynn L. *Current Biography, 55,* 47–51.
Bagley, P.M., Smith, A., & Priede, I.G. (1994). Tracking movements of deep demersal fishes in the Porcupine Seabight, North-East Atlantic Ocean. *Journal of the Marine Biological Association of the United Kingdom, 74,* 473–480.
Baker, J.D., Antonelis, G.A., Fowler, C.W., & York, A.E. (1995). Natal site fidelity in northern fur seals, *Callorhinus ursinus. Animal Behaviour, 50,* 237–247.
Baker, R.R. (1982). *Migration: Paths through time and space.* London: Hodder & Stoughton.
Batschelet, E. (1981). *Circular statistics in biology.* New York: Academic Press.
Belden, R.C., & Hagedorn, B.W. (1993). Feasibility of translocating panthers into northern Florida. *Journal of Wildlife Management, 57,* 388–397.
Benhamou, S. (1990). An analysis of movements of the wood mouse *Apodemus sylvaticus* in its home range. *Behavioural Processes, 22,* 235–250.
Benhamou, S., Bovet, P., & Poucet, B. (1995). A model for place navigation in mammals. *Journal of Theoretical Biology, 173,* 163–178.
Benhamou, S., & Séguinot, V. (1995). How to find one's way in the labyrinth of path integration models. *Journal of Theoretical Biology, 174,* 463–466.
Berthold, P. (Ed.) (1991). *Orientation in birds.* Basle: Birkhäuser.
Bovet, J. (1968). Trails of deer mice (*Peromyscus maniculatus*) traveling on the snow while homing. *Journal of Mammalogy, 49,* 713–725.
Bovet, J. (1980). Homing behavior and orientation in the red-backed vole, *Clethrionomys gapperi. Canadian Journal of Zoology, 58,* 754–760.
Bovet, J. (1984). Strategies of homing behavior in the red squirrel, *Tamiasciurus hudsonicus. Behavioral Ecology and Sociobiology, 16,* 81–88.
Bovet, J. (1992). Mammals. In F. Papi (Ed.), *Animal homing.* London: Chapman & Hall.
Bovet, J. (1995). Homing in red squirrels (*Tamiasciurus hudsonicus*): The importance of going straight. *Ethology, 101,* 1–9.
Bovet, J., & Bovet, P. (1993). Computer-simulations of rodent homing behavior, using a probabilistic model. *Journal of Theoretical Biology, 161,* 145–156.
Bramanti, M., Dall'Antonia, L., & Papi, F. (1988). A new technique to monitor the flight path of birds. *Journal of Experimental Biology, 134,* 467–472.
Brown, C.G. (1992). Movement and migration patterns of mule deer in southeastern Idaho. *Journal of Wildlife Management, 56,* 246–253.
Buchler, E.R. (1976). A chemiluminescent tag for tracking bats and other small nocturnal animals. *Journal of Mammalogy, 57,* 173–176.
Bünning, E. (1973). *The physiological clock* (3rd ed.). Berlin: Springer.
Cabrera, J., Schmidt-Koenig, K., & Watson, G.S. (1991). The statistical analysis of circular data. In P.P.G. Bateson & P.H. Klopfer (Eds.), *Perspectives in Ethology,* (Vol. 9). New York: Plenum.
Chelazzi, G. (1992). Invertebrates (excluding arthropods). In F. Papi (Ed.), *Animal homing.* London: Chapman & Hall.
Christian, S.F. (1994). Dispersal and other inter-group movements in badgers, *Meles meles. Zeitschrift für Säugetierkunde, 59,* 218–223.
Clapham, P.J., Baraff, L.S., Carlson, C.A., Christian, M.A., Mattila, D.K., Mayo, C.A., Murphy, M.A., & Pittman, S. (1993). Seasonal occurrence and annual return of humpback whales, *Megaptera novaeangliae,* in the southern Gulf of Maine. *Canadian Journal of Zoology, 71,* 440–443.
Clarke, M.F., Burke da Silva, K., Lair, H., Pocklington, R., Kramer, D.L., & McLaughlin, R.L. (1993). Site familiarity affects escape behaviour of the eastern chipmunk, *Tamias striatus. Oikos, 66,* 533–537.

Clevenger, A.P. (1993). Pine marten (*Martes martes* L.) home ranges and activity patterns on the island of Minorca, Spain. *Zeitschrift für Säugetierkunde, 58*, 137–143.

Collett, T.S. (1987). The use of visual landmarks by gerbils: Reaching a goal when landmarks are displaced. *Journal of Comparative Physiology A, 160*, 109–113.

Cowan, P.E. (1983). Exploration in small mammals: Ethology and ecology. In J. Archer & L.I.A. Birke (Eds.), *Exploration in animals and humans*. Wokingham, UK: Van Nostrand Reinhold.

Danilkin, A.A., Darman, Yu.A., & Minayev, A.N. (1992). The seasonal migrations of a Siberian roe deer population. *Revue d'Ecologie (La Terre et la Vie), 47*, 231–243.

Darwin, C. (1873). Perception in the lower animals. *Nature, 7*, 360.

DeLong, R.L., Stewart, B.S., & Hill, R.D. (1992). Documenting migrations of northern elephant seals using day length. *Marine Mammal Science, 8*, 155–159.

Doncaster, C.P., & MacDonald, D.W. (1996). Intraspecific variation in movement behaviour of foxes (*Vulpes vulpes*): A reply to White, Saunders and Harris. *Journal of Animal Ecology, 65*, 126–127.

Drake, V.A., & Gatehouse, A.G. (Eds.) (1995). *Insect migration: Tracking resources through space and time*. Cambridge: Cambridge University Press.

Drickamer, L.C., Lenington, S., Erhart, M., & Robinson, A.S. (1995). Trappability of wild house mice (*Mus domesticus*) in large outdoor pens: Implication for models of t-complex gene frequency. *American Midland Naturalist, 133*, 283–289.

Duplantier, J.M., Cassaing, J., Orsini, P., & Croset, H. (1984). Utilisation de poudres fluorescentes pour l'analyse de déplacements de petits rongeurs dans la nature. *Mammalia, 48*, 293–298.

Durner, G.M., & Amstrup, S.C. (1995). Movements of a polar bear from Northern Alaska to Northern Greenland. *Arctic, 48*, 338–341.

Ellis-Quinn, B.A., & Simon, C.A. (1991). Lizard homing behavior: The role of the parietal eye during displacement and radio-tracking, and time-compensated celestial orientation in the lizard *Sceloporus jarrovi*. *Behavioral Ecology and Sociobiology, 28*, 397–407.

Etienne, A.S. (1992). Navigation of a small mammal by dead reckoning and local cues. *Current Directions in Psychological Science, 1*, 48–52.

Etienne, A.S., Joris-Lambert, S., Dahn-Hurni, C., & Reverdin, B. (1995). Optimizing visual landmarks: Two- and three-dimensional minimal landscapes. *Animal Behaviour, 49*, 165–179.

Etienne, A.S., Maurer, R., & Séguinot, V. (1996). Path integration in mammals and its interaction with visual landmarks. *Journal of Experimental Biology, 199*, 201–209.

Fabre, J.H. (1879/1882). *Souvenirs entomologiques* (1st and 2nd series). Paris: Delagrave. (Edition consulted: 1989, Paris: Laffont.)

Fancy, S.G., Pank, L.F., Whitten, K.R., & Regelin, W.L. (1989). Seasonal movements of caribou in arctic Alaska as determined by satellite. *Canadian Journal of Zoology, 67*, 644–650.

Fielden, L.J. (1991). Home range and movements of the Namib Desert golden mole, *Eremitalpa granti namibensis* (Chrysochloridae). *Journal of Zoology, 223*, 675–686.

Gallistel, C.R. (1990). *The organization of learning*. Cambridge, MA: MIT Press.

Ganzhorn, J.U., & Burkhardt, J.F. (1991). Pigeon homing: New airbag experiments to assess the role of olfactory information for pigeon navigation. *Behavioral Ecology and Sociobiology, 29*, 69–75.

Gautestad, A.O., & Mysterud, I. (1995). The home range ghost. *Oikos, 74*, 195–204.

Gehrt, S.D., & Fritzell, E.K. (1996). Sex-biased response of raccoons (*Procyon lotor*) to live traps. *American Midland Naturalist, 135*, 23–32.

Gese, E.M., & Mech, L.D. (1991). Dispersal of wolves (*Canis lupus*) in northeastern Minnesota, 1969–1989. *Canadian Journal of Zoology, 69*, 2946–2955.

Goodyear, J.D. (1993). A sonic/radio tag for monitoring dive depths and underwater movements of whales. *Journal of Wildlife Management, 57*, 503–513.

Gorman, M.L., & Zubaid, A. (1993). A comparative study of the ecology of woodmice *Apodemus sylvaticus* in two contrasting habitats: deciduous woodland and maritime sand-dunes. *Journal of Zoology, 229*, 385–396.

Grigg, G.C., Pople, A.R., & Beard, L.A. (1995). Movements of feral camels in central Australia determined by satellite telemetry. *Journal of Arid Environments, 31*, 459–469.

Gwinner, E. (1996). Circadian and circannual programmes in avian migration. *Journal of Experimental Biology, 199*, 39–48.

Hailey, A., & Coulson, I.M. (1996). Differential scaling of home-range area to daily movement distance in two African tortoises. *Canadian Journal of Zoology, 74*, 97–102.

Hansen, B. (1995). The latest word in ursine accessories. *Telonics Quarterly* [TM], *8*, 1–2.

Harris, S., Cresswell, W.J., Forde, P.G., Trewhella, W.J., Woollard, T., & Wray, S. (1990). Home range analysis using radio-tracking data: A review of problems and techniques particularly as applied to the study of mammals. *Mammal Review, 20*, 97–123.

Harrison, D.J. (1992). Dispersal characteristics of juvenile coyotes in Maine. *Journal of Wildlife Management, 56*, 128–138.

Helbig, A.J. (1991). Experimental and analytical techniques used in bird orientation research. In P. Berthold (Ed.), *Orientation in birds*. Basle: Birkhäuser.

Helbig, A.J. (1996). Genetic basis, mode of inheritance and evolutionary changes of migratory directions in palaearctic warblers (Aves: Sylvidae). *Journal of Experimental Biology, 199*, 49–55.

Herz, R.S., Zanette, L., & Sherry, D.F. (1994). Spatial cues for cache retrieval by black-capped chickadees. *Animal Behaviour, 48*, 343–351.

Jacobs, L.F. (1992). Memory for cache locations in Merriam's kangaroo rats. *Animal Behaviour, 43*, 585–593.

Jamon, M. (1994). An analysis of trail-following behaviour in the wood mouse, *Apodemus sylvaticus*. *Animal Behaviour, 47*, 1127–1134.

Jewell, P.A. (1966). The concept of home range in mammals. *Symposia of the Zoological Society of London, 18*, 85–109.

Joslin, J.K. (1977). Rodent long distance orientation ("homing"). *Advances in Ecological Research, 10*, 63–89.

Ketterson, E.D., & Nolan, V., Jr. (1990). Site attachment and site fidelity in migratory birds: Experimental evidence from the field and analogies from neurobiology. In E. Gwinner (Ed.), *Bird migration: Physiology and ecophysiology*. Berlin: Springer.

Key, G.E., & Woods, R.D. (1996). Spool-and-line studies on the behavioural ecology of rats (*Rattus* spp.) in the Galapagos Islands. *Canadian Journal of Zoology, 74*, 733–737.

Khan, J.A. (1992). Efficiency of Wonder trap against roof rat, *Rattus rattus* L. *Applied Animal Behaviour Science, 34*, 175–180.

Kramer, G. (1953). Wird die Sonnenhöhe bei der Heimfindeorientierung verwertet? *Journal für Ornithologie, 94*, 201–219.

Krebs, C.J., Kenney, A.J., & Singleton, G.R. (1995). Movements of feral house mice in agricultural landscapes. *Australian Journal of Zoology, 43*, 293–302.

Lair, H. (1987). Estimating the location of the focal center in red squirrel home ranges. *Ecology, 68*, 1092–1101.

Larsen, K.W., & Boutin, S. (1994). Movements, survival, and settlement of red squirrel (*Tamiasciurus hudsonicus*) offspring. *Ecology, 75*, 214–223.

Lawrence, D.H. (1928). *Lady Chatterley's Lover*. Firenze: Orioli. (Edition consulted: 1983, New York: Bantam Classic; the sentence quoted here is on p.45.)

LeBoeuf, B.J., Crocker, D.E., Blackwell, S.B., Morris, P.A., & Thorson, P.H. (1993). Sex differences in diving and foraging behaviour of northern elephant seals. *Symposia of the Zoological Society of London, 66,* 149–178.

Lemen C.A., & Freeman, P.W. (1985). Tracking mammals with fluorescent pigments: A new technique. *Journal of Mammalogy, 66,* 134–136.

Mardia, K.V. (1972). *Statistics of directional data.* London: Academic Press.

Mather, J.G., & Baker, R.R. (1981). Magnetic sense of direction in woodmice for route-based navigation. *Nature, 291,* 152–155.

Maurer, R., & Séguinot, V. (1995). What is modelling for? A critical review of the models of path integration. *Journal of Theoretical Biology, 175,* 457–475.

McMillan, B.R., & Kaufman, D.W. (1995). Travel path characteristics for free-living white-footed mice (*Peromyscus leucopus*). *Canadian Journal of Zoology, 73,* 1474–1478.

McNaughton, B.L., Barnes, C.A., Gerrard, J.L., Gothard, K., Jung, M.W., Knierim, J.J., Kudrimoti, H., Qin, Y., Skaggs, W.E., Suster, M., & Weaver, K.L. (1996). Deciphering the hippocampal polyglot: the hippocampus as a path integration system. *Journal of Experimental Biology, 199,* 173–185.

Mech, L.D. (1983). *Handbook of animal radio-tracking.* Minneapolis, MN: University of Minnesota Press.

Miles, M.A., de Souza, A.A., & Povoa, M.M. (1981). Mammal tracking and nest location in Brazilian forest with an improved spool-and-line device. *Journal of Zoology, 195,* 331–347.

Morgan, C.L. (1894). The homing of limpets. *Nature, 51,* 127.

Nelson, M.E. (1994). Migration bearing and distance memory by translocated white-tailed deer, *Odocoileus virginianus*. *Canadian Field-Naturalist, 108,* 74–76.

Papi, F. (Ed.) (1992a). *Animal homing.* London: Chapman & Hall.

Papi, F. (1992b). General aspects. In F. Papi (Ed.), *Animal homing.* London: Chapman & Hall.

Papi, F., & Wallraff, H.G. (1992). Birds. In F. Papi (Ed.), *Animal homing.* London: Chapman & Hall.

Perdeck, A.C. (1958). Two types of orientation in migrating starlings, *Sturnus vulgaris* L., and chaffinches, *Fringilla coelebs* L., as revealed by displacement experiments. *Ardea, 46,* 1–37.

Pezzack, D., & Duggan, D.R. (1986). Evidence of migration and homing of lobster (*Homarus americanus*) on the Scotian shelf. *Canadian Journal of Fisheries and Aquatic Sciences, 43,* 2206–2211.

Piéron, H. (1909). Contribution à la biologie de la patelle et de la calyptrée. Le sens du retour et la mémoire topographique. *Archives de Zoologie Expérimentale et Générale* (5e série), *1,* 18–29.

Poucet, B. (1993). Spatial cognitive maps in animals: New hypotheses on their structure and neural mechanisms. *Psychological Review, 100,* 163–182.

Rado, R., Wollberg, Z., & Terkel, J. (1992). Dispersal of young mole rats (*Spalax ehrenbergi*) from the natal burrow. *Journal of Mammalogy, 73,* 885–890.

Ramsay, M.A., & Andriashek, D.S. (1986). Long distance route orientation of female polar bears (*Ursus maritimus*) in spring. *Journal of Zoology, 208,* 63–72.

Rempel, R.S., Rodgers, A.R., & Abraham, K.F. (1995). Performance of a GPS animal location system under boreal forest canopy. *Journal of Wildlife Management, 59,* 543–551.

Riley, J.R., Smith, A.D., Reynolds, D.R., Edwards, A.S., Osborne, J.L., Williams, I.H., Carreck, N.L., & Poppy, G.M. (1996). Tracking bees with harmonic radar. *Nature, 379,* 29–30.

Ross, P.I., & Jalkotzy, M.G. (1992). Characteristics of a hunted population of cougars in southwestern Alberta. *Journal of Wildlife Management, 56,* 417–426.

Salsbury, C.M., & Armitage, K.B. (1994). Home-range size and exploratory excursions of adult, male yellow-bellied marmots. *Journal of Mammalogy, 75,* 648–656.

Samuel, M.D., & Fuller, M.R. (1994). Wildlife radiotelemetry. In T.A. Bookhout (Ed.),

Research and management techniques for wildlife and habitats (5th ed.). Bethesda, MD: The Wildlife Society.
Schmidt-Koenig, K. (1961). Die Sonne als Kompass im Heim-Orientierungssystem der Brieftauben. *Zeitschrift für Tierpsychologie, 18,* 221–244.
Schmidt-Koenig, K., Ganzhorn, J.U., & Ranvaud, R. (1991). The sun compass. In P. Berthold (Ed.), *Orientation in birds.* Basle: Birkhäuser.
Schnute, J.T., & Groot, K. (1992). Statistical analysis of animal orientation data. *Animal Behaviour, 43,* 15–33.
Schwartz, C.C., & Franzmann, A.W. (1992). Dispersal and survival of subadult black bears from the Kenai Peninsula, Alaska. *Journal of Wildlife Management, 56,* 426–431.
Sherry, D.F., & Duff, S.J. (1996). Behavioural and neural bases of orientation in food-storing birds. *Journal of Experimental Biology, 199,* 165–171.
Sinclair, A.R.E. (1983). The function of distance movements in vertebrates. In I.R. Swingland & P.J. Greenwood (Eds.), *The ecology of animal movement.* Oxford: Clarendon Press.
Sinsch, U. (1987). Orientation behaviour of toads (*Bufo bufo*) displaced from the breeding site. *Journal of Comparative Physiology A, 161,* 715–727.
Smith, J.L.D. (1993). The role of dispersal in structuring the Chitwan tiger population. *Behaviour, 124,* 165–195.
Sokolov, V.E., Rodionov, V.A., Sukhov, V.P., Kudryashov, V.S., & Kuznetzov, M.S. (1977). [A radiotelemetrical study of diurnal activity in *Castor fiber*]. *Zoologicheskii Zhurnal, 56,* 1372–1380. (In Russian, with English summary)
Stamps, J. (1995). Motor learning and the value of familiar space. *American Naturalist, 146,* 41–58.
Stapp, P., Young, J.K., VandenWoude, S., & Van Horne, B. (1994). An evaluation of the pathological effects of fluorescent powder on deer mice (*Peromyscus maniculatus*). *Journal of Mammalogy, 75,* 704–709.
Stenseth, N.C., & Lidicker, W.Z., Jr. (Eds.) (1992). *Animal dispersal: Small mammals as a model.* London: Chapman & Hall.
Thomson, P.C., Rose, K., & Kok, N.E. (1992). The behavioural ecology of dingoes in north-western Australia. VI. Temporary extraterritorial movements and dispersal. *Wildlife Research, 19,* 585–595.
Wagner, G. (1970). Verfolgung von Brieftauben im Helicopter. *Revue suisse de Zoologie, 77,* 29–60.
Wallace, A.R. (1873). Perception and instinct in the lower animals. *Nature, 8,* 65–66.
Wallraff, H.G. (1993). Correct and false olfactory orientation of homing pigeons as depending on geographical relationships between release site and home site. *Behavioral Ecology and Sociobiology, 32,* 147–155.
Wehner, R. (1992). Arthropods. In F. Papi (Ed.), *Animal homing.* London: Chapman & Hall.
White, G.C., & Garrott, R.A. (1990). *Analysis of wildlife radio-tracking data.* San Diego, CA: Academic Press.
White, P.C.L., Saunders, G., & Harris, S. (1996). Spatio-temporal patterns of home range use by foxes (*Vulpes vulpes*) in urban environments. *Journal of Animal Ecology, 65,* 121–125.
Wilkinson, D.H. (1952). The random element in bird "navigation". *Journal of Experimental Biology, 29,* 532–560.
Wiltschko, R., & Wiltschko, W. (1995). *Magnetic orientation in animals.* Berlin: Springer.
Wiltschko, W., & Balda, R.P. (1989). Sun compass orientation in seed-caching scrub jays (*Aphelocoma coerulescens*). *Journal of Comparative Physiology A, 164,* 717–721.
Wiltschko, W., & Wiltschko, R. (1991). Magnetic orientation and celestial cues in migratory orientation. In P. Berthold (Ed.), *Orientation in birds.* Basle: Birkhäuser.
Winter, J.D. (1983). Underwater biotelemetry. In L.A. Nielsen & D.L. Johnson (Eds.), *Fisheries techniques.* Bethesda, MD: American Fisheries Society.

Author Index

Aadland, J., 113
Abdulla, F.A., 173, 174, 203
Able, K.P., 241, 252, 260
Abler, R., 2
Abraham, K.F., 244
Abraham, W.C., 106, 108
Abu-Bakra, M.A.J., 174
Adams, J.S., 2
Aggleton, J.P., 211, 217, 223
Aigner, T.G., 174
Aihara, N., 228
Al-Khamees, K., 225
Alberoni, M., 38
Albert, P., 161, 163, 170
Albin, R.W., 165
Alerstam, T., 246, 258
Alexander, D., 43
Alexinsky, T., 165
Alivisatos, B., 39
Allen, Y., 190, 198, 199, 208, 214, 215, 227, 228
Alyan, S., 155, 251
Amaral, D.G., 223
Ammassari-Teule, M., 66, 78, 93, 94, 106, 110, 114, 115, 117–119
Amos, J.N., 245
Amstrup, S.C., 256
Anderson, D.J., 174
Anderson, E., 148, 172
Anderson, G.M., 93, 96, 121, 126
Anderson, J.C., 245
Anderson, P.K., 255
Andrews, M.W., 113, 114
Andriashek, D.S., 246
Angelucci, L., 91, 111, 112, 118, 120, 125, 128
Anisman, H., 93, 114, 126, 165
Annett, L.E., 228
Antonelis, G.A., 256
Appleyard, D., 2
Arber, M., 113
Archer, T., 112
Arendash, G.W., 112
Arendt, T., 190
Arguin, H., 35
Armitage, K.B., 257
Arolfo, M.P., 156, 158, 159, 164, 165, 177
Arthur, G.A., 22
Ashikhmina, O.V., 116
Assal, G., 46
Attree, E.A., 225, 226
Augerinos, G., 91, 95, 98, 99, 103, 104, 113, 125, 126, 127
Awh, E., 38
Ayres, A.J., 16

Baade, L.E., 23
Backer Cave, C., 49, 50
Backes, R.C., 176
Baddeley, A.D., 34, 35, 36, 37, 39, 40
Bagley, P.M., 244
Bailey, D.W., 101, 113
Baisden, R.H., 99
Baker, H.F., 223
Baker, J.C., 153, 154, 156, 157, 177, 178
Baker, J.D., 256
Baker, R.R., 255, 259
Balda, R.P., 105, 114, 251
Bandera, R., 35
Bannerman, D.M., 172, 173
Baraff, L.S., 256
Baranowski, J.R., 107
Bare, D.J., 157
Barnes, C.A., 130, 146, 150, 254
Barnett, A.M., 99, 100, 107, 176
Baron, S.P., 91, 96, 118, 127, 128, 213, 214
Barta, A., 167
Barton, A., 40
Barton, M., 21
Bartus, R.T., 190
Basso, A., 37
Basso, G., 27
Batschelet, E., 262, 263
Batson, J.D., 116
Bättig, K., 124, 196
Baudry, M., 148, 172
Beard, L.A., 256
Beatty, W.W., 93, 96, 104, 105, 109, 111, 113, 116, 117, 124, 126
Beaumont, J.G., 13, 19
Beck, C.H.M., 118, 119, 121, 125
Becker, J.T., 106–108, 110, 115, 126, 171, 200, 217, 222
Beer, B., 190
Beerten, A., 39
Belden, R.C., 256
Belleville, S., 35
Bellugi, U., 29
Belyavin, M., 40
Benhamou, S., 150, 177, 246, 251, 254
Beninger, R.J., 107, 108, 116, 190, 193, 217
Bennett, G.W., 91, 94, 109
Bennett, M.C., 166
Benson, F., 42
Benton, A.L., 18, 23
Bereacochea, D.J., 193, 211, 222
Berg, B., 105
Berlyne, D. E., 60, 61

Berman, R.F., 111, 119, 125, 126
Bernstein, D., 114
Bernstein, I.G., 112, 114
Berry, A.J., 245
Berthold, P., 251
Berthoz, A., 6
Best, M.R., 116, 121
Best, P.J., 107, 118, 125
Bhavnani, G., 21
Biegler, R., 198
Bierley, R.A., 105, 109, 117, 127, 128
Birch, D., 196
Bishop, T.W., 126
Bisiach, E., 25, 47
Bitterman, M.E., 112
Bjerkemo, M., 92, 95
Bjorklund, A., 190, 228
Bjornsson, T.D., 94, 114
Black, A.H., 91, 95, 98, 99, 103, 104, 113, 125, 126, 127
Blackwell, S.B., 256
Blank, G.S., 153
Blozovski, D., 174
Bodnoff, S.R., 166
Boegman, R.J., 107, 108, 116
Boetz, M.I., 156
Bolhuis, J.J., 129
Boller, F., 35, 36
Bolles, R.C., 90, 121, 125
Bolson, B., 64, 65
Bolton, R.L., 67
Bond, A.B., 93, 114
Book, A., 46
Bostock, E., 91, 93, 100, 111
Bottini, G., 42
Boutin, S., 257
Bouzouba, L., 66, 68
Bovet, J., 246, 247, 251, 252, 256, 259–261, 263, 264
Bovet, P., 177, 251, 264
Boyd, T.M., 47
Bradbury, E., 227
Bramanti, M., 247
Brandeis, R., 112, 146
Brandner, C., 161, 162, 174
Brandys, Y., 146
Bravo, H., 206
Bredart, S., 39
Bressi, S., 35, 38
Brewer, B., 29
Brierley, R.A., 124, 126
Brioni, J.D., 164, 165
Brokofsky, S., 105
Brooks, D.N., 36
Brooks, L.R., 36
Brown, C.G., 256
Brown, C.R., 89, 94, 101, 118

AUTHOR INDEX 271

Brown, M.F., 87, 91, 96–99, 101, 102, 106, 107, 109, 116, 121, 124, 126, 200, 201, 204, 213
Brown, P., 203
Brown, W.L., 88
Brundin, P., 203
Bruto, V., 93, 114, 126
Buchler, E.R., 246
Buhot, M.-C., 66, 71, 73, 74–78, 80, 81, 89, 101
Bulgarelli, E., 47
Bunch, S.T., 228
Bunning, E., 264
Bunsey, M., 105
Burchinal, M., 150, 208
Bures, J., 95, 101, 102, 105, 116–118, 121, 124, 126, 129, 130, 155, 156, 158, 159, 164, 177, 202, 204, 207
Buresova, O., 92, 94, 95, 101, 102, 105, 117, 118, 121, 126, 129, 130, 158, 164, 204, 207
Burke da Silva, K., 251
Burkhardt, J.F., 259
Burmeister, S., 112
Burwell, R., 150, 208
Bussey, T.J., 223
Butcher, S.P., 172, 173
Butelman, E.R., 2
Butters, N., 22
Byatt, G., 100, 107
Byrne, R.W., 42, 43, 44, 47

Cabrera, J., 262
Cain, D.P., 172, 173
Calaminici, M.-R., 173, 174, 203
Callahan, M.J., 174
Caltagirone, C., 38
Calvanio, R., 37, 40, 42
Campbell, K., 190
Cannon, R.L., 99
Cantone, G., 35
Cappa, S.F., 38, 42
Caprioli, A., 91, 94, 111, 112, 114, 118, 120, 125, 128
Caramazza, A., 39
Carbone, C.P., 126
Carlson, C.A., 256
Carreck, N.L., 246
Cartwright, B.A., 98, 158, 177
Carvell, S.E., 206
Cassaing, J., 245
Cassel, J.-C., 172, 178, 179
Castro, C.A., 164, 195, 166, 167, 222
Cave, C.B., 224
Chaix, K., 66, 68
Chapillon, P., 111, 162, 170
Chapuis, N., 66, 68
Cheal, M.-L., 67
Chelazzi, G., 259
Chen, L.L., 155
Cheng, K., 6, 70
Chevalley, A.-F., 161, 163, 168
Chew, G.L., 154, 156, 157, 177, 178
Chew, L., 114
Chiacchio, L., 36
Chollet, F., 224
Christian, M.A., 256
Christian, S.F., 257
Cipolotti, L., 38
Clapham, P.J., 256
Clarke, M.F., 251
Clarke, S., 46

Clevenger, A.P., 250
Clouse, B.A., 104, 105, 116, 124, 126
Cocchetto, D.M., 94, 114
Cockburn, J., 20, 21
Cohen, D.J., 93, 96, 121, 126
Cohen, N.J., 190, 204, 222–224
Cole, B.J., 162, 166, 201, 210
Collett, T.S., 98, 158, 177, 251
Collette, F., 39
Collison, C., 92, 95, 99, 102, 119, 125, 200, 204, 208
Colombo, P.J., 108
Coltheart, M., 33
Conner, R.L., 92
Conrad, C.D., 166
Contant, B., 150, 174
Contant-Astrom, B., 152
Conway, D.H., 196, 200, 213
Cook, D., 202
Cook, J.M., 245
Cook, R.G., 93, 96, 106, 114
Corman, C.D., 75
Cornell, E.H., 103, 128
Corwin, J.V., 202, 203
Couillard, N.L., 95, 105, 113
Coulson, I.M., 245
Couvillon, P.A., 112
Cowan, P.E., 61, 255
Cowey, A., 26
Coyette, F., 35
Cramer, C.P., 110
Crannell, C.W., 89
Cresswell, W.J., 250
Crocker, D.E., 256
Croset, H., 246
Crowne, D.P., 203
Crutcher, K.A., 110
Culver, C.M., 19
Cummings, J.L., 42
Cunitz, A.R., 222

D'Amato, F.R., 117, 128
Dachir, S., 93
Dahn-Hurni, C., 251
Dale, R.H.I., 8, 91, 93, 99, 101, 103, 104, 110, 114, 116–119, 121, 124, 125, 127, 128
Dall'Antonia, L., 247
Dallal, N.L., 99
Dallas, M., 49
Dalrymple-Alford, J.C., 100, 107, 193, 210, 222, 229
Danilkin, A.A., 256
Danion, J.M., 51
Darman, Y.A., 256
Darwin, C., 247, 251, 259
Dashiell, J.F., 89
Davey, M.J., 225, 226
Davidoff, J.B., 13, 38
Davies, A.D.M., 42, 46
Davies, J., 97, 98, 115, 127
Davis, H.P., 107, 108, 191, 193
Davis, J.L., 91, 105, 108, 114, 126
Davis, R.E., 174
Davis, S., 172, 173
De Haan, E.H.F., 33, 41
De Kloet, E.R., 166
De Luca, G., 36
De Michele, G., 35
De Renzi, E., 28, 34, 35, 37, 38, 41, 42, 44
de Souza, A.A., 245
De Tribolet, N., 46

Deacon, R.M.J., 90, 131
Dean, P., 196
Dean, R.L., 190
DeBoissiere, T., 114
Decker, M.W., 167, 174
Degos, J-D., 35, 36
Degueldre, C., 39
Dehane, S., 25
Dekeyne, A., 165
Del Valle, R., 101, 121
Delfiore, G., 39
Dell, P.A., 225
Della Sala, S., 34, 35, 36, 38, 39
DeLoache, J., 113
DeLong, R.L., 247
Dember, W.N., 90
Dennis, W., 90
Dermen, D., 15
Devan, B.D., 153
Deweer, B., 35, 36
Deyo, R.A., 92
Diamond, A., 8
Diamond, D.M., 166
Dickenson, S.L., 91, 94, 109
Dickinson, A.K., 161
Diez-Chamizo, V., 100
DiMattia, B.D., 180
DiMattia, B.V., 110, 202, 222
Dimen, K.R., 94
Dinkla, A., 47
Doherty, D., 64
Doncaster, C.P., 250
Donohoe, T., 90, 125
Dopkins, S., 48
Dostrovsky, J., 2, 6
Dou, H., 110, 111
Douglas, R.J., 8, 77, 90, 101, 121, 127
Downs, R.M., 2
Doyle, J., 164
Drake, V.A., 258
Drickamer, L.C., 242
Dryver, E., 95, 112
Dudchenko, P., 159
Duff, S.J., 251
Duggan, D.R., 258
Dunbar, G.L., 126
Dunlap, W.P., 191
Dunnett, S.B., 174, 190, 217, 227–229
Duplantier, J.M., 246
Durkin, T.P., 90, 91, 107–109, 119, 126, 128, 131
Durner, G.M., 256
Durup, H., 227
Durup, M., 66, 68, 70, 83
Dutrieux, G., 118
Dyck, R.H., 146, 156, 157, 197

Ebendahl, T., 228
Ebendal, T., 206
Eckerman, D., 92, 94, 95, 115, 117
Edberg, J.A., 97, 99, 107, 109, 200, 204, 213
Edwards, A.S., 246
Eichenbaum, H., 152, 171, 178, 179, 190, 196, 200, 204, 222–224
Eilan, N., 29
Einon, D., 91, 93, 111, 113, 118, 124, 125, 226
Ekstrom, R.B., 15
Elithorn, A., 17
Elkins, K.M., 95, 105, 113
Ellen, P., 63, 64, 100

AUTHOR INDEX

Ellis-Quinn, B.A., 264
Elsmore, T.F., 104, 112, 114
Ennaceur, A., 110
Ergis, A-M., 35, 36
Erhart, M., 242
Ermakova, I.V., 121, 129
Etienne, A.S., 154, 251
Evans, A.C., 39
Evenden, J.L., 115
Everitt, B.J., 162, 166, 174, 201, 210, 217

Fabre, J.H., 247, 251, 259
Fagioli, S., 93
Faglioni, P., 35, 38, 42
Fairbairn, A.F., 49
Fancy, S.G., 244
Farah, M.J., 37, 40, 42
Farmer, E., 40
Farnsworth, G., 202
Fawcett, J.W., 229
Fazio, F., 38
Feldon, J., 211
Fenton, A.A., 156, 177
Ferguson, G.A., 15
Fibiger, H.C., 107, 108
Fielden, L.J., 246
File, S.E., 191
Fine, A., 228
Finger, S., 224, 226
FitzGerald, R.W., 175
Fleming, P., 197, 206, 209, 220
Fleshner, M., 166
Fletcher, A., 197, 216, 217
Fone, K.C.F., 91, 94, 109
Forde, P.G., 250
Foreman, N., 16, 75, 77, 78, 80, 90, 93, 94, 98, 99, 100, 108–110, 113, 118, 119, 121, 124, 125, 127, 128
Foster, T.C., 167
Fowler, C.W., 256
Frackowiak, R.S.J., 39, 224
Franceschi, M., 38
Francis, D.D. 165
Franck, G., 39
Franzmann, A.W., 256
Frassinetti, F., 27
Fredriksson, A., 112
Freeman, P.W., 246
French, C.C., 19
French, J.W., 15
Fricker, R.A., 190
Friston, K.J., 224
Frith, C.D., 39
Fritzell, E.K., 242
Fuller, G.B., 23
Fuller, M.R., 243
Furukawa, S., 109, 111
Furuya, Y., 217
Fussinger, M., 202, 203
Fuster, J.M., 39

Gabriel, M., 102
Gaffan, D., 222, 224
Gaffan, E.A., 97, 98, 115, 222
Gage, F.H., 211, 216
Gage, P.D., 107
Galea, L.A., 167, 170, 176
Galey, D., 90, 91, 107, 109, 119, 126, 128
Gallagher, M., 91, 93, 100, 111, 150, 161, 167
Gallagher, M., 208

Gallistel, C.R., 6, 70, 99, 154, 251, 254
Ganzhorn, J.U., 259, 262
Gardner, H., 14
Gärling, T., 46
Garrott, R.A., 243
Garrud, P., 160
Garvey, K.J., 95, 105, 113
Gatehouse, A.G., 258
Gathercole, S.E., 34, 37
Gaulin, S.J.C., 175
Gautestad, A.O., 250
Gazzaniga, M.S., 26, 27
Gehrt, S.D., 242
Geminiani, G., 42
Gentry, G., 88
Gerbrandt, L.K., 99
Gerrard, J.L., 254
Gese, E.M., 257
Ghent, L., 4, 18, 43
Ghirardi, O., 111, 112, 118, 120, 125, 128
Gilbert, M.E., 99, 117
Gilleland, K.R., 116
Gillett, R., 95, 109, 110, 113
Giraudo, M.-D., 47
Gisquet-Verrier, P., 165
Giuliani, A., 91, 111, 118, 120, 125, 128
Givens, B., 174
Glanzer, M., 61, 222
Glassman, R.B., 95, 105, 113
Godding, P.R., 93
Gold, P.E., 167
Goldmeier, E., 48
Gollin, E.S., 16
Golub, L., 114, 115
Gonzalez, M.I., 96
Good, M.A., 172, 173
Goodale, M.A., 93, 110
Goodlett, C.R., 154, 163
Goodridge, J.P., 155, 159
Goodyear, J.D., 244
Gorman, M.L., 248
Gothard, K., 254
Gottschaldt, K., 36
Gould, E., 166
Gould, P., 2
Grabowski, M., 203, 228, 229
Grange, D., 51
Granon, S., 71, 73, 74, 150, 180
Grassi, F., 38
Gray, J.A., 62, 190, 193, 197–199, 203, 204, 206, 208–212, 214–220, 225, 227, 228, 230
Green, R.J., 107, 108
Green, S., 201, 208
Grigg, G.C., 256
Grigoryan, G.A., 211
Grobety, M.-C., 94, 101, 116, 125, 159, 160, 177, 178
Groot, K., 262
Grossi, D., 35, 36
Guajardo, G., 206
Guic-Robles, E., 206
Gwinner, E., 253

Habib, M., 42
Hadley, D.M., 36
Hagan, J.J., 161, 162, 173, 174, 208, 217
Hagedorn, B.W., 256
Haig, K.A., 110, 115
Hailey, A., 245
Hain, J.D., 23

Hall, C.S., 60
Hall, G.S., 1
Hall, J., 119, 125, 126
Halligan, P.W., 20, 26
Hambrecht, K.L., 108
Hamilton, W.J., 124
Hammond, K.M., 37, 40, 42
Hamsher, K. de S., 18
Handelmann, G.E., 106–108, 110, 115, 126, 171, 200, 217, 222
Hanley, J.R., 35, 36, 38, 40, 42, 46, 47
Hansel, M.C., 115
Hansen, B., 243
Harbaugh, R.E., 174
Harbluk, J.L., 49
Hardy, W.T., 102
Hargreaves, E.L., 95, 121, 176
Harley, C., 88
Harley, C.W., 159
Harman, H.H., 15
Harper, D.N., 193, 222
Harris, L.J., 8
Harris, S., 250
Harrison, D.J., 256
Harrison, S., 224
Hart, R.H., 101, 113
Hasher, L., 48
Hatanaka, H., 106
Hay, D.C., 33, 41
Healy, S.D., 113
Heaton, R.K., 23
Hebb, D.O., 1
Heilman, K.M., 13, 19, 25
Hein, A., 5
Helbig, A.J., 241, 253
Held, R., 5
Hemmings, R., 16
Henderson, C., 111
Herrmann, T., 64, 65
Herz, R.S., 251
Hess, C., 174
Heth, C.D., 103, 128
Hewitt, M.J., 91, 94, 109
Higashida, A., 108, 114, 119
Hill, R.D., 247
Hilton, S.C., 114
Hirsch, R., 104
Hirst, W., 48, 49
Hitch, G., 34
Hodges, H., 146, 175, 190, 193, 195, 197–199, 201, 203–206, 208–212, 214–220, 223, 225, 227, 228, 230
Hoffman, G.E., 175
Hoffman, N., 104, 105, 126
Hoffman, H.J., 106, 110, 124
Holmes, G., 3
Homuta, L., 158
Honig, W.K., 100, 107, 114, 146
Hooper, H.E., 16
Horner, J., 88, 101
Hoyles, E.S., 159
Huggins, C.K., 87, 102
Humphreys, A.G., 166
Humphreys, G.W., 33, 42, 43, 46
Hunt, P.R., 211, 217, 223
Hurly, T.A., 113
Huston, J.P., 206
Hutt, M.L., 23
Hynes, C.A., 159

Ijaz, S., 114
Ikegami, S., 106
Ilersich, T.J., 116

AUTHOR INDEX 273

Ingle, D., 67, 71, 72
Ingram, D.K., 114
Innes, D., 175
Innis, N.K., 99, 103, 116, 117, 119, 121, 125, 127, 128
Introini-Collison, I.B., 167
Isaacson, R.L., 77
Itoh, M., 90, 115
Itoh, T., 90, 115
Iversen, S.D., 173
Ivy, G., 99
Iwamoto, E.T., 90, 92, 103, 106, 107, 109
Iwasaki, T., 109, 111
Izquierdo, I., 164, 165

Jacobs, J.H., 196
Jacobs, L.F., 175, 251
Jacoby, L.L., 49
Jaffard, R., 90, 91, 107, 109, 119, 126, 128, 131, 193, 211, 222
Jalkotzy, M.G., 256
James, M., 19, 38
Jamon, M., 250
Jander, R., 185, 251
Janis, L.S., 126
Jarrard, L.E., 107, 109, 164, 171, 172, 178, 179, 191, 200, 201, 213, 217, 228
Jastrow, J., 1
Jeffrey, T.E., 16
Jenkins, M.W., 206
Jerusalinsky, D., 164, 165
Jewell, P.A., 258
Jhamandas, K., 107, 108, 116
Job, R., 33
Johansson, B.B., 203, 228, 229
Johnson, C.T., 164, 197, 209, 216
Johnson, K.L., 23
Johnston, C.S., 47
Johnston, T.D., 71
Jones, C.H., 107, 113
Jones, D., 17
Jones, G., 48, 49
Jones, S., 109, 110, 113
Jonides, J., 38
Joravsky, D., 5
Joris-Lambert, S., 251
Joslin, J.K., 260
Joubert, A., 67
Jung, M.W., 254
Juraska, J.M., 110, 111

Kadar, T., 93, 112
Kalish, D., 88, 89
Kametani, H., 92, 111, 125
Kamil, A.C., 93, 103, 105, 114, 115, 124, 127, 208
Kaplan, S., 88
Kasal, K.L., 95, 105, 113
Kaufman, D.W., 250
Kavaliers, M., 167, 170, 176
Kawamura, H., 106
Keith, A.B., 217
Keith, J.R., 153, 176
Kelche, C., 210, 229
Kelsey, J.E., 173, 217, 223
Kenealy, P.M., 13
Kenney, A.J., 250
Kerbusch, J.M.L., 131, 195
Kerkhoff, G., 21
Kermisch, M.G., 104, 114, 116
Kerr, M., 17
Kershaw, T., 190, 197, 198, 199,

206, 208, 209, 214, 216, 217–220, 227, 228
Keseberg, U., 107, 109
Kesler, M., 105
Kesner, R.P., 6, 195, 110, 180, 202, 222
Kessler, J., 94, 130
Ketterson, E.D., 251
Key, G.E., 245
Khan, J.A., 242
Kimble, D., 113, 176
King, R.A., 91, 93, 100, 111
King, V.R., 202, 203
Kinoshita, M., 113
Kinsbourne, M., 38
Kinsora, J.J., 174
Kirk, S.A., 16
Kirk, W.D., 16
Kiyota, Y., 193
Kjellstrand, P., 92, 95
Klein, D.B., 1
Kluge, P.B., 112, 114
Knierim, J.J., 155, 157, 254
Kobayashi, S., 92, 111, 125
Koeppe, R.A., 38
Koide, H.M., 228
Koide, K., 228
Kok, N.E., 257
Kolb, B.E., 180, 225
Komisaruk, B.R., 206
Kopelman, M.D., 49
Kovner, R., 48
Kraemer, P.J., 99, 117
Kramer, D.L., 251
Kramer, G., 254
Krebs, C.J., 250
Krebs, J.K., 114
Krechevsky, I., 128
Krekule, I., 105, 116, 118, 124, 158, 164
Kritchevsky, M., 29
Kromer, L.F., 223
Kubie, J.L., 112
Kudrimoti, H.S., 155, 254
Kudryashov, V.S., 243
Kumazaki, M., 228
Kunka, M.G., 112
Kupke, T., 17
Kuznetzov, M.S., 243

Lachman, S.J., 89, 94, 101, 118
Ladavas, E., 26, 27
Lair, H., 247, 251
Laird, J.T., 23
Lalonde, R., 156
Lamb, M.R., 93, 114
Landis, T., 42
Landry, B.A., 173
Lanke, J., 92, 95
Lantos, P.L., 190, 198, 199, 208, 214, 227, 228
Larkfors, L., 228
Larsen, K.W., 257
Larsen, T., 222
Lasalle, J.M., 90, 96, 111, 114, 127
Lashley, K.S., 5
Lavenex, P., 103, 150, 157, 159, 160, 177, 178
Lawrence, D.H., 256
Laws, K.R., 28
Le Moal, M., 91, 93
Le Peillet, E., 193, 204, 206, 209, 210, 216–219, 227, 230
Lebessi, A., 204

LeBoeuf, B.J., 256
Lebrun, C., 131
Lee, D., 17
Lee, H., 88
Lee, R.K.K., 165
Lehman, J.C., 166
Leis, T., 94, 95, 108, 117, 118, 121, 126, 127
Lemen, C.A., 246
Lenington, S., 242
Lenzi, G., 38
Leret, M.L., 96
Lescaudron, L., 193, 211, 222
Levere, T.E., 112
Levin, E.D., 110
Levine, D.N., 37, 40, 42
Levitsky, D., 105
Levy, A., 112, 114
Lewis, R., 17
Lezak, M.D., 13, 16, 23
Lichtman, A.H., 94
Lidicker, W.Z., 255
Lieberman, K., 40
Likert, R., 16
Lincoln, N., 21
Lindner, M.D., 174
Linggard, R.C., 153, 154, 156, 157, 177, 178
Lipp, H.-P., 159, 160, 177, 178
Lippa, A.S., 190
Logie, R.H., 34, 35, 38, 40, 41
Loh, E.A., 118, 119, 121, 125
London, E.D., 114
Long, J.M., 164, 197, 209, 216
Loseva, E.V., 121, 129
Love, S., 225
Lucchessi, H., 91, 98, 101
Luine, V.N., 111, 112, 166
Lukaszewska, I., 67, 77
Luria, A.R., 4
Luxen, A., 39
Luzzatti, C., 25
Lydon, R.G., 108, 109
Lyeth, B.G., 108
Lynch, G., 148, 172
Lynch, K., 2

Mabry, T.R., 167
MacAndrew, S.B.G., 48, 49
MacDonald, C., 49, 50
MacDonald, D.W., 250
MacDonald, P., 103, 108
Macgillivray, M., 119, 125
Mackintosh, N.J., 89, 100, 103, 128, 161
Maclean, C.J., 222, 223
Magni, S., 105, 116, 118, 124
Maho, C., 115, 117, 118, 119
Maier, N.R.F., 62, 88
Maier, S.F., 164, 165, 166, 167
Maisch, B., 78
Majchrzak, M.J., 174
Maki, R.H., 113
Maki, W.S., 104, 105, 116, 124, 126
Mallet, P.E., 190, 193, 217
Mandel, R.J., 211, 216
Mandler, G., 49
Mansson, L., 92, 95
Maquet, P., 39
Marchbanks, R., 190, 227
Mardia, K.V., 262
Marighetto, A., 90, 91, 107, 109, 119, 126, 128, 131
Markowitsch, H.J., 94, 130

274 AUTHOR INDEX

Markowska, A., 77, 105, 117
Markowska, A.L., 164, 174, 197, 209, 216
Markus, E.J., 155
Marquardt, C., 21
Marshall, J.C., 25, 26
Martin, B.R., 94
Martin, G.M., 159
Martinez, C., 112, 166
Marzolf, D., 113
Mason, S.E., 106, 108
Masuda, Y., 90, 115
Mather, J.G., 259
Matthews, B., 40
Mattila, D.K., 256
Mattson, N.B., 228, 229
Maurer, R., 154, 251, 254
Mayer, E., 229
Mayes, A.R., 48, 49, 50, 51, 52
Mayo, C.A., 256
Mazmanian, D.S., 98, 116
McBride, S.A., 104, 114
McCarthy, J.J., 16
McCarthy, M.M., 25
McCarthy, R.A., 28, 29
McCarty, R., 167
McDonald, R.J., 161, 172, 180, 201
McEwan, B.S., 112, 157, 166
McGaugh, J.L., 161, 167, 172
McGhee, R., 107, 113
McLachlan, D.R., 49
McLaughlin, R.L., 251
McLean, A.P., 193
McMillan, B.R., 250
McNamara, R.K., 146, 167, 189, 209
McNaughton, B.L., 155, 157, 254
McVety, K.M., 153
Mead, A., 115
Meaney, M.J., 166
Means, L.W., 102, 154, 163
Mech, L.D., 243, 256
Meck, W.H., 99, 100, 107, 108, 176
Medina, J.H., 164, 165
Meldrum, B.S., 190, 193, 197, 203, 204, 206, 209, 210, 216–220, 225, 227, 230
Menzel, C.R., 67
Menzel, E.W., 67, 113
Mesulam, M.M., 25
Meudell, P.R., 48, 49, 50
Meyer, E., 39
Meyer, R.C., 202, 203
Micheau, J., 109
Midgley, G., 111
Miles, M.A., 245
Miller, D.B., 94, 114
Miller, G.A., 105
Miller, L.L., 94, 114
Milner, B., 35, 36, 38, 48, 50, 51, 52
Minayev, A.N., 256
Minoshima, S., 38
Mirmiran, M., 96, 112
Mishkin, M., 40, 174
Misslin, R., 67
Mitchell, D., 101, 121
Mitchell, S.N., 190, 211, 214
Mittelman, G., 152, 163
Miyamoto, M., 193
Mizukawa, K., 228
Mizumori, S.J.Y., 104, 114, 116

Moghaddam, M., 177, 180, 202
Mohammed, A.K., 228
Molina-Holgado, M., 96
Money, J., 19, 43
Mooney, C.M., 15
Morgan, C.L., 251
Morris, P.A., 256
Morris, R.G., 37, 39, 41
Morris, R.G.M., 103, 108, 145, 146, 148, 150–152, 154, 160–162, 164, 166, 171–174, 178, 179, 196, 198, 200, 208, 217, 258
Morrow, L., 47
Morton, N., 37, 41
Muggia, S., 39
Muir, J.L., 174
Mulder, T., 47
Muller, J., 111
Muller, R U., 2
Mumby, D.G., 115, 217, 223
Mundy, W.R., 90, 92, 103, 106, 107, 109
Murai, S., 90, 115
Murphy, M.A., 256
Murray, C.L., 107, 108
Mysterud, I., 250

Nadel, L., 6, 34, 61, 62, 96, 103, 108, 116, 145, 146, 173, 196, 198, 200
Nagai, M., 228
Nagaoka, A., 193
Nageli, H.H., 124
Nakajima, S., 108, 109
Needham, G., 115
Neisser, U., 178, 179
Nelson, A., 205, 210, 223
Nelson, M.E., 262
Nerad, L., 156, 158, 159, 177, 207
Nesbit, J.C., 111
Netto, C.A., 193, 197, 204, 206, 209, 210, 216, 217, 227, 230
Newcombe, F., 4, 33, 41
Nichelli, P., 35, 37, 38
Nicolle, M.M., 167
Nihonmatsu, I., 106
Nikkah, G., 190, 228
Nishino, H., 228
Nolan, V., 251
Nonneman, A.J., 154, 163
Norman, D.A., 34
Novak, J.M., 105
Nunn, J.A., 193, 197, 204, 209, 210, 217, 223, 225, 230

O'Keefe, J., 2, 6, 61, 62, 96, 103, 116, 121, 145, 160, 173, 196, 198, 200, 213
O'Mara, S.M., 6
O'Steen, W.K., 157
Odashima, J., 90, 115
Ogawa, N., 108, 114, 119
Ohta, H., 113
Oitzl, M.S., 95, 121, 126, 13, 166, 204
Okaichi, H., 96, 107, 109, 118, 121, 126, 127
Oliverio, A., 117, 128
Olton, D.S., 6, 8, 87–97, 99, 102–104, 106–108, 110, 113, 114–119, 123–126, 197, 200, 204, 208, 209, 216, 217, 222, 223
Orenstein, D., 91, 118, 126
Orsini, A., 35

Orsini, P., 246
Osani, M., 92, 111, 125
Osborne, J.L., 246
Oshima, Y., 96, 107, 109, 118, 121, 126, 127
Ossenkopp, K.P., 95, 121, 167, 170, 176
Ostergaard, A.L., 38
Osterreith, P.A., 23
Otto, B., 94, 130
Otto, T., 190, 204, 222, 223, 224

Packard, M.G., 90, 91, 100, 161, 172
Pacteau, C., 226
Pailhous, J., 47
Paillard, J., 180
Palfrey, R., 158, 177
Pallage, V., 91, 94, 95, 108, 117, 118, 121, 126
Pallis, C., 42
Palmer, E.P., 42
Pank, L.F., 244
Papagno, C., 34, 36
Papas, B.C., 107
Papi, F., 241, 247, 252, 253, 255, 260, 261
Pappas, B.C., 208, 217
Parko, E.M., 64
Patel, H., 116
Patel, S.N., 228
Patterson, A., 42
Paulesu, E., 39
Paylor, R., 163
Pearson, D.E., 93, 96, 121, 126
Pearson, N.A., 35, 36, 38, 40, 47
Peele, D.B., 91, 96, 118, 127, 128, 213, 214
Pei, G., 206
Pellegrino, G. di, 27
Pelleymounter, M.A., 161
Perani, D., 38
Perdeck, A.C., 262
Peretz, I., 35
Petri, H.L., 153
Petrides, M., 39
Petrie, B.F., 163, 173
Petrinovich, L., 90, 121, 125
Pezzack, D., 258
Pfister, J.P., 110
Phelps, M.T., 115
Phillips, A.G., 115, 217, 223
Phillips, D.L., 116
Piaget, J., 61
Pickering, A.D., 48, 49, 50
Pico, R.M., 91, 99, 105, 108, 114, 126, 127
Picq, J.L., 107, 111, 113
Pieron, H., 252
Pigott, S., 36, 38
Pinel, J.P.J., 217, 223
Pittman, S., 256
Pocklington, R., 251
Pondel, M., 99
Pople, A.R., 256
Poppy, G.M., 246
Porto Saito, M.I., 71
Posner, M.I., 25
Potegal, M., 89, 116
Poucet, B., 6, 50, 64–68, 70, 71, 75–78, 80, 81, 83, 89, 91, 98, 101, 146, 150, 177, 178, 180, 251
Povoa, M.M., 245
Press, G.A., 223
Previdi, P., 35, 38

AUTHOR INDEX 275

Priede, I.G., 244
Pulsinelli, W.A., 107

Qin, Y., 254
Quasha, W.H., 16

Rabin, P., 16, 37, 38
Radek, R.J., 174
Rado, R., 256
Raffaele, K.C., 217, 223
Rage, P., 66
Ramacci, M.T., 91, 111, 112, 118, 120, 125, 128
Ramsay, M.A., 172, 173, 246
Ramsay, S.C., 224
Ranvaud, R., 262
Raphelson, A.C., 77
Raskin, L.A., 93, 96, 111, 121, 126
Ratcliff, G., 4, 47
Rauch, S.L., 111
Rawlins, J.N.P., 90, 115, 131, 160–162, 208, 211, 217, 222, 223
Reeder, T.M., 174
Reep, R.L., 202, 203
Regehr, J.C., 173
Regelin, W.L., 244
Reinstein, D.K., 114
Rempel, R.S., 244
Renner, M.J., 61, 228
Reverdin, B., 251
Reynolds, D.R., 246
Reynolds, M.A., 114
Richards, R.W., 101, 113
Riddoch, M.J., 33, 42, 43, 46
Ridley, R.M., 223
Riley, D.A., 96, 106
Riley, J.R., 246
Rish, P.A., 97, 99, 107, 109, 200, 204, 213
Ritchie, F.B., 88, 89
Rittenhouse, L.R., 101, 113
Rixen, G.J., 117, 127, 128
Rizzo, L., 51
Robbins, T.W., 115, 162, 166, 174, 201, 210, 217
Roberts, W.A., 91, 93, 94, 98, 103–106, 114–119, 125, 127
Robertson, I.H., 25
Robinson, A.S., 242
Robinson, C., 93
Robinson, C., 93, 118, 119, 126
Robinson, N., 114
Robinson, S.E., 108
Rodgers, A.R., 244
Rodionov, V.A., 243
Rodriguez, M., 111
Rogers, L.L., 246
Rogers, M.J.C., 13
Rohmer, G., 51
Roitblat, H.L., 8, 99, 104, 114, 115, 117
Rolls, E.T., 6
Rose, D., 8
Rose, F.D., 225, 226
Rose, G.M., 166
Rose, K., 257
Rosenzweig, M.R., 104, 114, 116, 228
Ross, E.D., 37, 38
Ross, P.I., 256
Rossi-Arnaud, C., 66, 78, 93
Rossier, J., 160
Rothblat, L.A., 223

Roullet, P., 90, 96, 111, 114, 127, 162, 170
Roy, E.J., 166
Rudy, J.W., 79, 104, 161, 163–167, 170, 171
Rush, J.R., 93, 109
Russell, J.T., 4, 5
Ryabinskaya, E.A., 116

Saghal, A., 217
Saito, H., 90, 115
Saksida, L., 167
Sakurai, T., 228
Salamone, J.D., 173
Salmon, E., 39
Salsbury, C.M., 257
Salway, A.F., 40
Samuel, M.D., 243
Samuelson, R.J., 87, 90, 92, 93, 95, 116
Sanberg, P.R., 112
Sandkuhler, J., 78
Sansone, M., 117, 128
Sartori, G., 33
Saucier, D., 172, 173
Saucy, F., 154
Saunders, G., 250
Sautter, S.W., 47
Savage, J., 113
Save, E., 66, 71, 73–78, 80, 81, 177, 180
Sawrey, D.K., 176
Schacter, D.L., 34, 49
Schadler, M., 1
Schallert, T., 174
Schenk, F., 94, 101, 103, 116, 125, 150, 152, 156–164, 168, 171, 174, 176–179
Schmidt, W.J., 107, 109
Schmidt-Koenig, K., 262, 264
Schnute, J.T., 262
Schugens, M.M., 190
Schumacher, E.H., 38
Schwarting, R.K.W., 206
Schwartz, C.C., 256
Seamans, J.K., 115
Seckl, J.R., 166
Segu, L., 66
Seguinot, V., 251, 254
Seiterle, D., 159
Selden, N.R.W., 162, 166, 210, 210
Semenov, L.V., 155
Semmes, J., 4, 18, 43
Sengstock, G.J., 112
Seron, X., 33, 35, 44, 53
Seymoure, P., 110, 111
Shafer, J.N., 75
Shallice, T., 33, 34, 37, 38, 39
Shanks, N., 165
Shavalia, D.A., 96, 104, 105
Shaw, C., 211
Shaywitz, B.A., 93, 96, 121, 126
Sherry, D.F., 175, 251
Shimamura, A.P., 49
Shoqeirat, M.A., 49, 50, 51, 52
Shors, T.J., 95, 112
Shuaib, A., 114
Siegel, A.W., 1
Silbermann, M., 112
Simon, C.A., 264
Simon, H., 91, 93
Simpson, J., 173
Sinclair, A.R.E., 255
Sinden, J.D., 173, 174, 190, 197–

199, 203, 206, 208, 209, 211, 212, 214–217, 220, 225–228
Singleton, G.R., 250
Sinsch, U., 245
Sirigu, A., 42
Sirinathsinghji, D.J., 190
Skaggs, W.E., 254
Skelton, R.K., 146, 167
Skelton, R.W., 189, 209
Slobin, P., 100, 104, 127
Small, M., 33, 41
Smeele, M., 115
Smith, A., 244
Smith, A.D., 246
Smith, A.R., 159
Smith, B.A., 98
Smith, E.E., 38
Smith, J.L.D., 257
Smith, L.E., 115
Smith, M.L., 48, 50, 51, 52
Smith, M.Y., 161
Smith, R.A., 108
Soderstrom, S., 206
Soffie, M., 66
Sokolov, E.N., 62
Sokolov, V.E., 243
Sorenson, J.C., 228, 229
Soteres, B.J., 63
Sowinski, P., 193, 197, 204–206, 209, 210, 216–220, 223, 227, 230
Spalding, J.M.K., 46
Speakman, A., 196
Spencer, R.L., 157
Spetch, M.L., 94, 95, 100, 114
Spinnler, H., 35, 36, 37, 39
Spreen, O., 16, 18, 19, 21, 23
Squire, L.R., 49, 50, 171, 214, 223, 224
Staddon, J.E.R., 99
Stadler Morris, S., 161, 163, 170
Stamps, J., 251
Stanton, M.E., 107, 108
Stapp, P., 246
Stea, D., 2
Stein, C.S., 71
Stenseth, N.C., 255
Stephenson, J.D., 173, 174, 203
Sterio, D., 100
Sterzi, R., 42, 47
Stevens, R., 93, 94, 96, 108, 110, 118, 119, 125, 126
Stewart, B.S., 247
Stewart, C., 152, 196, 200
Stewart, C.A., 146, 150, 171, 178, 179
Stiles-Davis, J., 29
Stoerig, P., 26
Stokes, K.A., 107, 118, 121, 125
Strauss, E., 16, 19, 21, 23
Street, R.F., 15
Strupp, B.J., 105
Sukhov, V.P., 243
Suster, M., 254
Sutherland, N.S., 61, 101, 103, 104, 128
Sutherland, R.J., 79, 146, 153, 154, 156, 157, 171, 173, 177, 178, 180, 197
Suzuki, S., 91, 95, 98, 99, 103, 104, 113, 125–127
Suzuki, W.A., 223
Szuran, T., 166

Taghzouti, K., 91, 93
Takai, R.M., 111

AUTHOR INDEX

Takei, N., 106
Tako, A., 193, 211, 222
Talland, G.A., 17
Taube, J.S., 155, 159
Taylor, B., 115
Taylor, L.B., 23
Tees, R.C., 111
Terkel, J., 256
Testa, T.J., 165
Teuber, H-L., 4, 18, 43
Thal, L.J., 211, 216
Tham, W., 114, 115
Thifault, St., 156
Thinus-Blanc, C., 5, 6, 64, 66–78, 80, 81, 83, 87, 89, 91, 96, 98, 101, 104, 109, 116, 177
Thomson, P.C., 257
Thorpe, W.H., 61
Thorson, P.H., 256
Thrasher, S., 214
Thurstone, L.L., 16
Timberlake, W., 92, 115, 116
Timothy, C.J., 223
Toates, F., 90, 125
Tokrud, P.A., 203
Tolman, E.C., 5, 8, 88, 89, 145, 195, 199
Tomie, J., 107, 113, 130, 209
Tomie, J.-A., 173, 177
Tomlinson, W.T., 71
Tomonaga, M., 113
Toniolo, G., 94, 95, 108, 117, 118, 121, 126, 127
Tonkiss, J., 211
Toumane, A., 90, 91, 107, 108, 109, 119, 126, 128
Towle, A., 191
Tozzi, A., 66, 78
Trewhella, W.J., 250
Trojano, L., 36
Tromp, E., 47
Troster, A.I., 117, 127, 128
Truncer, P.C., 127
Turnbull, O.H., 28
Turner, J.J., 211, 212, 214
Tweedie, F., 164, 171, 173, 174, 178, 179

Ueki, S., 217
Ugawa, Y., 92, 111, 125
Ungerleider, S., 40
Uphold, J.D., 102
Uster, H.J., 124

Vaher, P.R., 166
Valdiviesco, C., 206
Valenstein, E., 13, 19
Valentino, M.L., 154, 163
Vallar, G., 34, 35, 36, 37, 38, 47
Valouskova, V., 121, 129
Van de Poll, N.E., 111

Van der Linden, M., 33, 34, 35, 36, 39, 44, 51, 53
Van der Staay, F.J., 131, 195
Van Gool, W.A., 96, 112
Van Haaren, F., 96, 111, 112
Van Horne, B., 246
Van Luijtelaar, E.L.J.M., 131, 195
Van Velderhuizen, N., 114
Vanden Woude, S., 246
Vargas, H., 217, 223
Varney, N.R., 18
Vauclair, J., 67
Verna, A., 193, 211, 222
Villa, P., 42
Villegas, M., 112, 166
Volpe, B.T., 48, 49, 107, 108, 191, 193
Von Culin, J.E., 97, 99, 107, 109, 200, 204, 213

Wages, C., 63, 64
Wagner, G., 246
Walker, A., 112
Walker, D.L., 174
Walker, H.T., 43
Wallace, A.R., 251
Waller, S.B., 114
Wallraff, H.G., 252, 260, 261
Warach, J., 42
Ward, L., 106, 108
Warren, D.A., 164–167
Warrington, E.K., 16, 19, 37, 38, 42, 46
Wasserstein, J., 15
Wathen, C.N., 98, 106, 117
Watson, C.D., 91, 94, 109
Watson, G.S., 262
Watson, R.T., 25
Watt, K.E.F., 124
Watts, J., 93, 118, 119, 126
Weaver, K.L., 254
Wechsler, D., 14
Wehner, R., 241, 252, 259, 262
Weiller, C.R., 224
Weinstein, S., 4, 18, 43
Weltzl, H., 166
Werffeli, P., 150, 174
Wertlieb, D., 8
Werz, M.A., 91, 92, 95, 96, 118, 119, 124, 125, 208
West, J.R., 154, 163
Wheeler, E.A., 106
Whishaw, I.Q., 107, 113, 130, 150, 152, 153, 163, 172, 173, 176–180, 209, 217, 225, 228
White, G.C., 243
White, N.M., 90, 91, 100, 161, 172, 180, 201
White, P.C.L., 250
White, W., 92, 115, 116
Whiteley, A.M., 42, 46

Whiting, S., 21
Whitten, K.R., 242
Wiedmann, K.D., 36
Wilkie, D.M., 94, 95, 100, 104, 107, 111, 113–115, 127, 158, 177
Wilkinson, D.H., 260
Will, B., 91, 94, 95, 108, 117, 118, 121, 126, 127, 229
Williams, C.L., 99, 100, 107, 108, 176
Williams, I.H., 246
Wilson, B.A., 20
Wilson, J.T.L., 35, 36, 40, 51
Wilson, M.A., 157
Wiltschko, R., 262
Wiltschko, W., 251, 262
Wilz, K.J., 67
Winblad, B., 228
Winocur, G., 91, 93, 118, 127
Winter, J.D., 243
Wirsching, B.A., 107, 108, 116
Wise, R.J.S., 224
Wishart, T.B., 114
Wollberg, Z., 256
Wood, E.R., 217, 223
Woodruff, M.L., 99
Woods, R.D., 245
Woollard, T., 250
Wouters, M., 111
Wray, S., 250
Wurtman, R.J., 114

Xavier, G.F., 71

Yamamoto, T., 217
Yashpal, K., 167
Yatsugi, S., 217
Yau, J.L.W., 166
Yehuda, S., 146
Yoerg, S.L., 93, 115, 124, 127, 208
York, A.E., 256
Young, A.W., 33, 35, 36, 38, 40, 41, 47
Young, J.K., 246

Zaborowski, J.A., 225
Zacks, R.T., 48
Zahalka, A., 95, 121, 126, 130, 164, 204
Zaharia, M.D., 165
Zanette, L., 251
Zangwill, O.L., 42, 46
Zanobio, M.E., 37
Zimmer, J., 228, 229
Zimmerman, M., 78
Zimmermann, E., 166
Zola-Morgan, S., 171, 223
Zoladek, L., 103
Zubaid, A., 248
Zuffi, M., 39

Subject Index

Age differences, 8, 87, 111–112, 119, 125, 150–152, 157, 162–163, 166–170, 174, 180
Algorithmic strategies, 96, 98–99, 102, 111, 113–114, 116–131
Allocentric learning, 50, 71, 84, 195–202, 206, 211, 226, 228–230
Alzheimer's disease, 8, 35–36, 38, 39, 190
Amnesia, 48–50, 52
Animal cognition, 59–86
Anterograde amnesia, 48, 223
Aquatic RAM, 121, 128–131
Arena maze, *see* Open field
Associative place learning, 99, 189, 195–202, 214, 229
Attention, 25–26, 28, 36, 48, 160

Block design test, 14, 22
Block tapping test, 35–37, 40–41
Brooks matrix task, 36–37

Capture-marking-recapture (CMR), 241
Caudate nucleus, 108, 172, 202–203
Central executive, 34, 39–41, 43, 46–47
Cerebellum, 156
Cholinergic system, 50, 108–110, 126, 173–175, 180, 189–190, 197, 203, 211, 214–215, 226–229
Circular statistics, 261–263
Closure flexibility (concealed figures) test, 16–17
Cognitive map, 5, 42–43, 61–62, 84, 97–98, 117, 145, 158, 177–178, 195–197, 199–202, 207, 217, 230, 254–255
Cognitive neuropsychology, 24–25, 33–58
Compass orientation, 251, 262–264
Construction tests, 4, 22–23
Containment, 96–97, 125
Context memory, 48–52
Copying tests, 4, 11, 23
Core area, 248
Corticosterone, 166
Course reversal, 254

Dead reckoning, 154, 254
Dentate gyrus, 166, 218, 220
Disability, 3
Dishabituation paradigm, 60–61, 66–80
Dispersal, 239, 255

Displacement experiments, 259–262, 264
Dorsal noradrenergic bundle (DNAB), 166, 201–202, 211
Drawing tests, 4, 11, 23

Early experience, 110–111, 125
Egocentric learning, 89, 146, 163, 195, 202–203, 228–229
Elevated plus maze, 191–192, 226
Enriched conditions (EC), 225–226, 228–229
Environmental enrichment, 224–226, 228–230
Excursion, 239, 256–258
Exploration, 6, 59–86
Extramaze cues, 99–102, 111, 113, 127, 131, 189, 195–202, 206, 211, 226, 228–229

Figure-ground test, 16
Fluorescent powders, 245
Food caches, 251
Footprints, 246
Foraging, 87, 90, 105, 113–116, 124, 131
Fornix, 152, 172, 178, 196, 201, 204, 211, 229
Frames of reference, 26–27

Genetically based orientation, 253
Glucocorticoid, 166
Gollin incomplete figures test, 16
Gottschaldt's embedded figure test, 16–17, 36
Grid-based navigation, 255
Guidance cue, 162, 172, 174, 181, 197

Habituation, 60, 66–80
Hippocampus, 6, 48, 50, 62, 76–82, 94, 108, 110, 126, 145, 162, 166–168, 171–175, 178–180, 190, 193–201, 204, 209–210, 218–224, 226–228
Hole board, *see* Platform task
Home range, 7, 175, 239, 247–251
Homing behaviour, 7, 239, 252–264
Homing board task, 150–151, 159, 168, 170, 176–177
Huntington's chorea, 190

Innateness, 5

Intelligence, 4, 14–15
Internal model, 60, 66–67
Internal representation, 60, 66
Intracerebral transplants, 226–230
Intramaze cues, 99–102, 127, 131, 189, 195–202, 214, 219, 226, 229
Ischaemia, 114, 190, 193–195, 204–207, 210, 216–223, 227

Korsakoff's disease, 8, 50, 190

Learning set, 163–164
Location memory task, 51
Long-distance travels, 7, 239–241, 251–264
Long-term memory, 4, 35–37, 48–53, 87, 164–165, 190–195, 211, 213–219, 254–255
Lose-shift strategy, 219
Luminescent markers, 246

Magnetic compass, 262
Maier three-table maze, 62–65
Map-and-compass, 254
Map-based navigation, 255
Map-reading test, 43–44
Matching to sample, 131
Memory capacity, 105
Memory consolidation, 164–165
Memory persistence, 105
Mental rotation task, 36–37, 41
Migration, 7, 239, 256
Missing stimulus, 75–77
Morris water task, *see* Water maze
Motivation, 159, 208–210

Neophobia, 82
Network map, 43–46
NMDA receptors, 108, 148, 172, 180
Noradrenergic system, 166, 201–202, 211

Object assembly test, 14
Olfactory cues, 71, 76, 83, 102–103, 150, 170, 204–205, 230
Ontogeny, 108, 147, 168–170
Open field, 60, 68–75, 191–194
Operant learning, 103–104
Orientation, 260–264
Orienting response, 62, 77
Overshadowing, 161–163, 168, 170, 173–174, 180

Parietal cortex, 6, 25, 38, 202–203

277

SUBJECT INDEX

Parkinson's disease, 8, 190
Path integration, 154–156, 158–159, 177, 254
Pattern span test, 35–36, 40, 51
Phonological loop, 34–35, 37–38, 40–41, 52
Picture completion test, 15
Pilotage, 254
Place learning, 146, 150–151, 156–174, 177–179, 189, 196–197, 213–217
Platform task, 191, 195, 130–131
Primacy effect, 164–165, 222
Proactive interference, 104, 107–108, 126, 193, 211
Procedural tasks, 213–222
Proprioceptive cues, 146, 195, 202–203, 228–229
Prospective memory, 105
Psychometric tests, 14–23

Radar, 246
Radial arm maze (RAM), 6–8, 59, 87–143, 190–196, 198–201, 204–205, 208, 210–217, 222, 226, 228, 230
Radial water maze, see Aquatic RAM
Radio-tracking, 242–245
Random search, 99, 253
Rapid acquisition tasks, 213–222
Recency effect, 105, 164–165, 222
Recording devices, 247
Recovery of function, 224–230
Reference memory, 87, 106–114, 128, 171, 190–195, 198–201, 213–219, 230
Relocation memory task, 51
Retroactive interference, 105, 211
Retrograde amnesia, 48, 181
Retrospective memory, 105

Reversal learning, 107, 163–164, 173
Reversible inactivation technique, 77–82
Reward type, 91, 94, 119, 124
Right-left orientation test, 18–19
Road-map test, 43–44
Route reversal, 254
Route-based orientation, 254

Satellites, 244
Schizophrenia, 51
Search strategies, 96–103, 116–131, 152, 198–199, 207–211
Sedentariness, 239, 251
Sensorimotor cues, 146, 152–153, 195, 202–203, 228–229
Serial position, 104–105, 193, 220, 222
Sex differences, 111, 175–176
Short-distance travels, 247–251
Short-term memory, 4, 34–38, 46, 87, 164, 179, 190–195, 254
Site fidelity, 239, 251
Spatial information processing, 70
Spatial novelty, 61, 66–80
Spatial representation, 60, 66–67, 70, 75, 81, 83
Species differences, 112–114
Spontaneous alternation task, see T-maze, Y-maze
Spool and line devices, 245
Star compass, 262
Statistical analysis, 261–263
Strain differences, 112–114
Stress, 150, 165–167, 173, 209–210
Striatum, 89, 172, 180, 190, 201–202, 228
Sun compass, 176, 262
Supervisory attentional system (SAS), 34
Surveillance, 246–247

Systematic search, 253

T-maze, 61, 90, 121, 146, 191–192, 202, 204, 211, 217, 223, 226
Tactile cues, 205–207, 230
Three-door runway, 192–193, 199, 217–222, 228
Topographical memory, 4, 41–47, 52–53
Trail following, 253
Triangulation, 243

Unilateral neglect, 12, 20, 25–26, 47

Vector map, 43–46
Ventrolateral cortex, 203
Vestibular system, 84, 154–156
Virtual reality, 2–3
Visual imagery, 29, 36, 41
Visual inference, 13, 16
Visual object and space perception battery (VSOP), 19–20
Visual organisation, 13, 15–17
Visual scanning, 13, 17
Visual-spatial neglect, 4, 5–27
Visuospatial sketchpad, 34–35, 37–41, 52

Water maze (WM), 7, 59, 113, 145–188, 190–195, 197, 201–219, 221, 223, 228, 230
Win-shift strategy, 114–116, 154
Win-stay design, 100, 115
Win-stay strategy, 154, 163
Working memory, 34–41, 43, 46, 52, 87, 96, 106–114, 128, 131, 155, 171, 190–195, 198–201, 204–205, 208, 210, 213–224, 228, 230

Y-maze, 61, 90, 190–192, 202, 214, 216–217, 223

Printed in Great Britain
by Amazon